OPTICS

OPTICS

BY

W. H. A. FINCHAM, M.Phil., F.Inst.P., F.S.M.C.,

FORMERLY LECTURER APPLIED OPTICS DEPARTMENT NORTHAMPTON POLYTECHNIC
(NOW THE CITY UNIVERSITY), LONDON

and

M. H. FREEMAN, B.Sc., M.Phil., M.Inst.P.,

CHIEF PHYSICIST, PILKINGTON P-E LTD, ST ASAPH, NORTH WALES

EIGHTH EDITION

BUTTERWORTHS

ENGLAND: BUTTERWORTH & CO. (PUBLISHERS) LTD.
LONDON: 88 Kingsway, WC2B 6AB

AUSTRALIA: BUTTERWORTHS PTY. LTD.
SYDNEY: 586 Pacific Highway, 2067
MELBOURNE: 343 Little Collins Street, 3000
BRISBANE: 240 Queen Street, 4000

CANADA: BUTTERWORTH & CO. (CANADA) LTD.
TORONTO: 14 Curity Avenue, 374

NEW ZEALAND: BUTTERWORTHS OF NEW ZEALAND LTD.
WELLINGTON: 26–28 Waring Taylor Street, 1

SOUTH AFRICA: BUTTERWORTH & CO. (SOUTH AFRICA) (PTY.) LTD.
DURBAN: 152–154 Gale Street

©

W. H. A. Fincham and M. H. Freeman
1974

First Edition	... ...	1934
Second Edition	... ...	1936
Third Edition	... ...	1939
Fourth Edition	... ...	1942
Fifth Edition	... ...	1945
,, ,, *(Reprinted)*	... ...	1947
,, ,, *(Reprinted)*	... ...	1949
Sixth Edition	... ...	1951
,, ,, *(Reprinted)*	... ...	1954
,, ,, *(Reprinted)*	... ...	1956
,, ,, *(Reprinted)*	... ...	1959
Seventh Edition ...	... ...	1965
,, ,, *(Reprinted)*	... ...	1969
,, ,, *(Reprinted)*	... ...	1972
Eighth Edition	... ...	1974

Suggested UDC Number: 535

ISBN: 0 407 93421 9

Contents

Preface to Eighth Edition

In preparing this new edition the opportunity has been taken to involve a second author and to bring up to date some sections of the book. In particular, the chapters on physical optics have been largely rewritten to include lasers and the interference and diffraction effects of coherent light. Ideas of multiple-beam interference and a more extensive treatment of optical thin films are incorporated.

Various sections on photometry and optical materials have been rewritten. The inclusion of colour plates is intended to give the student a more confident introduction to physical optics. For these and other plates and diagrams the authors are indebted to a number of sources and full acknowledgement is made where appropriate.

London   W. H. A. Fincham
North Wales   M. H. Freeman

Preface to First Edition

During recent years considerable progress has been made in bringing the teaching of optics in line with practical requirements, and a number of text books dealing with Applied Optics have been published. None of these, however, caters for the elementary student, who has no previous knowledge of the subject and who, moreover, is frequently at the same time studying the mathematics required.

This book, which is based on lectures given in the Applied Optics Department of the Northampton Polytechnic Institute, London, is intended to cover the work required by a student up to the stage at which he commences to specialise in such subjects as ophthalmic optics, optical instruments and lens design. It includes also the work required by students of Light for the Intermediate examinations of the Universities.

The first eleven chapters deal with elementary geometrical optics, Chapters XII to XVI with physical optics, and the last three with geometrical optics of a rather more advanced character.

The system of nomenclature and sign convention adopted is that in use at the Imperial College of Science and the Northampton Polytechnic Institute, London. The sign convention is founded on the requirement that a converging lens shall have a positive focal length measured from the lens to the second principal focus. This is easily understood by the elementary student and is the convention commonly used throughout the optical industry. In ophthalmic optics – the most extensive branch of optical work at the present time – lenses are always expressed in terms of focal power; this idea has been introduced quite early and used throughout the work.

The solution of exercises plays an important part in the study of a subject such as Optics, and it is hoped that the extensive set of exercises with answers will be found useful. Typical examples from the examination papers of the London University, the Worshipful Company of Spectacle Makers and the British Optical Association are included by permission of these bodies.

My best thanks are due to my colleagues Messrs. H. T. Davey and E. F. Fincham for their valuable assistance in the preparation of the diagrams, which, together with the photographs, have been specially made for the book. I wish particularly to express my indebtedness to Mr. H. H. Emsley, Head of the Applied Optics Department, Northampton Polytechnic Institute, for reading the manuscript and for his very valuable help and suggestions given during the whole of the work.

LONDON, January, 1934   W.H.A.F.

I – The Propagation of Light

1. Introductory

The branch of Science known as **optics** is concerned with the study of light and vision; it may also embrace the study of other radiations closely allied to light.

Light is a form of radiant energy, being similar in its essential nature to other forms, such as heat and electrical (wireless) radiation. It is radiated out through space at an enormous speed by certain bodies called **luminous bodies.** Most of these are capable of emitting light by virtue of their exceedingly high temperature, as a result of which their constituent atoms are in a condition of considerable agitation, the effects of which are transmitted outwards from the body in all directions through its surroundings.

A poker, when cold, emits no radiation that we can detect. When gradually heated, it sets up a disturbance in the form of vibrations or waves in the surrounding medium, which radiate outwards at a speed of 186,000 miles per second; at some distance away we can "feel" the effect as heat, i.e., we detect this form of radiation by our sense of touch. As the temperature rises and the vibrations become more rapid, .the poker is seen to become red; the radiation is such that we can "see" its effects; it is in the form of light. We detect this form of radiation by our sense of sight, the "end organ" called the eye acting as detector. With further rise in temperature the poker passes to a "yellow" and then to a "white" heat.

The exact nature of light and kindred radiations, and the manner of their propagation through space, are not completely known; but a clear working idea of these "wave motions" can be derived from the ripples set up on the surface of water into which a stone is dropped. The important characteristics of the disturbance are: the **velocity** with which it travels outwards, the **frequency** of the rise and fall of the undulations, and the distance between successive wave crests, called the **wave-length**. Clearly the velocity is equal to the product of the wave-length and the frequency or number of vibrations arriving at a given point in unit time.

In the case of light, heat and electrical radiations, the velocity has been found by experiment (Chapter XII) to be 186,000 miles or 300 million metres per second. As we should expect from the poker experiment, the frequency of the vibrations is greater for light than for heat, and greater for some colours of light than others. The wave-length is consequently longer for heat than light and longer for red light (which appeared first) than for other colours. The figures for a few selected radiations are given in the table.

Radiation.	Velocity cm. per sec.	Frequency Billions per sec.*	Wave-Length cm.
Wireless	3×10^{10}	10^{-6}	3×10^4
Heat (infra-red)	3×10^{10}	30	10^3
Light ⎧ Red	3×10^{10}	395	$7 \cdot 59 \times 10^{-5}$
⎨ Yellow	3×10^{10}	509	$5 \cdot 89 \times 10^{-5}$
⎩ Violet	3×10^{10}	764	$3 \cdot 93 \times 10^{-5}$
Ultra-violet	3×10^{10}	1000	$3 \cdot 0 \times 10^{-5}$
X-Rays	3×10^{10}	3×10^6	10^{-8}

*Refers to British billion = 10^{12}. The American billion = 10^9.

In the table, only one typical value is given for each group of radiations. Actually, each group covers a wide range, and a more complete statement will be found in Fig. 145.

Of the many luminous sources such as the sun, the stars, various lamps, etc., the sun is our greatest source of light (and of heat), its brilliance being unapproached by anything else. Light gives rise to the sense of sight when it enters the eye, and in addition produces other effects such as the chemical change in the photographic plate or film, the bleaching of dyes, the pigmentation of the skin, and so on.

2. Rectilinear. Propagation of Light

Any space through which light travels is an **optical medium.** Most optical media have the same properties in all directions and are said to be **isotropic**—although there are a few optical substances, notably certain crystals, where this is not the case and in which particular phenomena arise (Chapter XVI). Most media, moreover, possess these same properties throughout their mass, so that they are **homogeneous.**

When light starts from a point source B (Fig. 1) in an isotropic medium, it spreads out uniformly at the same speed in all directions; the position it occupies at a given moment will be a sphere having the source at its centre. Such imaginary spherical surfaces will be called **light fronts** or **wave fronts.** In the case of the water ripples, the disturbance is propagated in one plane only and the wave fronts are *circular.*

Huygens (1629–1695), who is usually considered as the founder of the wave theory of light, assumed that any point on a wave surface behaved as a new source from which spherical "secondary waves" or "wavelets" spread out in a forward direction. The surface DAE, which

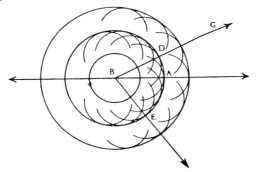

Fig. 1.–Wave front. Huygens' Principle

at any instant touches or envelops all these wavelets, forms a new wave front, and it is evident that while the light remains in the same isotropic medium, each new wave front will be a sphere with its centre at B.

Huygens' Principle assumes that the wavelets travel out in a forward direction only, and that the effect of each wavelet is limited to that part which touches the enveloping new wave front. The first of these assumptions may be explained by the fact that the points from which the wavelets arise are not independent sources, but are set in motion as the result of a wave front coming from the original source; while Fresnel (1788–1827) has shown (Chapter XV) by considering the "interference" that takes place between the wavelets, that the second assumption is approximately correct.

From Fig. 1 it is evident that the path, such as BDG, travelled by any part of the disturbance, is a straight line perpendicular to the wave front. We speak of such a line along which light travels as a **ray.** Huygens' construction for the new wave front enables us to find the change in

3

the form or direction of the wave front on passing into a new medium, or on reflection at a surface, and will be used in this way in subsequent chapters.

A luminous source, such as B (Fig. 2), will be seen only if light from it enters the eye. Thus an eye placed in the region between C and K, or between D and L, of the two screens CD and KL will see B, because the light reaches these regions without interruption. An eye placed between *c* and *d* will not see the object B, because the opaque obstacle *a b* shields the region *c a b d*, which is said to be in **shadow.** B will be visible, however, to an eye at *e*, just beyond the limits of the shadow region.

If we have two opaque screens at KL and MN, with circular apertures at *k l* and *p q* respectively, the screen MN will be illuminated over a circular patch *m n*, but will be dark elsewhere, and an eye placed at the aperture *p q* will not see the source B.

The fact that B is invisible to an eye placed between *c d* and visible to one placed within the region *k m n l* shows, as was also seen from Huygens' principle, that the light effect at a given point is due sensibly to that portion of the light that has travelled along the straight line joining the point to the source.

If the conditions within the regions *c d* and *m n* were to be studied more closely, it would be found that the patch *m n* is not uniformly illuminated, nor is the patch *c d* completely dark and separated from the surrounding brightness on the screen CD by a sharp line of demarcation, especially if *a b* and *k l* are small. The light waves, in passing the edges of these apertures, bend round into the space behind them in the same way as water waves may be seen curving round the end of a breakwater,

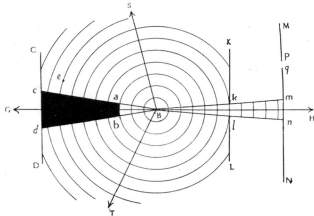

Fig. 2.—Rectilinear propagation of light.
Aperture at *kl*; Obstacle at *ab*

or as sound waves penetrate into the region behind a building. Light waves are so very small however, that this bending or **diffraction,** as it is called, takes place to an extent so minute, that special precautions have to be taken to observe it. (See Chapter XV).

3. Pencils and Beams

The light from a luminous point, or from any one point on a large source or illuminated object, after passing through a limiting aperture such as k l (Fig. 2) constitutes a **pencil.** The aperture may be simply an opening in an opaque screen or the edges of a lens, mirror, etc. The pencil increases in width as the distance from the source increases, and the light is said to be **divergent.** Under certain circumstances, however, as for example, after the light has passed through a convex lens, the pencil may be modified in such a way that its width is gradually decreasing; the light is then said to be **convergent,** and it converges to a point or **focus.** This focus will be the **image** of the object point from which the light started. Beyond the focus the pencil again diverges, its width still being limited by the original aperture. When the object point or the focus is at a great distance as compared with the width of the aperture

Fig. 3.—Pencils showing light fronts

the edges of the pencil will be practically parallel. Divergent, convergent and parallel pencils, with the corresponding wave fronts, are illustrated in Fig. 3. The ray passing through the centre of the aperture, will be the **typical direction** or **chief ray** of the pencil.

The collection of pencils arising from an extended source or object (Fig. 4a) constitutes a **beam.** The edges of the beam may be divergent or convergent, irrespective of the divergence or convergence of the pencils constituting the beam. Thus, if light from the sun passes through

Fig. 4.—Pencils and beams

an aperture (Fig. 4b) the individual pencils from each point on the sun will be parallel, but the pencils are not parallel to each other, and the edges of the beam are divergent. Similarly, the beam from a lens (Fig. 4c) may be divergent, while its constituent pencils are convergent. *The terms divergent, convergent, or parallel light refer to the form of the pencils and not to that of the beam.*

In Fig. 5 BEF is the section of a pencil diverging from a luminous

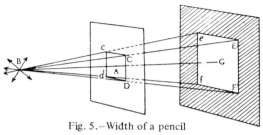

Fig. 5.—Width of a pencil

point B, and limited by an aperture CD, EF being the width of the cross section of the pencil in a plane parallel to the plane of the aperture. In the similar triangles CDB and EFB

$$\frac{EF}{CD} = \frac{BG}{BA} \qquad \dots \qquad \dots \qquad \dots \qquad (1.01)$$

As the light spreads out uniformly in all directions,

$$\frac{\text{area EF fe}}{\text{area CD dc}} = \left(\frac{BG}{BA}\right)^2 \dots \qquad \dots \qquad \dots \qquad (1.02)$$

that is, *the area of the cross section of a pencil varies as the square of the distance from the source.*

Since the amount of light in a pencil depends only upon the amount given out by the source, the amount falling upon a unit area of an illuminated surface—the **illumination**—varies inversely as the area of the cross section of the pencil; hence, *the illumination of a surface placed perpendicularly to the direction in which the light is travelling varies inversely as the square of the distance of the surface from the source.* This is the well-known **Law of Inverse Squares,** which is of great importance in questions of light measurement, illumination, etc.

4. Vergence

For a given aperture, the divergence or convergence of a pencil and the curvature of the wave front at the aperture decrease as the distance of the luminous point or of the focus from the aperture is increased. As the effect of a lens or curved mirror is to change the divergence or convergence of light, it is necessary to have some means of expressing the amount of divergence or convergence of a pencil in any given position. This is done by defining the divergence or convergence of a pencil at any particular position as the reciprocal of the distance *from* that position *to* the luminous point or the focus. It will be seen in Chapter V that this quantity also represents the curvature of the wave front. The general term **vergence** has been suggested to denote either divergence or convergence, the difference being shown by a difference of sign. The unit of vergence is the **dioptre**, the vergence of a pencil one metre from the luminous point or focus.

5. The Pinhole Camera

We have an interesting verification of the law of rectilinear propagation in the pinhole camera. The light from each point on an illuminated object, on passing through a small aperture in an opaque screen, gives rise to a narrow pencil, and if the light is received on a second screen at some distance from and parallel to the plane of the aperture, each pencil produces a patch of light of the same shape as the aperture. As the light travels in straight lines, the patches of light on the screen occupy similar relative positions to those of the corresponding points on the object, and the illuminated area of the screen is similar in shape to the original object, but inverted (Fig. 6a). If the aperture is made quite small, the individual patches of light will overlap to only a small extent, and a fairly well defined picture or image of the object is formed (Fig. 6b). A pinhole camera may be constructed by mounting in one side of a box, a thin sheet of metal with a small needle hole in it, and in the opposite side of the box, a photographic plate or ground glass screen on which the image is formed, or the metal plate with the small aperture may be mounted in an ordinary camera in place of the lens. As will be seen from

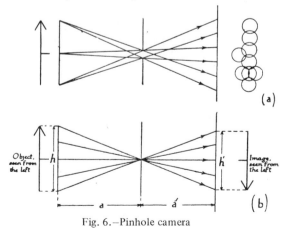

Fig. 6.—Pinhole camera

Fig 6*b*, the size of the inverted image of any object will depend on the distances of object and image from the aperture, thus

$$\frac{h'}{h} = \frac{a'}{a}$$

The degree of sharpness of the pinhole image can never be of a very high order, because if the diameter of the aperture is reduced beyond a certain amount, additional blurring becomes evident, due to diffraction effects. Also, the illumination of the image is very low as compared with that formed by a lens, owing to the very small aperture used. The pinhole camera is occasionally used in practice, and possesses certain advantages over a lens. The image is free from distortion, the depth of focus is very great, i.e., images of objects at greatly varying distances are reasonably sharp in one plane, and much greater angles of view are obtainable than are possible with a lens.

6. Shadows and Eclipses

The properties of shadows may easily be deduced from the law of rectilinear propagation. If the source is a point, the boundary of the shadow is sharply defined (neglecting diffraction) and its section, as received on a screen perpendicular to the typical direction of the light, will be the same shape as the opaque obstacle would appear when viewed from the position of the source. Thus the shadow of a sphere will have a circular section, while that of a flat circular body will be circular or elliptical according to how it is tilted. The size of the shadow in any position may be determined in the same way as the cross section of a pencil (§ 3).

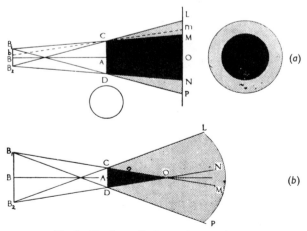

Fig. 7.—Shadows. Umbra and penumbra

Usually, the light will be coming from an extended source and we must consider the shadow as formed by light from different points on the source. In Fig. 7a B_1B_2 is a circular source, CD a circular opaque object and LP a screen. No light from the source can enter the space CDNM, and on the screen there is a circular area of diameter MN in total shadow; this total shadow is known as the **umbra**. Surrounding the umbra is a space, CLM, DNP, in partial shadow, known as the **penumbra.** In the penumbra the illumination gradually increases from total darkness at the edges of the umbra to the full illumination at the outer edges of the shadow; for example, the point m is receiving light from the portion B_1b of the source. The shadow received on the screen consists, then, of a central dark portion surrounded by an ill-defined shaded edge. The diameters of the umbral and penumbral portions of the shadow received on a screen, may be found as follows:

In Fig. 7a from similar triangles

$$\left. \begin{aligned} LN = MP = \frac{CD \times BO}{BA} \\[2mm] LM = NP = \frac{B_1B_2 \times AO}{BA} \end{aligned} \right\} \qquad \ldots \qquad \ldots \quad (1.03)$$

When the source is larger than the opaque object (Fig. 7b), the umbra is in the form of a cone having the object as its base and at a certain distance from the object the umbra disappears, and the shadow is wholly penumbral. This may readily be seen in the case of the shadows of objects in sunlight. As the sun subtends an angle at the earth of 32 minutes, the length of the umbra in sunlight is 105 times the diameter of the opaque object.

In Fig. 7b,

$$\frac{AO}{BO} = \frac{CD}{B_1B_2}$$

$$AO = \frac{CD \times BA}{B_1B_2 - CD} \qquad \ldots \qquad \ldots \qquad \ldots \quad (1.04)$$

Eclipses of the sun and of the moon are interesting examples of shadows.

Eclipse of the Sun. At certain times the moon comes between the sun and the earth and as the distance between the moon and the earth is less than the length of the umbra of the moon's shadow, the shadow on the earth's surface will consist of an umbral portion surrounded by a penumbra. An observer in the area covered by the umbra will see the

9

moon completely covering the sun and the eclipse will be total; in the penumbra the eclipse will be partial, only a portion of the sun being covered.

Eclipse of the Moon. When the earth is directly between the sun and the moon, the shadow of the earth is seen projected upon the moon's surface, and the eclipse will be total when the moon lies wholly within the umbra. When the eclipse is partial the penumbra of the shadow may be seen as a partial darkening of the moon's surface outside the complete shadow.

Shadows play an important part in our visual interpretation of the forms of solid objects; for example, a distant white sphere, evenly illuminated, is indistinguishable from a circular flat surface, while an illumination giving a shadow on some part of it at once gives it its solid appearance. Much can be done in disguising the form of objects by darkening various parts to create a false impression of shadows.

7. The Nature of White Light

In the poker experiment, it was seen that the light emitted changed from red, which appeared first, to yellow and finally to white when the temperature was sufficiently high. The nature of white light was investigated by Sir Isaac Newton (1642–1727) in his classical experiments in 1666. On passing a narrow beam of sunlight through a glass prism, Newton found that the patch of light received on a screen was broadened out into a band of colours, in which he recognised seven distinct colours in the following order: red, orange, yellow, green, blue, indigo and violet, each colour gradually shading into the next. This coloured band is called a **spectrum** and the white light is said to be **dispersed.** Recombining the various coloured beams by passing them through a similar prism with its base in the opposite direction to the first, or through a convex lens, Newton found that the light was again white: he therefore concluded that white light is composed of a mixture of light of the seven spectrum colours.

It is now known that each of these colours consists of vibrations of a given range of frequency, red having the lowest frequency and therefore the longest wave-length, and violet the highest frequency and the shortest wave-length, the other colours occupying intermediate positions in the scale of frequencies according to their position in the spectrum. Any solid body, when raised to a "white" heat, is emitting vibrations of all these frequencies and, in addition, a certain amount which lie outside the range of frequencies to which the eye is sensitive. Of these, those of shorter wave-length than the violet are known as the **ultra-violet** radiations and those of longer wave-length than the red, as the **infra-red** (see Chapter XIII).

1. The velocity of light *in vacuo* is 3×10^{10} cm. per second. What will be the wavelengths corresponding to the following frequencies?

Orange	450 billions per second.
Green	569 billions per second.
Blue	696 billions per second.

2. Describe an experiment to show that light travels in approximately straight lines.

3. A person holding a tube 6 inches long and 1 inch in diameter in front of his eye just sees the whole of a tree through the tube. What is the apparent (angular) height of the tree, and what is its distance if its actual height is 40 feet?

4. Explain the terms, convergent, divergent and parallel light. Illustrate your answer with diagrams showing the form of the wave fronts in each case.

5. A pencil of light diverges from a point source through a rectangular aperture 3 cm. × 4 cm. at 30 cm. from the source. Find the area of the patch of light on a screen parallel to, and 120 cm. from the aperture.

6. A pencil of light on passing through a lens of 40 mm. diameter converges to a point 60 cm. from the lens. Find the areas of the cross sections of the pencil at 20, 50 and 75 cm. from the lens.

7. What are meant by pencils and beams of light? Illustrate your answer with diagrams.

A spherical source of 5 cm. diameter is placed 30 cm. from a circular aperture of 8 cm. diameter in an opaque screen; find the size and nature of the patch of light on a white screen 50 cm. from and parallel to the plane of the aperture.

8. A small source of light is 30 cm. from a rectangular aperture 18 cm. × 8 cm. A screen is placed at 105 cm. from and parallel to the plane of the aperture; what will be the area of the illuminated patch on the screen?

A second source is placed 8 cm. from the first, and in a line parallel with the shorter side of the aperture. Find the areas of full and partial illumination.

9. What is the height of a tower that casts a shadow 65 feet 3 inches in length on the ground, the shadow of the observer who is 6 feet high being at the same time 7 feet 6 inches in length?

10. A man 5 feet 11 inches in height is standing at a distance of 4 feet 6 inches from a street lamp. What will be the length of the man's shadow, if the lamp be 9 feet above the roadway?

11. A circular opaque disc of 2 feet diameter is 6 feet from an arc lamp; find the size of the shadow on a screen 20 feet from and parallel to the disc. What will be the form and size of the shadow if the arc be surrounded by a diffusing globe of 1 foot 6 inches diameter?

12. Explain carefully the formation of the image in a pinhole camera. How does the character and size of the image depend on:
 (*a*) the size of the aperture.
 (*b*) the shape of the aperture,
 (*c*) the distance of the aperture from the screen,
 (*d*) the distance from the aperture to the object.

13. If the distance between an object and its image formed by a pinhole be 5 feet, what will be the position of the pinhole for the image to be one-tenth the size of the object?

14. A pinhole camera produces an image 2·25 inches diameter of a circular object,

and when the screen is withdrawn 3 inches further from the pinhole, the diameter of the image increases to 2·75 inches. What was the original distance between the pinhole and the screen? (S.M.C.)

15. If the sun subtends an angle of 32 minutes at the earth, what must be the distance between the aperture and the screen of a pinhole camera in order that the image of the sun shall have a diameter of $\frac{1}{2}$ inch?

16. Explain the formation of an image in a pinhole camera. Sunlight is reflected from a small plane mirror about 2 mm. square and the reflected light falls normally on a white screen. Describe and explain the patch of light formed on the screen when this is (a) a few inches from the mirror. (b) about 10 feet from the mirror. (B.O.A.)

17. A circular opaque object 3 inches diameter is placed 12 inches from and parallel to a circular source of 5 inches diameter. Find the nature and size of the shadow on a screen perpendicular to the line passing through the centres of the source and the object and 3 feet from the object.

18. Explain with diagrams the way in which total and partial eclipses of (a) the sun and (b) the moon are produced.

19. Find the diameter of the umbra and penumbra of the earth at the distance of the moon (240,000 miles) taking the earth's diameter as 8,000 miles, and the visual angle subtended by the sun from the earth as 32 minutes. (S.M.C.) (Assume the light from the sun to be parallel light).

20. Why are shadows much sharper in the case of an arc lamp without a surrounding diffusing globe than with one? Explain with a diagram.

21. A dark room 10 feet square, with white walls, has a small hole in one wall. The image of a man 6 feet high outside the room is formed on the wall and is 5 inches high. How far away is the man? What will be the size of the image of a tree 50 feet high and 300 feet away?

22. An opaque circular disc is interposed between a luminous disc of larger diameter and a screen so that the shadow consists of a central umbra and surrounding penumbra. The discs and screen are parallel and the line joining the centres of the discs is perpendicular to them. Show that the width of the penumbra ring is independent of the size of the opaque disc. (S.M.C.)

II – The Behaviour of Light on Reaching a New Medium

8. Introduction

The light from a source continues to travel out until it meets the surface bounding some new medium. Here a number of effects are produced, which depend upon the nature of the two media and of the surface. In all cases some of the light is sent back or *reflected*, while the remainder passes into the new medium, where some of it will be transformed into some other form of energy, or *absorbed*, and some will be *transmitted*. These three modifications of the light always take place, but one often predominates over the others. It will generally be convenient to consider separately the light which is reflected at the surface and that which passes into the new medium, but it must be remembered that both effects are always taking place at a surface.

9. Reflection—Regular and Diffuse

When the surface is a polished one, practically the whole of the reflected light travels back in definite directions, as though coming from the source placed in some new position, the surface acting as an aperture in limiting the reflected beam (Fig. 8). An observer in this reflected beam sees the image of the original source, but is not conscious of the presence of the reflecting surface if this is perfectly polished. Such

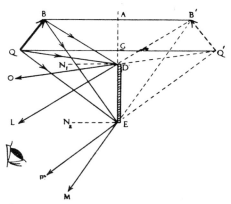

Fig. 8.–Regular reflection. Position of image formed in plane mirror

reflection is said to be **regular** or **specular.** When the surface is not polished, every irregularity of the surface will reflect the light in a different direction, and the light will not return as a definite beam. The surface itself will be visible from all directions, and the light may be considered as spreading out from every point on the surface. This reflection is said to be **diffuse** or **irregular,** although the actual reflection at each minute portion of the surface will be regular. All illuminated objects are visible by reason of the light irregularly reflected at their surfaces.

The amount of light reflected at a surface depends upon a number of factors. It will vary greatly with different media, as may be seen in the amount of light diffusely reflected by, say, white, grey and black paper, while the quantity reflected by a polished silver surface is much greater than that reflected from polished steel. The angle at which the incident light meets the reflecting surface may also have a considerable effect on the quantity of light reflected. This is very apparent when light is reflected from a polished glass surface at different angles.

In the case of transparent substances, such as glass, water, etc., the percentage of light reflected for any angle of incidence may be found from Fresnel's law (Chapter XVI). It will depend not only upon the angle of incidence, but also upon the difference in the refractive indices (§ 12) of the two media. For example, if some pieces of glass are placed in glycerine, or, better still in canada balsam, media which have almost the same refractive index as glass, practically no light is reflected at the surface of the glass, and the pieces are almost invisible. A similar effect is seen in the case of a grease spot on paper. The grease filling the pores of the paper has a refractive index more nearly that of the paper fibres than the air has, and so less light is reflected, and the spot appears darker than the surrounding paper.

The light reflected by some substances is coloured. This is due to the fact that, of the various colours contained in white light, some only are reflected, and the others are absorbed by the substance. Such reflection is said to be *selective*. A red object, for example, when illuminated with white light, reflects mainly the red, the other colours passing into the substance and being absorbed. Such an object, illuminated with green light, would appear practically black, as the green light contains no red.

10. The Law of Reflection

In Fig. 9 BA is a ray of light incident on a surface DE. The perpendicular to the surface at the point A is known as the **normal to the surface at the point of incidence,** and the plane containing the incident ray BA and the normal AN is the **plane of incidence.** The angle BAN, which the incident light makes with the normal is the **angle of incidence,** *i*. The light after reflection will travel along AF, and the angle FAN is the **angle of reflection,** *r*.

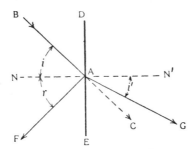

Fig. 9.—Reflection at a plane surface

The law of reflection, which may easily be found experimentally, has been known from very early times. It may be stated as follows:

The incident and reflected rays and the normal to the surface lie in the one plane and the incident and reflected rays make equal angles with the normal on opposite sides of it.

The way in which the light is reflected at a surface and thence the law of reflection, may be found by applying Huygens' principle. In Fig. 10 BCGA represents a parallel pencil of light incident on a plane reflecting surface AD, AG being the position of a plane wave front at a certain instant. Had there been no change in medium, the new wave front would, after a time, have reached the position HD. Assuming the speed of the light to be the same before and after reflection, in a given time each part of the reflected light will have travelled back from the surface a distance equal to that which it would have travelled forward, had there been no reflection. Each point on the surface between A and D becomes, in succession, the centre of a reflected wavelet as the incident wave front

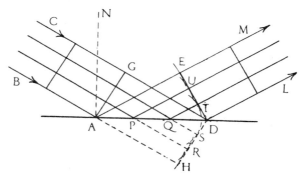

Fig. 10.—Reflection of plane wave at plane surface. Huygens' construction

meets it. The reflected wavelets arising, for example, at the points A, P and Q will have travelled out distances AE, PU and QT, equal to AH, PR and QS respectively, while the edge of the wave front at G is reaching the surface at D. From Huygens' principle, the envelope or common tangent, ED, to the wavelets will be the reflected wave front. It may easily be proved that the angles GAD and EDA are equal, and therefore the angle of incidence equals the angle of reflection.

11. Absorption and Transmission

The subsequent behaviour of the light that enters the new medium will depend upon the nature of the substance. In most cases the whole of the light, after travelling a short distance, is changed into some other form of energy, such as heat or chemical action, and the light is said to be absorbed. Substances through which light cannot pass are said to be **opaque.** The degree of absorption will depend upon both the nature of the substance and the thickness through which the light travels, as may be seen in the case of, for example, gold, which, in a thin film, will transmit a considerable amount of light.

When all colours are equally absorbed, as will usually be the case when the substance is of any considerable thickness, the absorption is said to be *general*; but with some substances, coloured glasses, dyes, etc., the absorption is *selective*; some colours are absorbed and others are transmitted. We shall deal with this again when considering transmission. Certain substances possess the property of absorbing energy of short wave-length, the ultra-violet and the violet and blue light, and emitting it as radiations of longer wave-length. This is known as **fluorescence** and is considered more fully in Chapter XI.

Some substances absorb very little light and practically all of it is transmitted; such substances are **transparent** and are of great importance in optical work. If the surfaces of such a transparent substance are polished, the light passes through as a definite beam. No substance is perfectly transparent and some part of the light is always absorbed; thus measurements on a good optical glass show that about $2\frac{1}{2}$ per cent. of the incident light is absorbed in passing through a thickness of 1 cm., while it is well known that practially no light reaches the bottom of a deep sea, although a small thickness of the water absorbs very little light.

Other substances, which contain fine particles of a different optical nature from the surrounding mass, or which are composed of minute particles of transparent material with polished faces, such as minute crystals, transmit the light but diffuse or scatter it on the way, and objects cannot be seen clearly through such substances. These media, such as opal glass, paraffin wax, tracing paper and cloth, smoke and fog, are said to be **translucent.** The same effect is obtained with a transparent substance having an unpolished surface, as in ground or "frosted' glass. Screens made of translucent substances are frequently used to diffuse the light from a source, or as screens on which to receive images.

As was previously mentioned, the absorption and therefore the transmission of certain media are *selective* with respect to the colour or wave-length of the light absorbed or transmitted. Thus, when white light is incident on a piece of, say, green glass, some of the light is reflected at the polished surface and this will still be white; of that which passes into the glass, only green light is transmitted, the other colours composing the white being absorbed. All coloured glasses, dyes, etc., behave in this way. A good white glass, whilst being transparent to all the colours of the visible spectrum, is quite opaque to radiations below a certain wave-length in the ultra-violet. Some of the glasses used for protective purposes in ophthalmic work are designed to be opaque to ultra-violet or infra-red, or both, while transparent to ordinary light.*

12. Refractive Index

The changes in the direction of the light that occur when light passes from one medium to another can be explained on the assumption that

*The action of filters is dealt with in Chapter XI.

the velocity of light is different in different media, Foucault, in 1850, determined experimentally the velocity of light, both in air, and in water, and found that the velocity was less in water than in air. This important result was in accordance with the wave theory explanation of the behaviour of the light on passing between these two media. The action of lenses and prisms is entirely due to this change of velocity in different media, and it is necessary to have some means of expressing the relative velocity of the light in any medium.

If we take the velocity of light *in vacuo* as a standard, then its relative velocity in any other medium may be expressed in terms of the ratio $\frac{\text{velocity }in\ vacuo}{\text{velocity in medium}}$. This ratio is known as the **absolute refractive index** and will be denoted by n. As the velocity in any transparent medium is less than that *in vacuo*, the value of the absolute refractive index of any medium is greater than unity and, with the exception of gases, will be between 1·33, the refractive index of water and 2·42, that of the diamond.

When the light passes from one medium to another, such as from air to glass, glass to water, etc., it is sometimes convenient to express the change in velocity as a ratio of the velocities in the two media; thus

$$\frac{\text{velocity in air}}{\text{velocity in glass}} = {}_an_g, \qquad \text{this is called the}$$

relative refractive index from air to glass. Since the absolute refractive index of air

$$n_a = \frac{\text{velocity }in\ vacuo}{\text{velocity in air}}$$

and that of glass

$$n_g = \frac{\text{velocity }in\ vacuo}{\text{velocity in glass}}$$

the relative refractive index from air to glass

$$_an_g = \frac{\text{velocity in air}}{\text{velocity in glass}} = \frac{n_g}{n_a}$$

Hence the relative refractive index from one medium to another is the ratio of the absolute refractive index of the second medium to that of the first. As the velocity of light in air is very nearly that *in vacuo*, the average absolute refractive index of air being 1·00029, the relative refractive index from air to any substance is usually given as the refractive index of that substance.

The terms "rare" and "dense" are frequently used in a comparative sense in referring to media of low and high refractive index respectively.

It will be seen, when we consider the action of a prism (§ 35), that the dispersion of white light, that takes place in Newton's experiment with the prism, is due to the refractive index being different for each of the colours composing the white light, the refractive index being highest for the violet and lowest for the red in all substances with a few exceptions. In stating any refractive index, therefore, it is necessary that the colour of the light shall be specified; this is considered more fully in Chapter XIII. Usually, a mean value of refractive index is taken as that for a particular yellow. This used to be the light emitted by incandescent sodium vapour. Then light from helium vapour was used as this allowed a more accurate measurement. Lately, yellow-green light from mercury vapour has been used although values for sodium light are still commonly quoted. The following are the mean refractive indices of some of the substances commonly used in optical work:

Water	1·3336
Canada Balsam	1·526
Cedar Wood Oil	1·510
Boro-silicate Crown Glass	1·509
Hard Crown Glass	1·5175
Spectacle Crown Glass	1·5230
Dense Flint Glass	1·6214
Double Extra Dense Flint Glass	1·7129
Quartz	$\left\{\begin{array}{l}1\cdot5442\\1\cdot5534\end{array}\right.$
Diamond	2·4173

13. Refraction

When light is incident obliquely on the surface between two media of different refractive index, its direction is changed on passing into the new medium, and it is said to be **refracted**. The new direction of the light may be found by applying Huygens' principle as follows: Fig. 11 represents a parallel pencil of light BADC incident on the surface AD between two media of refractive index n and n', n' being greater than n, AG is a plane wave front one edge of which has just reached the surface at A. Had there been no change in medium, the incident light would have travelled on with velocity V without change of direction, and the wave front after a certain time would have been in the position HD. As the incident wave front meets the surface, each point on the surface, such as A, P, Q, R and D, becomes in succession the centre of a system of wavelets travelling in the new medium. These wavelets travel in the second medium at a velocity V′, where $\dfrac{V'}{V} = \dfrac{n}{n'}$, and by the time the light at G has travelled to D in the first medium, the radii of the wavelets in the second medium will be

19

$$AE = AH \times \frac{n}{n'}, \quad PX = PS \times \frac{n}{n'}, \quad QY = QT \times \frac{n}{n'}, \text{ etc.}$$

The envelope of all these wavelets will be the refracted wave front ED. It will be found that the refracted wave front is still a plane surface and that when, as in Fig. 11, the light is passing from a rarer to a denser medium, it is refracted *towards the normal*. A similar construction for

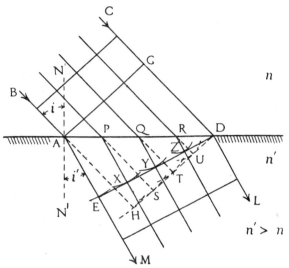

Fig. 11.–Refraction of plane wave front at plane surface. Huygens' construction

light passing from a denser to a rarer medium will show that in this case it is refracted *away from the normal*. The angle BAN is the angle of incidence i and the angle MAN′, the angle of refraction i'.

Law of Refraction.

From the definition of refractive index (§ 12).

$$n \text{ GD} = n' \text{ AE} \quad \text{(Fig. 11)}$$

$$n \frac{\text{GD}}{\text{AD}} = n' \frac{\text{AE}}{\text{AD}}$$

$$n \sin \text{DAG} = n' \sin \text{ADE}$$

but $\qquad\qquad \widehat{DAG} = i,$ and $\widehat{ADE} = i$

$$\therefore \ n \sin i = n' \sin i' \qquad \ldots \qquad \ldots \qquad \ldots \quad (2.01)$$

and $\qquad\qquad \dfrac{\sin i}{\sin i'} = \dfrac{n'}{n} \qquad \ldots \qquad \ldots \qquad \ldots \quad \ldots (2.01\,\text{A})$

$\dfrac{n'}{n}$ is the relative refractive index from the first to the second medium, and is a constant for these two media and for light of any one colour.

The law of refraction may be stated as follows: *The incident and refracted rays and the normal to the surface at the point of incidence lie in the one plane on opposite sides of the normal, and the ratio of the sine of the angle of incidence to the sine of the angle of refraction is a constant for any two media for light of any one colour.* This law, which was first stated, but in a somewhat different form from the above, by Willebrord Snell (1591–1626), is often known as *Snell's Law.**

The refraction at a plane surface and the law of refraction may be demonstrated by the following simple experiment. A block of glass having two parallel polished surfaces is placed on a sheet of paper with its polished surfaces vertical. Two pins A and B (Fig. 12) are placed a few inches apart to represent the path of a narrow beam of light incident

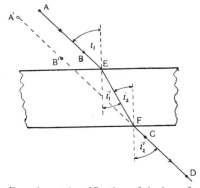

Fig. 12.—Experimental verification of the law of refraction

obliquely on one surface; on looking through the glass, the pins will be seen to occupy the apparent positions A′ B′. Two further pins C and D

*Snell stated the law of refraction as a constant ratio of the cosecants of the angles of incidence and refraction. The statement of the law as a constant ratio of the sines of the angles was first given by Descartes (1596–1650) in 1637.

are then placed in apparent line with A and B, as seen through the glass. The line AEFD then represents the path of the light between the points A and D. By repeating the experiment for a number of different angles of incidence and measuring these angles and the corresponding angles of refraction, the law of refraction may be experimentally verified and the refractive index of the glass obtained.

14. The Principle of Least Time

The laws of reflection and refraction may be summed up in a general law, which was first stated by Fermat (1601–1665), and is known as the **principle of least time**; this may be stated as follows: *The actual path*

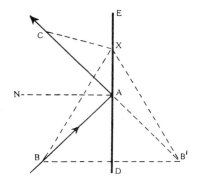

Fig. 13.—Path of least time on reflection

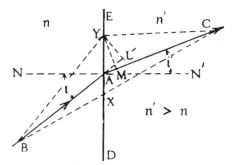

Fig. 14.—Principle of least time. Refraction at plane surface

travelled by light in going from one point to another is that which, under the given conditions, requires the least time. In certain cases, however,

22

when the bounding surface is curved, the time may be a maximum instead of a minimum, but must always be one or the other.

In Fig. 13 DE is a reflecting surface and BAC the path of the light reflected at the surface in travelling from B to C. It can easily be proved that the path BAC, which is in accordance with the law of reflection, is shorter than any other path such as BXC. Let BB' be a perpendicular to DE and DB' = BD, then B'A = BA and B'X = BX; therefore the path BAC = B'C, and the path BXC = B'X + XC. But in the triangle B'CX the sides B'X and XC are together greater than the side B'C; therefore BAC is the shortest path and hence the path of least time. The principle of least time, in the case of reflection at a plane surface, was discovered by Hero of Alexandria (150 B.C.)

In the case of refraction the path that the light travels in the least time will not be the shortest distance between the two points. In Fig. 14 DE represents the surface between two media of refractive index n and n', n' being greater than n, and B and C two points between which the light is travelling. Since the light is travelling more slowly in the second medium than in the first, the straight line BXC, although the shortest path between B and C, is not the path of least time, for a greater part of the path lies in the second medium than is the case with the path, say BAC. For some other path BYC, the time gained by the further shortening of the distance in the second medium may be more than offset by the increase in the distance travelled in the first medium, and it can be shown that the path BAC that is in accordance with the law of refraction is the path of least time between B and C.

In Fig. 14

$$\frac{\sin \text{NAB}}{\sin \text{N'AC}} = \frac{n'}{n} = \frac{V}{V'}$$

where V and V' are the velocities of the light in the first and second media respectively. The time taken by the light in travelling over the path BAC is

$$t = \frac{\text{BA}}{\text{V}} + \frac{\text{AC}}{\text{V'}}$$

and it is required to show that this is less than the time $\dfrac{\text{BY}}{\text{V}} + \dfrac{\text{YC}}{\text{V'}}$ taken over any other paths such as BYC. Drawing YL perpendicular to BA and YM perpendicular to AC, then $\stackrel{\frown}{\text{AYL}} = \stackrel{\frown}{\text{NAB}}$ and $\stackrel{\frown}{\text{AYM}} = \stackrel{\frown}{\text{N'AC}}$.

therefore,
$$\frac{\sin \text{ AYL}}{\sin \text{ AYM}} = \frac{n'}{n}$$

and
$$\frac{AL}{AM} = \frac{n'}{n} \text{ or } \frac{AL}{V} = \frac{AM}{V'}$$

$$\frac{BA}{V} + \frac{AC}{V'} = \frac{BA}{V} + \frac{AM + MC}{V'} = \frac{BL}{V} + \frac{MC}{V'}$$

But
$$\frac{BL}{V} < \frac{BY}{V} \text{ and } \frac{MC}{V'} < \frac{YC}{V'}$$

therefore
$$\frac{BA}{V} + \frac{AC}{V'} < \frac{BY}{V} + \frac{YC}{V'}$$

15. Optical Length

From the definition of the refractive index of a medium, $n = \frac{V_0}{V}$ where V_0 is the velocity of light *in vacuo* and V the velocity of light in the medium concerned of refractive index n.

Therefore, $V = \frac{V_0}{n}$.

The time taken by light in travelling a distance l *in vacuo* $= \frac{l}{V_0}$, while the time taken in travelling a distance l in medium of refractive index $n = \frac{l}{V} = \frac{nl}{V_0}$.

When the light is travelling through a number of different media, if l_1, l_2, l_3, etc. are the distances in the separate media of refractive index n_1, n_2, n_3, etc. respectively, the total time taken to travel this distance will be

$$t = \frac{1}{V_0}(n_1 l_1 + n_2 l_2 + n_3 l_3 + \dots\dots) = \frac{1}{V_0}\Sigma nl.$$

nl is termed the **optical length** of a path l in medium of refractive index n. Paths of equal optical length are paths of equal time of travel.

The optical length of path is of considerable importance in dealing with the interference of light. (Chapter XIV).

16. The Fundamental Laws of Geometrical Optics

The study of optics may conveniently be divided into two parts, (1) that dealing with the physical nature of light, usually called **physical optics** and (2) that concerned chiefly with the way in which the directions in which the light travels are modified by mirrors, prisms and lenses. This is known as **geometrical optics.**

Geometrical optics is based on certain well-known facts as to the nature of light, and from the preceding work the fundamental laws of geometrical optics can now be stated. Ignoring diffraction effects due to the wave nature of light we have:

(1) Neighbouring rays of light are independent of one another.
(2) The propagation of light is rectilinear, i.e., light travels in straight lines.
(3) Law of reflection.
(4) Law of refraction.

Physical optics is considered in Chapters XII to XVI, while the remainder of the work is concerned chiefly with geometrical optics based on the above laws.

CHAPTER II

EXERCISES

1. State the law of reflection and show how is may be verified experimentally.

2. Describe Huygens' principle, and show how it may be used to explain the laws of reflection and refraction, on the wave theory.

3. Using Huygens' principle construct the reflected wave front when a parallel pencil of light is incident on a plane surface (a) normally, (b) at 30°, (c) at 70°.

4. Describe and give reasons for the appearances of the following substances:
(a) Grey paper; (b) Finely crushed glass;
(c) Paraffin wax, solid and molten.

5. Why is a lump of sugar practically opaque, while the small separate crystals of which the lump is composed are themselves transparent?

6. Explain what is meant by the refractive index of a substance. Why is it necessary to specify the colour of the light referred to in accurately stating a refractive index?

7. The velocity of light *in vacuo* is 300 million metres per second. Find the velocity (a) in water of refractive index 1·33; (b) in oil of refractive index 1·45; and (c) in dense flint glass of refractive index 1·65. (S.M.C.)

8. The velocity of yellow light in carbon disulphide is $183·8 \times 10^6$ metres per sec., in ether it is $221·5 \times 10^6$ metres per sec. Find the relative refractive index (a) for refraction from ether into carbon disulphide, (b) from carbon disulphide into ether.

9. The refractive indices of a few substances are given.
Water 1·33. Dense Flint Glass 1·65.
Crown Glass 1·50. Diamond 2·42.
Find the relative refractive index for the case of light being refracted from:
(a) Water into crown glass.
(b) Water into diamond.
(c) Diamond into water.
(d) Dense flint glass into air.
(e) Crown glass into dense flint glass.

10. Given that the speed of light in air is 3×10^{10} cm. per second and in water $2 \cdot 25 \times 10^{10}$ cm. per second, construct the emergent wave front from a small source situated in the water 10 cm. below the surface, and from this determine the position of the image.

11. State the law of refraction, and describe a method of measuring the refractive index of a block of glass having plane parallel faces.

12. Show that if a ray of light passes from a medium in which its velocity is V_1 to another in which it is V_2, $V_2 \sin i = V_1 \sin i'$ where i and i' are the angles of incidence and refraction respectively. What is the ratio $\dfrac{V_1}{V_2}$ called? (S.M.C.)

13. A poster has red letters printed on white paper; describe and explain its appearance when illuminated (a) with red light and (b) with green light.

14. Give diagrams showing the change in direction of the wave fronts of a parallel pencil on passing from air to water when the pencil is incident at (a) 30°, (b) 60° to the normal $\left(n \text{ of water} = \dfrac{4}{3} \right)$

15. Explain why objects cannot be seen through a fog, although the water particles of which fog is composed are transparent.

16. A parallel pencil of light is incident at 45° to the upper surface of oil ($n = 1 \cdot 47$) floating on water ($n = 1 \cdot 33$). Trace the path of the light in the oil and the water.

17. A parallel pencil is incident at an angle of 35° on the plane surface of a block of glass of refractive index $1 \cdot 523$. Find the angle between the light reflected from the surface and that refracted into the glass.

18. A white stone lies on the bottom of a pond. Its edges are generally observed to be fringed with colour, blue and orange. Explain this and state which is the blue edge. (L.U.)

19. State the fundamental laws of geometrical optics and describe an experimental verification of each.

20. What happens when some irregular fragments of glass are immersed in a liquid of the same refractive index? How may this device be employed to measure the refractive index of such fragments? (S.M.C.)

21. A ray incident at any point at an angle of incidence of 60° enters a glass sphere of refractive index $\sqrt{3}$ and is reflected and refracted at the further surface of the sphere. Prove that the angle between the reflected and refracted rays at this surface is a right angle. Trace the subsequent paths of the reflected ray.

III – Reflection at a Plane Surface

17. Images.—Virtual and Real

On looking into a plane reflecting surface one sees, apparently behind the surface, images of any objects that are in front of the surface. Some of the light from each object point is reflected at the surface, and enters the eye as though it were coming from points behind the surface. Apparent images of this kind from which the light is diverging are termed **virtual images.** When, by reflection or refraction at a curved surface, the light from object points is made to converge again through points, an image is formed that actually exists at the position to which the light converges. Such images, which can be received on a screen, as with the image formed by the lens of the camera or projection lantern, are termed **real images.**

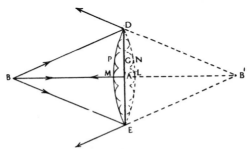

Fig. 15.—Image formed by reflection at a plane surface. Huygens' construction

18. Position of Image formed by Plane Mirror

The position of the image of any object before a plane reflecting surface can be determined from the law of reflection, or it may be found by using Huygens' construction to determine the form of the reflected wave front. Thus, in Fig. 15, B is an object point in front of a plane mirror DE. If the mirror had not been present, DLE would be the form of a wave front at a particular instant. As, however, the incident wave front meets the mirror, each point on the mirror becomes the centre of a series of wavelets travelling back at the same speed as the incident light was travelling forward. Thus the portion of the incident wave front meeting the mirror at A gives rise to wavelets that travel back to M in the same time as the incident light would have travelled to L. Similarly, the wavelets originating at G travel back to P in the same time as the incident light would have travelled to N, and so on over the whole mirror surface. The reflected wave front DME will be the envelope of all the reflected wavelets corresponding to points on the incident wave front. It can easily be proved from the figure that DME is an exactly similar curve to DLE, but in the opposite direction, and the reflected light is, therefore, diverging as though from a virtual image point B', where AB' = BA.

The same result may be obtained from the law of reflection. An object BQ (Fig. 8) is placed in front of a plane mirror DE. Rays BD and BE are reflected along DL and EM as though from an image at B', while rays QD and QE are reflected along DO and EP as though from an image at Q', B'Q' is, therefore, the virtual image of BQ, formed by the mirror. From the law of reflection and the geometry of the figure, it may easily be proved that BB' and QQ' are perpendicular to the mirror surface and that AB' = BA, GQ' = QG.

Fig. 16.—Reversion of image formed by plane mirror

Hence *the image formed by a plane mirror is situated as far behind the mirror as the object is in front, and the line joining the object and image is perpendicular to the plane of the mirror surface.* It is a simple matter, therefore, to find graphically the position of the image formed by a plane mirror, and as the reflected light is travelling apparently from the image, the path of the light by which an eye in a given position sees the image can readily be found.

It follows that in shape and size, the image will be an exact reproduction of the object. The point of the object nearest the mirror is represented by the point of the image nearest the mirror, and the top and bottom of the image correspond with the top and bottom of the object; that is, the image is erect. The question of whether the image is or is not reversed left for right, or *reverted*, as it is sometimes called, depends on the position and posture of the observer in viewing both object and image. Thus, an observer at O_1 (Fig. 16), in viewing the object BQ directly, sees B to his right; on turning round to view the image B′ Q′, B′ the image of B is now to his left. If the observer is at O_3 he will see both B and B′ to his left, but in this case he will see opposite sides of object and image; that is, if the side of the object nearer to him is called the front, he will see the back in the image. Similarly, if the light is turned through 90° by reflection, as is frequently done in optical instruments, an observer at O_2 (Fig. 17) will see the image reverted as compared with the object seen directly from O_1. Every case may be easily solved by a graphical construction of the position of the image, and the student should construct the images formed with the mirror in a number of different positions and compare the results with the images actually seen in a mirror in the various positions.

An application of the image formed in a plane mirror is frequently used in sight-testing. The test is usually made with the letter chart at a distance of six metres from the patient, but where this distance is not available, the patient is placed three metres from a plane mirror and the chart is placed beside or above him facing the mirror. The patient, therefore, sees a virtual image of the chart three metres behind the

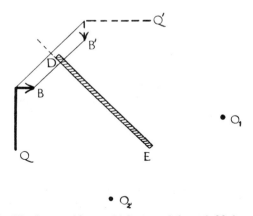

Fig. 17.—Reverted image. Light turned through 90 degrees

mirror, that is, six metres away, The letters on the chart will, of course, have to be reversed left for right if they are to appear the right way round in the image.

19. The Field of View of a Plane Mirror

As may be readily seen by experiment, the extent of image in a plane mirror that can be seen by an eye in any position will depend on the size of the mirror and the position of the eye. This extent of image seen is termed the **field of view** of the mirror.

If an eye at O (Fig. 18) is looking into a plane mirror DE, DO and EO will be the limiting directions along which the eye can see images reflected in the mirror. The reflected rays DO and EO correspond to incident rays from the object points Q and R respectively, and QR is, therefore, the extent of object in this plane of which the image, Q'R', can be seen from the position O. It will be seen that the field of view is

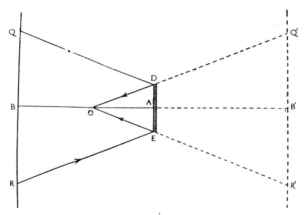

Fig. 18.—Field of view of a plane mirror

exactly the same as would be obtained if an eye situated at O was looking at an object occupying the position of the image, through an aperture in the position of, and the same size and shape as the mirror. From Fig. 18 we have

$$Q'R' = \frac{DE \times OB'}{OA} = QR$$

It will often be required in practice to find the least size and the position of a mirror in which the whole of a given object may be seen reflected from a given position. Such problems can be solved by treating the mirror as an aperture, and considering the image of the object formed behind the mirror. An example of this kind occurs in the case of the mirror used in conjunction with the sight-testing chart, § 18.

20. Deviation Produced by a Mirror

A plane reflecting surface is frequently used to produce a change in direction of a beam of light. The angle through which the beam is turned—**the angle of deviation**—will clearly depend on the angle of incidence at the mirror. In Fig. 9 a beam incident in the direction BA on a mirror DE is reflected along AF. Then $\widehat{\text{CAF}}$ is the angle of deviation and $\widehat{\text{CAF}} = 180° - 2i$.

21. Reflection from a Rotating Mirror

It can be shown that if a mirror is rotated through a given angle, a beam reflected from the mirror will be rotated through twice this angle, that is, the beam rotates twice as rapidly as the mirror. A beam incident

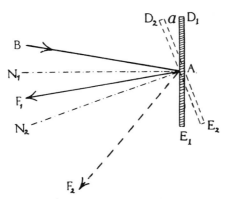

Fig. 19.—Rotating mirror. Reflected beam turned through twice the angle the mirror turns through

in the direction BA (Fig. 19) on the mirror D_1E_1, will be reflected in the direction AF_1. On the mirror being turned through an angle a to D_2E_2 the beam is reflected in the direction AF_2. Since the mirror and the normal turn through the angle a, the angle of incidence and, therefore, the angle of reflection are each increased by a, and the angle between the directions of the incident and reflected beams by $2a$. Hence $\widehat{F_1AF_2} = 2a$.

Considerable use of this fact is made in the measurement of small angular deflections. A beam of light, reflected from a mirror attached to the moving part of the apparatus, takes the place of the index or pointer, and this optical pointer turns through twice the angle the moving part turns through; also, the beam may be made of any length without adding weight to the apparatus, and thus quite small angular movements can be read with considerable accuracy. This method is frequently used in reading the small deflections of a galvanometer. A small light mirror is attached to the moving magnet, and a narrow beam of light is focused, after reflection at the mirror, in a small patch on a

Fig. 20.–Rotating mirror galvanometer

scale placed at some distance from the galvanometer. Another arrangement shown in Fig. 20 is also often used; a telescope is focused on the image of a scale formed by reflection at the mirror attached to the moving part. As the mirror turns, the image of the scale moves across the field of the telescope, and the reading is taken by means of a cross line in the telescope.

22. Deviation on Reflection at Two Mirrors in Succession

When a ray incident in a plane containing the normals of the two mirrors is reflected from two mirrors in succession, it undergoes a total deviation that depends only on the angle between the mirrors. The total deviation produced is, therefore, independent of the angle of incidence at the first mirror, and a rotation of the mirrors about an axis perpendicular to the plane containing their normals, and keeping the angle between them constant, produces no change in the final direction of the reflected light.

In Fig. 21 the total deviation of the ray

$$= C_1 \widehat{A_1 A_2} + C_2 \widehat{A_2 F}$$

$$= 2(\widehat{EA_1 A_2} + \widehat{EA_2 A_1}) = 2\widehat{A_1 EH}.$$

Therefore, total deviation $= 2 (180° - a)$ where a is the angle between the mirrors.

Successive reflection from two mirrors is used in all possible cases where considerable accuracy is required in the direction of a reflected beam. Providing the angle between the reflectors is accurately fixed, and this is frequently done by making them the surfaces of a prism

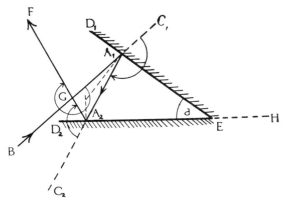

Fig. 21.—Deviation at two mirrors

(Chapter IV), the correct placing of the mirrors with respect to the incident light need not be made with any great accuracy.

A further advantage of the use of two reflectors is that the incident and reflected beams may be separated by any distance, thus making it possible to put lenses, prisms or other reflectors in either of the beams.

23. Multiple Images formed on Reflection at Two Mirrors

If an object is situated between two plane mirrors, more than the two direct images will be formed, because some of the light reflected at one mirror may be again reflected at the other mirror. The number of such images that will be seen will depend on the angle between the mirrors, the position of the object, and, to a certain extent, on the position of the eye viewing the images. The positions of the images may be found by using the graphical construction of § 18, and considering any image so found as acting as an object for the mirror of which it is in front.

Thus light from a luminous point B (Fig. 22) is reflected at the mirror D_1E giving rise to the image B'_1. Some of this reflected light reaches the mirror D_2E where it is again reflected, and as the incident light is travelling as though from B'_1, the reflected light is coming apparently from the image $B'_{1.2}$. In the same way, the light that is first reflected at D_2E travels as though from an image at B'_2, and on being reflected again at D_1E, it is coming apparently from the image $B'_{2.1}$. As both the images $B'_{1.2}$ and $B'_{2.1}$ lie behind the mirrors, the light coming as though from

33

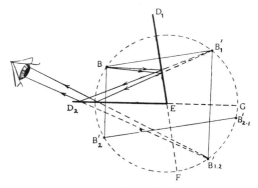

Fig. 22.—Multiple images formed by reflection at two mirrors

them does not meet the mirrors, and no further images are formed in this case. The object and each of the images will lie on a circle having its centre at E, the intersection of the mirrors. The path of the light by which an eye at O, looking into the mirror D_2E sees the image $B'_{1\cdot2}$ is shown in the figure.

The following facts concerning the number of images formed should be verified by constructing the images for a number of cases with different angles between the mirrors and different positions of the object. The results may also easily be verified by experiment.

Calling the angle D_1ED_2 between the mirrors a, the angle D_1EB, x and the angle BED_2, y, the total number of images in the series that starts with light reflected at D_1E, that is, the images B'_1, $B'_{1\cdot2}$, $B'_{1\cdot2\cdot1}$, etc., is given by the integer next higher than $\dfrac{180° - x}{a}$. The total number of images in the other series, that is, the images B'_2, $B'_{2\cdot1}$, $B'_{2\cdot1\cdot2}$, etc., is given by the integer next higher than $\dfrac{180° - y}{a}$.

The exception to this rule is when the angle a is contained in $(180° - x)$ or $(180° - y)$ an exact whole number of times. The number of images in the series, for which this is so, is given by the actual integer obtained by the division, the last image of the series in this case falling on the projection of the mirror, EG or EF. When the angle a is divisible an integral number of times into $180°$, the last images of each series coincide, and therefore the total number of images is $2\left(\dfrac{180°}{a}\right) - 1$.

A well-known application of the multiple images formed with inclined mirrors is the instrument known as the **Kaleidoscope**. In its simplest form it consists of two long strips of plane mirror, usually inclined to each other at an angle of 60° and mounted in a tube. An object is placed between the mirrors at one end and, on looking into the other end of the tube, one sees a symmetrical pattern, the sectors of which are formed by the object and its various images. If the angle between the mirror is 60° a six-sided pattern is formed, if the angle is 45° an eight-sided pattern, and so on. The kaleidoscope, while being a well-known toy, is also used commercially in various branches of design work requiring symmetrical patterns, as in the design of pottery, wallpaper, etc.

24. Two Parallel Plane Mirrors

If an object B (Fig. 23) is placed between two plane mirrors, that are facing each other and parallel, the light will be reflected backwards and forwards between the mirrors giving rise to the images B'_1, $B'_{1\cdot2}$, etc. behind the mirror D_1E_1 and the images B'_2, $B'_{2\cdot1}$, etc. behind the mirror D_2E_2, each image acting as an object for the mirror it is facing. It is clear that as the mirrors are parallel, the images and the object will be on a straight line and

$$A_1B'_1 = BA_1 \qquad A_2B'_2 = BA_2$$

$$A_1B'_{1\cdot2} = B'_2A_1 \qquad A_2B'_{2\cdot1} = B'_1A_2$$

$$A_1B'_{1\cdot2\cdot1} = B'_{2\cdot1}A_1 \qquad A_2B'_{2\cdot1\cdot2} = B'_{1\cdot2}A_2$$

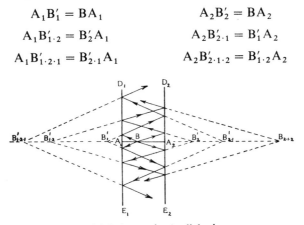

Fig. 23.—Multiple images in parallel mirrors

Theoretically, the number of images will be infinite, but, as only a certain fraction of the incident light is reflected each time, each successive image is fainter than the preceding one, and the number of images seen will depend on the brightness of the original object.

25. Reflection as a Means of Erecting an Inverted Image

Reflection from a number of plane surfaces is often used as a means of erecting the inverted real image formed by a lens; the commonest application of this is in the prism binocular. Each half of the binocular consists essentially of a telescope of the astronomical type, which as explained in Chapter X produces an image that is inverted in all directions. In the telescope, the light is reflected at two pairs of plane reflectors, the surfaces in each pair being at right-angles, and the common edge of the surfaces in the one pair being perpendicular to the common edge of the other pair. After these four reflections, the light

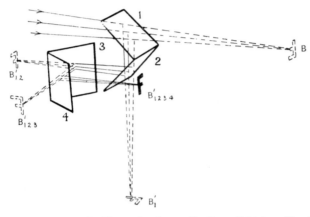

Fig. 24.—Complete reversal of image by four reflections, light travelling in original direction. Prism binocular principle

is travelling parallel to its original direction and the image has been completely reversed as shown in Fig. 24.

B is the real inverted image formed by convergent light from the first lens—the objective—of the telescope. After reflection at surface 1, the light is converging towards the image B'_1, and after a further reflection at surface 2, towards the image $B'_{1.2}$. It will be seen that the image has now been reversed up and down as compared with the original image B. The light is then intercepted by a second pair of reflecting surfaces, 3 and 4, having their common edge perpendicular to that of the first pair. These two further reflections give rise in turn to the images $B'_{1.2.3}$ and $B'_{1.2.3.4}$, and a left for right reversal is produced. The result of the four reflections is, therefore, a reversal of the image in two perpendicular planes, and in this way the inversion of the original image formed by the lens is corrected.

An additional advantage is obtained by the "folding-up" of the light path causing a considerable reduction in the length of the instrument. In the actual instrument, the reflectors consist of two 90° prisms.

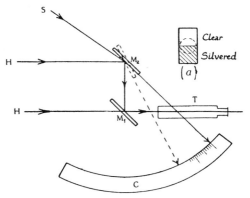

Fig. 25.—Principle of the sextant

26. The Sextant

Another instrument that depends on the successive reflection at two mirrors is the **Sextant** (Hadley 1682–1744), one of the most important nautical instruments. When it is required to measure the angle between two distant objects, as in surveying, the usual method is to employ some form of sighting device, such as a telescope, attached to a divided circle, and to measure the angle turned through by the telescope when directed to the two objects in turn. Such a method is, of course, impracticable when the measurement is to be made from a moving object, as, for example, in the measurement of the angle of elevation of the sun or a star from a moving ship, and it is for measurements of this kind that the sextant is commonly used.

The principle of the sextant is shown in Fig. 25, T is a telescope in front of which is fixed a mirror M_1 known as the *horizon glass*, half of which is silvered and the other half clear glass, as shown in Fig. 25a. The position of the telescope is such that it receives light from each half of this horizon glass. A second mirror M_2 known as the *index glass*, can be rotated and its position read by an index arm on a circular scale C. When the two mirrors are parallel, the telescope is receiving light from a distant object directly through the clear half of the horizon glass and also by reflection at the two mirrors, and a single image is seen in the telescope. The index attached to the index glass should then be reading zero. To measure the angle between two objects, say the sun and the horizon, the telescope is directed to the horizon, and the index mirror

37

turned until the light from the sun, after reflection from the two mirrors, also travels along the telescope axis. An image of the sun and of the horizon are then seen in coincidence in the telescope.

As the light is turned through twice the angle turned through by the mirror M_2, the angle SM_2H between the two objects, the angle of elevation of the sun, will be twice the angle turned through by the index glass from its zero position. To facilitate reading, the divisions on the scale are numbered with double their actual value; that is, half degrees are marked as degrees, and so the value of the angle SM_2H can be read directly. The scale is usually divided from 0° to 140° in 10′ divisions and is read by vernier to 10″.

CHAPTER III
EXERCISES

1. State the law of reflection. Show that the line joining an object and its image formed by a plane mirror is bisected at right angles by the mirror surface.

2. Describe, with a diagram, an experiment to prove that the image formed by a plane mirror is as far behind the mirror as the object is in front.

3. At 50 cm. from a plane mirror, 20 cm. by 15 cm. in size, and on a normal to its middle point, is situated a small source of light. What will be the size of the patch of reflected light received on a screen 2 metres from and parallel to the mirror?

4. A plane mirror 2 feet high is fixed on a wall of a room with its lower edge 3 ft. above the floor. If a man, whose eyes are 6 feet above the floor, stands directly in line with the centre of the mirror and 3 feet from it, what will be the length of floor that he can see by reflection?

5. A sight-testing chart, measuring 120 cm. by 80 cm., the longer dimension being vertical, is to be viewed by reflection in a plane mirror. Find the smallest size of mirror that can be used if the chart is 3·25 metres and the observer 2·75 metres from the mirror. If the observer's eye is 1·2 metres and the lower edge of the chart 1·5 metres above the ground, what must be the height of the bottom edge of the mirror?

6. Show that the length of the smallest mirror in which a person may see the whole of himself reflected is half the person's height.

7. On one wall of a room is a picture, 6 feet by 4 feet, and on the opposite wall 15 feet away is a plane mirror. What is the least size of the mirror in which a person 8 feet away may see the whole of the picture reflected?

8. Two plane mirrors are placed together with their edges touching so as to be approximately in the same plane, and an observer standing 8 feet away sees two images of his face which are just touching one another. If the width of his face is 6 inches, what is the angle between the mirrors? (S.M.C.)

9. A plane mirror is set up vertically facing a house 35 feet high and 25 feet wide and 100 feet away from it. What size and how high must the mirror be in order that a man standing 20 feet in front of it and whose eyes are 5 feet 3 inches above the ground may just be able to see the whole of the house reflected in the mirror? (S.M.C.)

10. What must be the minimum size of a rectangular plane mirror on one wall of a room which will enable a person standing in the middle of the room to see the whole of another person 5 feet 6 inches high and 2 feet wide entering a doorway behind him?

11. On one wall of a room 6 metres long hangs a picture of horizontal width AB = 24 cm. On the opposite wall hangs a plane mirror. Seated at a distance of 3 metres from the mirror, a man views a reflection of the picture; the side A of the picture is to his right and B to his left. If the separation of his eyes (P.D.) is 6 cm., find the least horizontal width of mirror in order that he may just see the total width of the picture with *each* eye.

If C is the middle point of AB, what portion or portions of the mirror width are required in order that the right eye can see only AC and the left eye only BC? (S.M.C.)

12. What will be the rotation of a mirror reflecting a spot of light on to a straight scale 1 metre from the mirror if the spot of light moves through 5 cm.?

13. Show that the images of an object placed between two inclined mirrors all lie on a circle.

14. Show how you would arrange an experiment to prove that when parallel light falls upon a plane mirror which is turned through an angle, the reflected beam turns through twice that angle. What conclusion can be drawn from this experiment? (S.M.C.)

15. Explain how an image is formed by reflection at a plane surface. In what respect do the object and image differ one from the other? How would you find by experiment the position of the image? (L.U.)

16. Show how by means of two plane mirrors a person standing in front of one of them can see the image of the back of his head. Trace the path of the light from the back of his head to his eye, and show clearly where the image is formed.

17. Account for the formation of multiple images by two plane mirrors hanging on adjacent walls of a room, and show in plan the object and images. (L.U.)

18. A kaleidoscope is made with three plane mirrors each 2½ inches wide, and a small square object is viewed by it. Draw a diagram showing the appearance that would be seen. (S.M.C.)

19. Two vertical plane mirrors are fixed at right angles to one another with their edges touching and a person walks past the combination on a straight path at 45° to each mirror. Show how his image moves. (S.M.C.)

20. It is required to turn a beam of light through 80° by means of mirrors; explain, giving diagrams, three ways in which this may be done. State the advantages and disadvantages of each method.

21. Explain the formation of an image by reflection in a plane mirror.

Two plane mirrors are placed so that each makes an angle of 45° with the other, and a luminous object is placed between them. Determine the position of all the reflected images. (B.O.A.)

22. Two plane mirrors are fixed together with their common edge vertical and their normals pointing North and North-East respectively; what deviation will a horizontal beam undergo after reflected at both mirrors? Find the final direction of a beam incident on the first mirror in a direction 15° North of East.

23. Two plane mirrors are placed 1 metre apart with faces parallel. An object placed between them is situated 40 cm. from one mirror; find the positions of the first six images formed by the mirrors.

24. Two mirrors are inclined at 35°. At what angle must a ray be incident on the first mirror so that after reflection at the two mirrors it returns along its own path?

25. Two plane mirrors are inclined at an angle of 60° to one another, and a small

object is placed between them and 1 inch from one mirror. Show on a diagram the position of each of the images and the path of the light by which an eye sees the last image formed. What practical application is made of this arrangement?

26. Prove that if two plane mirrors are inclined to one another at an angle x then any ray, after reflection once at each mirror, is turned through an angle of $(360 - 2x)$.

By treating the case of two plane mirrors inclined at $30°$, prove the following statement:

If two plane mirrors are inclined to one another at an acute angle x degrees, an incident ray parallel to one of them will retrace its original path after $\left(\dfrac{180}{x} - 1 \right)$ reflections at the two mirrors when this quantity is a whole number: and the incidence is normal at the $\left(\dfrac{90}{x} \right)$th reflection. (S.M.C.)

27. Explain with a diagram the principle of the sextant. How may the altitude of the sun above the horizon be found when the horizon is not visible?

28. Two objects 60 yards apart are observed by means of a sextant from a position on a line perpendicular to that joining the two objects. In order that the two objects may be seen in coincidence the rotating mirror of the sextant has to be rotated through $5°$; what is the distance away of the objects?

29. How would you arrange two mirrors so as to be able to see the side of your head when looking straight forward? Give a diagram showing the path of the light.

30. A small source of light S lies midway between the observer's eye E (considered as a point) and a plane mirror, the line joining the eye and source being perpendicular to the mirror and intersecting it at C. The mirror is tilted about the point C through a small angle, say $2°$. Find approximately the inclination to the line ESC at which the light now enters the eye, after reflection at the mirror. Draw a scale diagram showing the path of the light. (S.M.C.)

IV - Refraction
at a Plane Surface

27. Refraction by a Plane Parallel Plate

It follows from the law of refraction, that if a parallel pencil of light is incident normally on a plane surface separating two different media, the light will continue to travel in the new medium without change of direction. If a second surface is parallel to the first, the light reaches this normally, and again suffers no change in direction. Thus, light will be unchanged in direction on passing normally through a plane parallel plate of different refractive index from the surrounding medium, such, for example, as a parallel plate of glass in air.

When, however, the light is incident obliquely on such a plate, its direction will be changed both on entering and leaving and the directions can be found from the law of refraction. In Fig. 26 $D_1E_1D_2E_2$ is a plate having a refractive index $n'_1 = n_2$; the medium in front of the plate

Fig. 26.—Refraction at plane parallel plate

has a refractive index n_1, and that behind the plate a refractive index n'_2. Then, if BA_1 is the direction of the incident light, i_1 is the angle of incidence and i'_1 the angle of refraction at the first surface, i_2 and i'_2 are the angles of incidence and refraction at the second surface and

$$n'_1 \sin i'_1 = n_1 \sin i_1$$
$$n'_2 \sin i'_2 = n_2 \sin i_2$$

As the surfaces are parallel, $i_2 = i'_1$.

It follows that, when the media on the two sides of the plate are equal, that is, $n_1 = n'_2$, as in the case of a plate of glass in air, $i_2 = i'_2$, and the light emerges parallel to its original direction. It will however have suffered a lateral displacement GC, the extent of which will depend on the thickness and refractive index of the plate and on the angle of incidence at the first surface. As the light emerges from the plate parallel to its original direction, there will be no change in the apparent position of a *distant* object seen through the plate.

The apparent displacement of a *near* object or image, when it is viewed through a thick parallel plate of glass and the plate is rotated about an axis perpendicular to the line of sight, has been made use of in measuring the size of small objects—*parallel plate micrometer* (Clausen 1841)—and two such plates at an angle to one another were used by Helmholtz as the doubling device in his *ophthalmometer* for measuring the curvature of the cornea (1854).

Fig. 27.—Refraction by a series of parallel plates

28. Refraction by a Series of Parallel Plates

When light traverses a number of media in succession, as in Fig. 27,

$$n_1 \sin i_1 = n'_1 \sin i'_1$$
$$n_2 \sin i_2 = n'_2 \sin i'_2$$
$$n_3 \sin i_3 = n'_3 \sin i'_3$$
$$n_4 \sin i_4 = n'_4 \sin i'_4$$
$$\text{and } n'_1 = n_2, n'_2 = n_3, \text{ etc.}$$

If the surfaces are parallel, $i'_1 = i_2, i'_2 = i_3$, etc., and substituting these values in the above expressions, we have

$$n_1 \sin i_1 = n_2 \sin i_2 = n_3 \sin i_3 = n_4 \sin i_4 = n'_4 \sin i'_4$$

Thus it will be seen that the final direction of the light is the same as if it had undergone a single refraction at a surface separating the first and last media, and, if these are of equal refractive index, the final direction of the light is parallel to its original direction. There is, of course, a lateral displacement.

This fact, that intermediate parallel plates of different refractive index produce no change in the final direction of the light, is applied in certain cases of refractometry (§ 42), when a film of liquid is placed between a standard glass and the solid to be measured.

29. Reversibility of Optical Path

An important general law in optics is the **principle of the reversibility of the optical path.** It may easily be shown that if the light in Fig. 27 is reversed in direction, as, for example, by being reflected normally at a plane mirror, it will follow exactly the same path in a reversed direction. This is true of the light passing through any system, no matter how complicated, and problems may often conveniently be solved by considering the light as travelling in the opposite direction to which it is in the actual case.

30. Critical Angle and Total Reflection

Light on passing from a denser to a rarer medium is refracted away from the normal to the surface, and the angle of refraction in the rarer

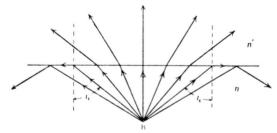

Fig. 28.—Refraction at a plane surface, $n' < n$. critical angle and total reflection

medium increases more rapidly than the angle of incidence. For a certain angle of incidence the angle of refraction will be 90°, that is, the refracted light will be just grazing the surface (Fig. 28). This angle of incidence is termed the **critical angle,** i_c, between the two media. Light meeting the surface at a greater angle of incidence than the critical angle cannot therefore enter the second medium, and the whole of the light will then be reflected at the surface. This is known as **total reflection.** The direction of this totally reflected light will be in accordance with the ordinary law of reflection. It must be remembered that some light is always reflected at the bounding surface between two media, and the amount reflected gradually increases as the angle of incidence increases, so that there is not a complete change in condition at the critical angle. The increase in the amount of the reflected light when the critical angle is exceeded is nevertheless considerable.

In Fig. 28 light is travelling from a medium of refractive index n to one of refractive index n', where n is greater than n', and

$$n \sin i = n' \sin i'$$

When i is the critical angle, i' is $90°$ and

$$n \sin i_c = n' \sin 90° = n'.$$

Hence
$$\sin i_c = \frac{n'}{n} \quad \dots \qquad \dots \qquad \dots \qquad \dots \quad (4.01)$$

The refractive indices and therefore the critical angle between two media will, of course, depend on the colour of the light. The critical angle will be smaller for blue light than for red, and in the case of incident white light, the red light may be passing into the second medium, while the other colours are totally reflected.

The following table gives the critical angle for yellow light of a few transparent substances when in contact with (a) air and (b) water.

	Refractive Index n_D	Critical Angle (in contact with Air)	Critical Angle (in contact with Water)
Water	1·3336	48° 34′	——
Crown Glass	1·5175	41° 14′	61° 32′
Light Flint Glass	1·5760	39° 23′	57° 46′
Dense Flint Glass	1·6214	38° 5′	55° 21′
Diamond	2·4173	24° 26′	33° 30′

The characteristic brilliancy of the diamond is due to its very high refractive index and therefore small critical angle. The stone is cut in such a way that a very large part of the light entering the stone is totally reflected and passes out through the *table*, the large plane surface of the stone.

Another interesting example of total reflection is seen in the case of the **mirage,** of which there are two distinct forms. One of these, which is much the commoner, is seen over land, and the other is occasionally seen at sea. The first form occurs in hot calm weather, especially over large expanses of sand, when inverted images of distant objects are seen reflected as though in a pool of water. This phenomenon is produced by the air close to the ground being hotter and therefore of lower refractive index than that a little higher up. Light travelling downwards from any point on an object is gradually refracted away from the normal, as it passes through air of decreasing refractive index, until near the ground it is totally reflected at a layer of low refractive index and enters the eye as though from an image below the surface of the ground. The reflection of the sky, trees, etc., gives the impression of the presence of a lake of water. This form of mirage may often be seen when motoring on an asphalte coated road, the surface of the road a hundred yards or so ahead appearing to be covered with water although actually dry.

The second type of mirage is seen at sea and most commonly in Arctic regions. In this case the air close to the sea is cold and therefore of comparatively high refractive index, becoming warmer and of lower refractive index higher up. Light travelling upwards from an object, such as a ship, is gradually refracted away from the normal until it is totally reflected downwards, and an eye receiving this reflected light sees an inverted image of the ship apparently floating in the sky.

Total reflection has many practical applications, both in replacing the reflection from a silvered surface over which, in many cases, it has definite advantages, and as the basis of methods of measuring refractive index. As these applications of total reflections in most cases involve the use of prisms they will be considered later.

31. Graphical Construction for Refraction

Problems involving the law of refraction are often conveniently solved by graphical construction, and several different constructions have been suggested. The following is easily carried out, and if carefully drawn will give sufficient accuracy for many practical problems.

Construction—With the point of incidence A (Fig. 29a) as centre, describe two circles having radii in the ratio of n to n'. The incident ray BA cuts the circle corresponding to medium n at B; from B draw a line parallel to the normal at the point of incidence, meeting the other circle at G. Join GA, then GA produced is the direction of the refracted ray.
Proof.

$$\frac{BM}{AB} = \sin i, \frac{GL}{AG} = \sin i'$$

$$GL = BM \quad \text{(from construction)}$$

45

Therefore

$$\frac{\sin i}{\sin i'} = \frac{BM}{AB} \bigg/ \frac{GL}{AG} = \frac{AG}{AB} = \frac{n'}{n}.$$

The same construction may be used to determine the critical angle between two media (Fig. 29b). In this case, as the refracted light in the medium of lower refractive index n' is grazing the surface, the construction line BG must be tangential to the smaller circle and i is then the critical angle. Sometimes, when using the construction for light travelling

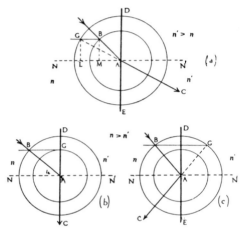

Fig. 29.—Graphical construction for refraction

from a denser to a rarer medium, it may be found that the construction line does not meet the smaller circle (Fig. 29c). This is a case of total reflection, and on producing the construction line to meet the larger circle again at G, GA produced is, of course, the direction of the reflected light.

The construction may be applied equally well to refraction at curved surfaces; in these cases the normal to the surface at the point of incidence will be a radius of the curve. A more convenient construction for curved surfaces is described in Chapter V.

32. Image Formed by Plane Refracting Surface

When all the rays of a pencil are diverging from or converging to one point, the pencil is said to be **homocentric.** In order that a perfect image may be formed, it is necessary that the homocentric pencils from points on the object shall still be homocentric after reflection or refraction. Except in reflection at a plane mirror, this ideal condition will never be realised and homocentric pencils in the object space are no longer homocentric in the image space. When however the pencils are very narrow, the rays in the image space may be considered, in many cases, to be diverging from or converging to a point with sufficient accuracy for a reasonably good image to be formed.

In Fig. 30, B represents a luminous point in a medium of refractive index n, which is separated from a second medium of index n', smaller than n, by the plane surface DE. Light diverges from B, and is refracted on passing into the second medium; the direction of the refracted rays

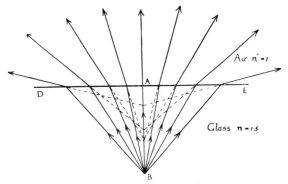

Fig. 30.–Light from object point in glass refracted on passage into air

Fig. 31.–Narrow pencil refracted from dense to rarer medium

may be calculated, or found by the graphical methods described above. It will be found that the refracted light is no longer travelling as though from a point. While all rays meeting the surface at equal angles travel,

47

after refraction, as though coming from a point on the normal to the surface through the object point, the greater the obliquity of the rays, the nearer to the surface will be the point from which the refracted rays are apparently coming.

As the pupil of the eye is small, the narrow pencils entering an eye, looking normally through the surface, will be reasonably homocentric and a fairly good image will be seen. A narrow pencil from B (Fig. 31) will, after refraction, appear to be coming from B', and as the light travelling along BA normally to the surface is unchanged in direction, B' must lie upon BA. As the angles are small,

$$\frac{\sin i'}{\sin i} = \frac{\tan i'}{\tan i} = \frac{AB}{AB'} = \frac{n}{n'}.$$

Calling AB and AB', l and l' respectively, we have

$$\frac{l'}{n'} = \frac{l}{n} \qquad \dots \qquad \dots \qquad \dots \qquad \dots \quad (4.02)$$

and when the second medium is air, $n' = 1$,

$$l' = \frac{l}{n}$$

Thus objects in a denser medium, when viewed from a rarer medium, appear closer to the surface than they actually are, a well-known appearance in the case of objects in water or bubbles in a block of glass.

33. Microscope Method of Determining Refractive Index

The principle explained in § 32 may be applied to the measurement of the refractive index of a transparent substance. A microscope, fitted with a scale and vernier for reading the position of the microscope body, is focused on an object in the usual way and the position of the microscope body is read on the scale. The glass or other substance to be measured, in the form of a block with plane parallel polished surfaces, is now placed on the stage in contact with the object, and the microscope is racked up until the object seen through the block is focused; the position is read on the scale. Finally the microscope is focused on a mark on the upper surface of the block, and a further reading taken. The difference between the first and third readings then gives the actual thickness of the block, l, and the difference between the second and third readings gives its apparent or *reduced* thickness, l'. Then the refractive index of the block is $n = \dfrac{l}{l'}$.

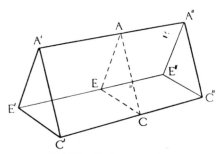

Fig. 32.—Refracting prism

PRISMS

34. Introduction

By a prism in optical work is meant a transparent substance bounded by plane polished surfaces inclined to one another. In the simplest form of prism, as used for refraction, only two surfaces, the **refracting surfaces,** need be considered, the light entering the prism at the first and leaving at the second. In the case however of the various reflecting prisms, there are a variety of forms having three or more polished surfaces at which the light is refracted or reflected.

Fig. 32 shows a simple refracting prism with refracting surfaces A'C'C''A'' and A'E'E''A''. The line of intersection of these two surfaces is the **refracting edge** or **apex** of the prism. Any section through the prism perpendicular to the edge, such as ACE, is a **principal section,** and the angle CAE of such a section is the **refracting** or **apical angle.**

35. Path of Light through a Prism

A ray BD (Fig. 33) incident in a principal section of a prism ACE will remain in the same plane after refraction, and its path may be traced by

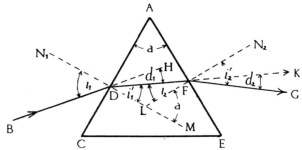

Fig. 33.—Path of light through a prism

49

considering the refraction at each surface in turn. We shall consider only the case of a prism of refractive index n surrounded by air. Then we have the following:—

at first surface, $\qquad \sin i'_1 = \dfrac{\sin i_1}{n} \qquad$ ... $\qquad$ ... $\qquad$ ... (4.03)

at second surface, $\qquad \sin i'_2 = n \sin i_2 \qquad$ ... $\qquad$ ... $\qquad$ ... (4.03A)

and from the geometry of the figure

$$i'_1 + i_2 = a \qquad \text{...} \qquad \text{...} \qquad \text{...} \qquad (4.04)$$

where a is the refracting angle of the prism.

The light is deviated through an angle d_1 at the first surface and d_2 at the second surface and from the figure

$$d_1 = i_1 - i'_1; \quad d_2 = i'_2 - i_2$$

then the total deviation

$$d = d_1 + d_2 = i_1 - i'_1 + i'_2 - i_2$$
$$= i_1 + i'_2 - a \text{ ...} \qquad \text{...} \qquad \text{...} \qquad \text{...} \qquad \text{...} \qquad (4.05)$$

It will be seen that when the prism is of higher refractive index than the surrounding medium, as with a prism in air, the light is always deviated away from the refracting edge.

Example: Refracting angle of prism 60°, $n = 1 \cdot 53$, angle of incidence at first surface, $i_1 = 45°$.

$$\sin i'_1 = \frac{\sin i_1}{n} = \frac{0 \cdot 7071}{1 \cdot 53} = 0 \cdot 4620$$

$$i'_1 = 27° \; 31'.$$

$$i_2 = a - i'_1 = 32° \; 29$$

$$\sin i'_2 = n \sin i_2 = 1 \cdot 53 \times 0 \cdot 5370 = 0 \cdot 8216$$

$$i'_2 = 55° \; 15'$$

Therefore

$$d = i_1 + i'_2 - a = 40° \; 15'.$$

36. Total Reflection in a Prism

In certain cases, the angle of incidence at the first surface of a prism will be such that the light meets the second surface at an angle greater than the critical angle, and is therefore totally reflected. This reflected light will meet a third surface of the prism and, if this is a plane polished surface, the light will emerge or again be totally reflected. Its direction after leaving the third surface may easily be found. It should be noted that total reflection will occur when the value obtained from expression (4.03A) is greater than unity.

For any given prism there will be a limiting angle of incidence at the first surface at which light can pass through the prism, and for a given glass there will be a maximum limiting value of the angle between the faces of the prism through which any light can pass. These limiting values may be found as follows: As the greatest possible angle of

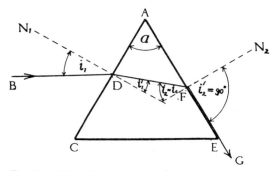

Fig. 34.–Refraction by a prism. Grazing emergence

emergence i'_2 is 90° (Fig. 34) i_2 is the critical angle i_C between the glass and air, then

$$i'_1 = a - i_c$$
$$\text{and } \sin i_1 = n \sin (a - i_c)$$

this value of i_1 gives the smallest angle of incidence for a ray that will pass through the prism; any ray closer to the normal will be totally reflected at the second surface. When the angle a is smaller than i_C the

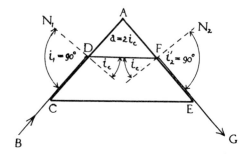

Fig. 35.–Maximum angle of refracting prism. $a = 2i_c$

limiting direction of the incident ray lies on the other side of the normal to that shown in Fig. 34. The greatest angle that a prism may have for light to pass through it will be such that the incident and emergent light is grazing the surface as in Fig. 35.

$$\text{Then } i_1 = i'_2 = 90°$$

$$\text{and } i'_1 = i_2 = i_C$$

$$\text{Therefore, } a = 2 \, i_C$$

Hence *no light can pass through a prism, the refracting angle of which is greater than twice the critical angle.* For a prism of crown glass in air, the refracting angle must not exceed about 82°.

37. Minimum Deviation

A special and important case of refraction by a prism occurs when the prism is in such a position that the light passes through it symmetrically, that is, when the angle of incidence at the first surface equals the angle of emergence at the second. It has been shown in § 35 that the deviation produced by a given prism depends on the incident angle of the light, and it may be shown experimentally or by tracing a number of rays through a given prism that, when the incident and emergent angles are equal, the deviation is a minimum. The prism is then said to be in its position of **minimum deviation.** An exact theoretical proof is beyond the scope of this book, but the student should calculate the deviation produced by a given prism for various angles of incidence and plot the deviations against the angles of incidence.

It will be found experimentally that there is only one position of a prism in which the deviation is a minimum, and this fact provides an experimental proof that minimum deviation only occurs when the angles of incidence and emergence are equal. For if these angles were not equal, from the principle of reversibility of path, there would be two positions of minimum deviation, the angle of incidence in the first position being the angle of emergence in the second and vice-versa.

In the position of minimum deviation, as the path of the light is symmetrical, we have

$$i_1 = i'_2$$

$$i'_1 = i_2 = a/2$$

$$d_1 = d_2 = d/2 = i'_2 - i_2 = i'_2 - a/2$$

Therefore
$$i'_2 = \frac{a}{2} + \frac{d}{2}$$

$$\sin i'_2 = n \sin i_2$$

Hence
$$\sin\left(\frac{a+d}{2}\right) = n \sin\left(\frac{a}{2}\right) \qquad \dots \qquad \dots \quad (4.06)$$

or
$$n = \frac{\sin\left(\frac{a+d}{2}\right)}{\sin\left(\frac{a}{2}\right)} \quad \dots \quad \dots \quad \dots \quad \dots (4.06\text{A})$$

Expression (4.06A) is of considerable importance as it is the basis of the standard method of determing the refractive index of glass. The sample of glass to be measured is made into a prism with refracting angle of about 60°, and the refracting angle and the angle of minimum deviation for light of a given colour are carefully measured. These measurements are made on the instrument known as the **spectrometer,** which is described in Chapter XIII.

Example: The same prism as above, $a = 60°$, $n = 1·53$.

When the prism is in the position of minimum deviation $i'_1 = i_2$

$$i'_1 + i_2 = a \text{ therefore } i'_1 = i_2 = 30°$$
$$\sin i_1 = n \sin i'_1 = 1·53 \times 0·5 = 0·765$$
$$i_1 = i'_2 = 49° 54'$$

Hence minimum deviation d

$$= i_1 + i'_2 - a = 39° 48'.$$

38. Normal Incidence and Emergence

When light enters the prism normally, $i_1 = 0°$, no deviation takes place at the first surface, and therefore $i_2 = a$. The total deviation produced

$$d = i'_2 - i_2 = i'_2 - a$$

or
$$i'_2 = a + d$$

Then as
$$\sin i'_2 = n \sin i_2$$
$$\sin (a + d) = n \sin a \quad \dots \quad \dots \quad \dots \quad (4.07)$$

As the path of the light is reversible, this expression applies equally to the case of light emerging normally at the second surface.

39. Ophthalmic Prisms

Prisms are frequently used in ophthalmic work for the measurement relief and correction of muscular defects of the eyes. These prisms are of comparatively small refracting angle, rarely exceeding 15°, and are generally placed normally to either the incident or emergent light. As for small angles the sines are very nearly proportional to the angles, expression (4.05) may, for *thin* prisms, be written

$$d = ni'_1 + ni_2 - a$$
$$= na - a$$
$$d = (n - 1)a \qquad \ldots \qquad \ldots \qquad \ldots \qquad \ldots \qquad (4.08)$$

Ophthalmic prisms are usually numbered in terms of the deviation produced; this may be expressed in degrees of deviation or, more conveniently in this class of work, in units which give the lateral displacement of a beam at a given distance. The unit generally adopted is the angle whose tangent is $\frac{1}{100}$, and the deviation is measured on a scale, the divisions of which are one hundredth of the distance between the scale and the prism. This angle is called the **prism dioptre** and is denoted by $\triangle$*.

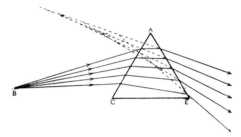

Fig. 36.—Light from a near object point refracted by prism

40. Images seen through Prisms

Different portions of the light diverging from a near object point B (Fig. 36) will meet the prism at different angles and, as the deviation produced by a given prism is dependent upon the angle of incidence, the various rays of the pencil will be deviated different amounts. The light leaving the prism is therefore no longer proceeding as though from a point, and a perfect image of the object point will not be formed. As, however, the pencils entering an eye are narrow, there will be little difference in the deviations of the rays in these narrow pencils, and the light entering the eye will be coming approximately from a point. This is particularly the case when the prism is in the position of minimum deviation, as changes of incident direction near the minimum deviation position produce small changes in deviation. As the light is deviated away from the apex of the prism, the eye sees the image apparently displaced towards the apex as in Fig. 37 and along a line perpendicular

*For further particulars of ophthalmic prisms see EMSLEY and SWAINE—
Ophthalmic Lenses.

to the refracting edge. This line, in the case of ophthalmic prisms, is known as the **base-apex line** of the prism.

On looking through a prism two other effects will usually be noticed. Unless the light from the object is monochromatic, that is, of one colour only, it will be dispersed by the prism into its component colours, the violet light being the most deviated and the red the least. Thus a small

Fig. 37.—Image seen through prism

white object appears, when seen through the prism, as a spectrum with the violet end the most deviated. A larger white object appears fringed with violet or blue on one edge and with red on the other, the colours from intermediate points on the object overlapping and giving white.

The other effect is seen on looking at an object such as a line parallel to the edge of the prism. Light from only one point on the line will be passing to the eye through a principal section of the prism and, as the effective refracting angle in an oblique section of the prism is greater than the angle in a principal section, the light in any oblique direction

Fig. 38.—Appearance of image of horizontal bar seen through prism

is deviated more than that in a principal section. Therefore if a horizontal line is viewed through a prism held with its principal section vertical and refracting edge upwards, the image will be displaced upwards, and distorted to a curve as shown in Fig. 38, the light from the ends of the line being deviated more than that from the centre.

41. Reflecting Prisms

The use of prisms for purposes of reflection often has definite advantages over the use of the silvered reflecting surface. The ordinary mirror silvered on the back surface cannot generally be used in optical instruments owing to the multiple reflections that take place between the two surfaces, while the mirror formed by silvering the front surface of a glass plate is delicate and because of its tendency to tarnish needs frequent resilvering. Front surface mirrors are now usually produced by aluminizing a glass surface. The aluminium surface is much more robust than the silver and does not tarnish. Stainless steel mirrors are also used, but the amount of light reflected is less than with silver or aluminium. When conditions allow of the use of total reflection a better reflection is obtained than with a silver surface, without the above drawbacks. The making of the reflecting surfaces in the form of a prism ensures that in those cases where two or more successive reflections are required the angle between the reflecting surfaces remains fixed. Even in those cases where total reflection is not possible and the prism surfaces are silvered, as in the end prisms of the rangefinder, the surfaces can be so arranged that there is no trouble caused by the reflections from the unsilvered surfaces.

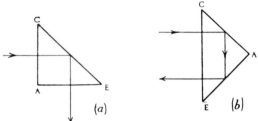

Fig. 39.—Totally reflecting prisms

Numerous forms of reflecting prisms are used in various optical instruments, and a few of the more important types will be considered here. In Fig. 39a ACE is the section of a prism having angles of 45° at C and E. Light falling normally on the surface AC is incident on the surface CE at an angle of 45° and, this being greater than the critical angle between glass and air, the light is totally reflected and emerges normally from the surface AE, having been deviated through 90°. The same form of prism is also used to deviate a beam through 180° by means of two reflections (Fig. 39b), the light being totally reflected at the faces AC and AE. Such prisms are used in the prism binocular, § 25.

A 90° reflecting prism may be employed as an **erecting prism,** as in Fig. 40. A prism of this form usually known as a **Dove prism** is sometimes used with the projection lantern for erecting the inverted image of an object that cannot be placed in the lantern in an inverted position. If the erecting prism is rotated about an axis passing through the faces

AC and AE and parallel to its hypotenuse face, the image will be rotated through twice the rotation of the prism.

Fig. 40.—Erecting or Dove prism

Fig. 41.—Pentagonal prism

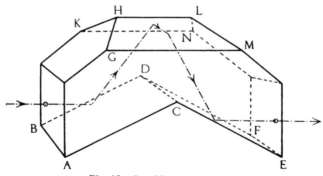

Fig. 42.—Roof Prism, Abbe form

When a constant deviation of 90° is required a prism of the form shown in Fig. 41 is often used. The reflecting surfaces CD and FE are at an angle of 45° to one another and the light is deviated through 270°, which gives the same result as a deviation of 90°. As the angle of incidence at the reflecting faces is less than the critical angle, these surfaces have to be silvered. This form of prism, which is known as the **pentagonal prism,** is used as the end prisms of the rangefinder and as one form of optical square used in surveying.

By a suitable arrangement of the surfaces, the complete reversal of an image, both up and down and left for right can be effected with a single prism. One such form is shown in Fig. 42. Reflection at the surfaces ABCD and CDEF gives a reversal of the image up and down, while reflection at the "roof" surfaces HGLM and HKLN produces a left for right reversal. Such prisms, of which there are many forms, are known as **roof prisms,** the two surfaces at the top of the prism illustrated in Fig. 42 forming the roof. The roof surfaces must always be perpendicular to one another.

42. Refractometers

The refractive index of a substance may be determined by measuring the critical angle between the substance and a glass of higher and known refractive index. This method is especially useful when only small samples of the substance to be measured are available, and with suitable

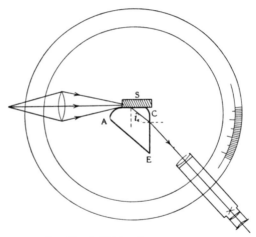

Fig. 43.—Pulfrich refractometer

instruments a high degree of accuracy can be obtained. Instruments for this purpose, of which there are a number of different forms, are known as **refractometers.**

In the **Pulfrich Refractometer** (Fig. 43) ACE is a prism of very dense flint glass, the refractive index n of which is accurately known. The specimen to be measured, S, is placed on the upper surface of the standard prism with a thin film of liquid, of a higher refractive index than the specimen, between the surfaces. (As was shown in § 28 this liquid has no effect on the final direction of the light.) A convergent beam of monochromatic light is incident horizontally on the end face of the test specimen close to the contact surface, and some of this light will enter the standard prism and emerge from the surface CE. The only light entering the standard prism will be that contained within the critical angle i_c between the substance being measured and the standard glass and if this angle can be measured the refractive index n' of the test specimen can be found, as $n' = n \sin i_c$.

The light emerging from the surface CE is focused by a telescope attached to a vertical divided circle, which turns about its centre at which the prism is mounted, the position of the telescope being read by a fixed vernier. The setting consists in bringing the cross-line or pointer

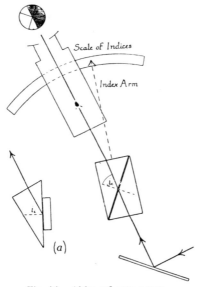

Scale of Indices

Index Arm

(a)

Fig. 44.—Abbe refractometer

of the telescope on to the edge of the "shadow," that is, the limiting direction of the light leaving the prism, and the refractive index corresponding to the particular angular reading of the telescope is found from prepared tables. The refractive index of a liquid may be determined by placing the liquid in a short length of glass tube cemented to the upper surface of the standard prism.

The **Abbe Refractometer** (Fig. 44) is based on the same principle. As used for the determination of the refractive index of liquids, two equal prisms of dense flint glass, having angles of 30°, 60° and 90°, are placed with their hypotenuse faces together and a thin film of the liquid to be measured between them. Light is reflected from a mirror on to the short face of the lower prism, and some of this light will emerge from the short face of the upper prism. The limiting direction of the light that can emerge will be that of the light which is at grazing incidence at the surface between the liquid and the upper prism, that is, the light that enters the upper prism at the critical angle between the glass and the liquid.

The light emerging from the prisms is received by a telescope, and the prisms are turned about a horizontal axis to bring the limiting direction —the "shadow" edge—on to the cross-line of the telescope. Attached to the prisms is an arm carrying an index that reads on a scale divided to give the refractive index of the substance tested. As the refractive index and therefore the critical angle varies with the colour of the light, the shadow edge would be coloured and badly defined when white light is used. This colour effect is neutralised by an equal but opposite dispersion given by two Amici direct-vision prisms (see Chapter XIII), which rotate in opposite directions about the telescope axis, the achromatised "shadow" edge corresponding to that for the yellow light.

The Abbe refractometer may be used to determine the refractive index of a solid by removing the lower prism and placing the substance to be measured on the hypotenuse face of the upper prism with a thin film of a high refractive index liquid such as monobromnaphthaline between the surfaces (Fig. 44a). The prism is then turned as before until the edge of the "shadow" comes on the cross-line in the telescope, when the scale reading gives the refractive index.

The refractometer may also be used to determine the refractive index of translucent substances, such as oils, butter and jam, and has important commercial applications in this way.

CHAPTER IV

EXERCISES

1. Describe and prove a graphical construction for refraction. Use the construction to find the direction of the refracted ray in glass ($n = 1 \cdot 53$) when the incident ray in air meets the surface at an angle of 45°.

2. Explain what is meant by critical angle and internal total reflection, and obtain the relation between the critical angle and the refractive indices of two media. Find by a graphical construction the critical angle between glass $n = 1 \cdot 67$ and water $n = 1 \cdot 33$.

3. A tank having a plane glass bottom 3 cm. thick is filled with water to a depth of 5 cm. Trace the path of a narrow parallel beam of light, showing the angles at each

surface, when the beam is (a) normal (b) at an angle of incidence of 45° at the upper surface of the water. (Refractive index of the glass 1·62).

4. A narrow beam of light is incident on one surface of a plane parallel plate of glass (n = 1·62) 5 cm. thick at an angle of 45° to the normal. Trace the path of the light through the plate and determine the lateral displacement of the beam on emergence.

What will be the direction of the emergent light if the second surface of the glass plate be immersed in oil of refractive index 1·49?

5. Light is diverging from a small object inside a block of glass towards a polished plane surface of the glass. Describe and show by means of a diagram what will happen to the light on reaching the surface.

6. Describe fully a method of determining the refractive index of a liquid contained in a tank with a plane glass bottom. Explain the principles upon which the method depends.

7. Explain how a microscope may be used to determine the refractive index of a sample of glass in the form of a parallel-sided slab. In a certain case the slab was 1¾ inches thick and made of glass of refractive index 1·6. Give figures for the settings of the microscope.

8. An object held against a plate glass mirror of refractive index 1·54 appears to be ⅜ inch from its image. What is the thickness of the mirror? (S.M.C.)

9 Explain the conditions under which total reflection occurs at the boundary between two transparent media. How would you make use of this phenomenon to measure refractive index? (L.U.)

10. The apparent thickness of a slab of glass of actual thickness t and refractive index n is $\frac{t}{n}$. Prove this statement (a) by considering the direction of a ray close to the normal and (b) by considering the change in curvature of the light front. Describe a practical application of this fact.

11. Find the critical angle for light passing from glass of refractive index 1·523 to (a) air, (b) water, (c) oil of refractive index 1·47.

12. A 4 inch cube of glass of refractive index 1·66 has a small object at its centre; show that if a disc of opaque paper exceeding a certain diameter be pasted on each face, it will not be possible to see the object from anywhere outside the cube. Find the minimum diameter cf disc required.

13. Rays of light are emitted upwards in all directions from a luminous point at the bottom of a trough containing liquid to a depth of 3 inches; the refractive index of the liquid is 1·25. Find the radius of the circle lying in the surface of the liquid with its centre vertically above the luminous point, outside which all rays will be totally reflected. (S.M.C.)

14. Explain with a diagram why an oar partly immersed in water appears to be bent.

15. An object placed 1 foot from a tank of water 1 foot wide and having thin plane glass sides is viewed through the tank, the line of sight being normal to the sides. Where will the object appear to be?

16. Show that the film of liquid placed between the substance to be measured and the prism of a refractometer has no effect on the reading of the instrument.

17. Given that the refractive indices of a liquid for red, yellow and blue light are 1·512, 1·532 and 1·543 respectively, find what will happen to a beam of white light travelling in the liquid and meeting the surface of the liquid separating it from air at the critical angle for yellow light.

18. A plate glass mirror 2 cm. thick is fixed in a vertical position. An object is placed 2 cm. in front of the glass surface and 4 cm. above the top edge of the mirror. Show by a full-size diagram how multiple images of the object are formed by the mirror. (Take the refractive index of the glass to be $\frac{3}{2}$.) (B.O.A.)

19. The refractive index of the air gradually increases towards the earth's surface. Show by means of a diagram how this will affect the apparent time of rising and setting of the sun or moon.

20. A rectangular trough has vertical sides of glass 1 inch thick and of refractive index 1·54. The bottom of the trough is covered by a layer of mercury above which is a layer of water 2 inches deep and of refractive index 1·33. A narrow beam of light strikes the surface of the water at an inclination of 45° and at a point 1½ inches from the one side of the trough, the beam being in a plane perpendicular to that side and turned towards it. Trace the course of the light after striking the surface of the water. (S.M.C.)

21. A strip of white paper is viewed through a glass prism. Explain carefully the character of the image seen, describing (a) its position and (b) the colours observed. Give a diagram. (B.O.A.)

22. A prism of 30° refracting angle is made of glass having a refractive index of 1·6. Find the angles of emergence and deviation for each of the following cases:

        (a) Angle of incidence at first face $= 24°\ 28'$
        (b) Angle of incidence at first face $= 53°\ 8'$
        (c) Incident light normal to first face.

23. A parallel beam of yellow light is incident at an angle of 40° to the normal to the first surface of a 60° prism of refractive index 1·53. Trace the path of the light through the prism and find the deviation produced on the beam. What will be the minimum deviation produced by this prism?

24. In measuring the refractive index of a prism for yellow light on a spectrometer, the refracting angle is found to be 60° 14' and the angle of minimum deviation 42° 26'. Find the refractive index.

25. A prism has a refracting angle of 60° and a refractive index of 1·5. Plot a graph showing the relation of the total deviation to the angle of incidence of the light on the first surface and find the angle of incidence for which the deviation is a minimum. Show that this angle of incidence is equal to the angle of emergence.

26. A parallel pencil of light is incident upon one face of a 30° prism at an angle of 54° with the normal and after refraction at this face the light reaches the second face, which is silvered, in such a way that it is reflected along its original path. Find the refractive index of the glass.

27. Light from a distant object is incident normally at the front surface of a prism of 3° angle, $n = 1·53$. Part of the light is reflected at the first surface and part at the second; find the angle between the two reflected beams leaving the prism.

28. A prism of refractive index 1·65 has one angle of 90° and another of 75° and a ray of light strikes the face between these two angles at an inclination of 65° to the normal on the side of the 90° angle. Trace its course through the prism and find the total deviation produced.

29. A parallel beam of light is incident at 30° to the normal to one face of an equilateral prism ($n = 1·53$), the three faces of which are polished. Trace the path of the light through the prism and find the total deviation produced. For what angle of incidence will the deviation be a minimum?

30. A narrow parallel beam of white light is incident at 40° to the normal to one face of a prism of 60° refracting angle made of glass having refractive indices 1·605 and 1·647 for red and blue light respectively. What will be the length of the spectrum on a screen 2 metres from the prism?

31. A narrow parallel beam of white light is incident at 50° on a prism of 30° refracting angle, $n = 1·53$. Some of the light is reflected at the first surface, and the remainder passes into the prism and is reflected at the second surface, which is silvered. Find the angle between the two beams on emergence from the prism and explain carefully any difference in their nature.

32. Show that if a prism of refractive index n is to give a minimum deviation d, its refracting angle a is given by

$$\tan \frac{a}{2} = \frac{\sin \frac{d}{2}}{n - \cos \frac{d}{2}}.$$

Hence find the refracting angle of a prism of refractive index 1·65 to give a minimum deviation of 60°. (S.M.C.)

33. A candle is mounted 1 foot away from and on the normal to one face of a thin prism of 1° refracting angle and refractive index 1·53. Find the positions of the successive reflected images from its front and back surfaces. (S.M.C.)

34. A plane mirror 2 feet 6 inches wide has its front edges bevelled, and an observer standing at 8 feet from the mirror sees three images of his eye, one in the centre of the mirror and the two others in the bevelled portions. If the refractive index of the glass of the mirror is 1·53, what is the angle at which the edges are bevelled? (S.M.C.)

35. The angle of a prism of nominally 90° is tested by directing a parallel beam of light on to the hypotenuse face of the prism and measuring the angle between the two emergent beams which have travelled through the prism in opposite directions. What will be the error in the angle of the prism if the angle between the emergent beams be 14 minutes? (n of prism 1·516).

36. Two isosceles right angle prisms ($n = 1·53$) are placed with their hypotenuse faces together. Find the limiting angle of incidence at the first face of one prism in order that light may pass through the two prisms when there is (a) air, (b) water between them.

37. Explain briefly the principle and optical construction of the Abbe refractometer. Give a diagram.

The right angle prisms of an Abbe refractometer are made of dense flint of $n = 1·75$, and the angle between the hypotenuse and the end face is 60°. Find the angle of emergence of the limiting ray from the last face of the upper prism block when testing substances with refractive indices (a) 1·35 (b) 1·50 (c) 1·65.

38. A thin prism is tested by observing through it a scale 80 cm. away, the scale divisions being one cm. apart. The deviation is found to be 1·75 scale divisions. Express the deviation in prism dioptres.

A ray of light falls normally on one face of this prism, is reflected by the second face, and emerges from the first face making an acute angle of 7·5 degrees with its original direction. Find the apical angle and refractive index of the prism. (S.M.C.)

39. A convergent pencil of light is intercepted before reaching its focus by a right-angled isosceles prism. The angle between the edges of the pencil is 18° and the central ray meets one of the short faces of the prism normally. What must be the minimum

refractive index of the prism in order that the whole of the pencil may be totally reflected at the hypotenuse face?

40. A pencil of parallel light is incident on the face PQ of a right-angled isosceles prism (right angle at P) in a direction parallel to the hypotenuse face QR. The width of the pencil is equal to one quarter of the length of face QR.

The prism is such as *just* to permit the whole width of the pencil to be reflected at face QR and then to emerge into air from face PR. Find the refractive index of the prism. (S.M.C.)

41. A ray of light is incident at grazing incidence (angle of incidence 90°) on the face AB of a right angle prism ABC (right angle at B) of refractive index 1·72; the ray travels in the direction from A to B. What happens to this ray at the face BC?

When a block of glass of refractive index n is placed in optical contact with the face AB, the same incident ray then emerges from face BC into air at an angle of 45° to the normal. Find the value of n.

Describe very briefly, with a sketch, an instrument in which this principle is employed. (S.M.C.)

42. A glass prism of refracting angle 72° and refractive index 1·66 is immersed in a liquid of refractive index 1·33. What is the angle of minimum deviation for a ray of light passing through the prism? (S.M.C.)

V – Curvature: Refraction at a Curved Surface

43. Curvature

Most of the curved surfaces used in optical work are of spherical form, although other forms of curved surfaces are sometimes used. Thus in ophthalmic work, cylindrical and toroidal curves are quite common, while the reflectors of headlamps and searchlights, and the mirror of the reflecting telescope, are of paraboloidal form. For the present, however, the consideration of curved surfaces will be confined to spherical surfaces.

The **curvature** of a surface may be defined as *the angle through which the surface turns in a unit length of arc.*

Let DAE (Fig. 45) be a portion of a circular track having its centre at C. An object travelling round the track in a clockwise direction will

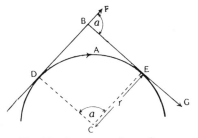

Fig. 45.—Curvature of a surface

be travelling, when at D, in the direction DF, and when it reaches E it will be travelling in the direction EG, having turned through the angle

65

FBE in a length of arc DE. As DF and EG are tangents to the curve, the angle FBE = the angle DCE = a; then from the definition

$$\text{Curvature} = \frac{a}{\text{arc DE}}; \quad a \text{ in radians} = \frac{\text{arc DE}}{r}.$$

Hence curvature $= \dfrac{1}{r} = R$.

The curvature of a surface may therefore be expressed as the *reciprocal of its radius of curvature.*

The unit of curvature is the **dioptre** (D or Dp.) which is the curvature of a curve having a radius of 1 metre.

$$\text{Curvature in Dioptres} = \frac{1}{r \text{ in metres}} = \frac{100}{r \text{ in cm.}} = \frac{39\cdot37}{r \text{ in inches}}$$

44. Measurement of Curvature

As the curves to be considered are usually very small portions of the complete sphere, a direct measurement of curvature will not be possible. The curvature of a spherical surface can, however, be determined from two measurements that can be readily made. In Fig. 46 DAE represents

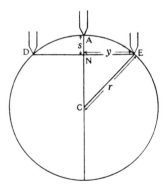

Fig. 46.–Measurement of curvature

a small portion of a spherical curve, having its centre of curvature at C. AN is known as the **sagitta** or **sag,** s, of the curve for a given chord $2y$, being so called from its resemblance to an arrow on a bow.

In the triangle ECN

$$r^2 = y^2 + (r-s)^2$$

$$r = \frac{y^2}{2s} + \frac{s}{2} \quad \dots \qquad \dots \qquad \dots \qquad \dots \quad (5.01)$$

When s is small compared with r, as it frequently is,

$$r = \frac{y^2}{2s} \text{ (approx.)}$$

and, Curvature, $R = \frac{2s}{y^2}$ (approx.)

Curvature is in dioptres when y and s are in metres.

For small portions of curves the curvature is approximately proportional to the sag, s.

The mechanical determination of the curvature of a surface by means of various forms of **spherometers** consists in measuring the sag s for a given chord $2y$ of the curve, when the radius or curvature may be found from expression (5.01). One of the simplest forms of spherometer is the **optician's lens measure** illustrated in Fig. 47. This instrument has two fixed points D and E, midway between which is the moving point A, the position of which is read by means of a series of levers actuating a pointer over a circular scale. The movement of the centre point from its zero position, when the three points are resting on a plane surface, to its position when the points rest on the curved surface being measured, gives the distance s, the distance between the two fixed points being $2y$. The scale of the lens measure is divided to give the focal power (*see* § 53) of the surface for the refractive index 1·523 of the glass of which spectacle lenses are usually made.

Front.   Interior (diagrammatic).
Fig. 47.—Optician's lens measure

As the three points of the lens measure are in a straight line the instrument may be used to measure the curvature of any meridian of cylindrical and toroidal surfaces. This arrangement, however, introduces errors in that it is impossible to ensure that the instrument is standing perpendicularly on the surface. To overcome this source of error the more accurate spherometers usually have the moving point at the centre of a ring, or equidistant from three fixed points that lie on a circle.

The usual form of simple spherometer is shown in Fig. 48. It consists of a metal stand with three pointed legs arranged at the corners of an equilateral triangle, the fourth leg is formed by a micrometer screw having a pointed end and passing through the stand equidistant from the three fixed legs. A divided head is attached to the screw to read fractions of a complete turn; thus if the pitch of the screw is $\frac{1}{2}$ mm. and the head is divided into one hundred equal divisions, a rotation of one division represents a movement, up or down, of $\frac{1}{200}$ mm. Complete rotations of the screw are read on a short length of scale divided in $\frac{1}{2}$ and 1 mm. divisions attached to the stand against the divided head. To measure a surface, the instrument is first placed on a plane surface, and the screw adjusted until all four points touch the surface; this position is best found by holding one of the fixed legs lightly between finger and thumb,

Fig. 48.—Spherometer

and noting the position of the screw where the slightest further movement of the screw downwards allows the instrument to rotate freely about the centre point. The mean of a number of settings gives the reading on the plane or the zero setting of the instrument. The instrument is then placed on the surface to be measured, and reset in the same manner until the four points touch the curved surface. The difference between the readings on the plane and curved surfaces gives the value s. The value y, the distance between either of the fixed points and the centre point is best found by measuring the distance between each of the three fixed points, and dividing the mean by $\sqrt{3}$.

A more accurate instrument is the **Abbe spherometer.** In this the surface to be measured rests on an accurately turned metal ring, a number of such rings being provided to accommodate surfaces of various diameters. Passing through the centre of the ring is a plunger, which is kept in contact with the surface by means of a counterbalance weight. A finely divided silver scale, attached to the plunger and observed with a microscope fitted with a micrometer eyepiece, enables one to read movements of the plunger to $\frac{1}{1000}$ mm. The displacement of the plunger from its position when in contact with a plane to that when in contact with the curve gives the value s, and the diameter of the ring is $2y$. As the edge of the ring has a certain thickness, its inner diameter must be taken when measuring convex curves and its outer diameter when measuring concave curves.

The spherometer with sharply pointed feet or a sharp-edged ring is very liable to error due to the points or edge wearing and thus altering the value of y; also there is always the danger of damage to a polished surface. To overcome these difficulties, Aldis has suggested the modification of fitting to the points or at three equally spaced positions around the ring, small polished steel spheres. With the **Aldis spherometer** readings are taken in the usual way, and a correction for the small spheres is applied as follows:

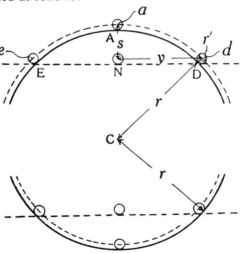

Fig. 49.—Principle of Aldis spherometer

In Fig. 49 the positions of two of the fixed feet and the movable centre foot are shown as resting on a plane surface END and on a curved surface EAD. If y is taken as the distance between the centres of the outer and the centre spheres and s, as before is the displacement of the centre leg, then, as may be seen from the figure, expression (5.01) gives the radius of the curve *ead* passing through the centres of the small spheres. The radius of the surface is therefore found by subtracting the radius r' of the small spheres from the calculated value in the case of a convex curve and adding r' to the calculated value for a concave curve.

45. Curvature of Wave Fronts

The effects of mirrors, prisms and lenses on the light may be studied by considering the changes produced in the curvature of the wave fronts (§ 2). For example, Fig. 50 shows the action of a convex lens in changing the light diverging from an object point B into light converging to a focus B'. The effect of the lens has been to change the curvature of the wave front, and if the amount of this change for a given lens and the curvature of the incident wave front at the lens is known, the new curvature and therefore the position of the focus of the emergent light is easily found.

The curvature of a wave front is expressed in the same way as that of a surface. Thus if l is the distance of the object point B from the lens, then $\dfrac{1}{l}$ will be the curvature of the incident wave front at the lens, and similarly if l' is the distance of the focus B', $\dfrac{1}{l'}$ is the curvature of the emergent wave front at the lens.

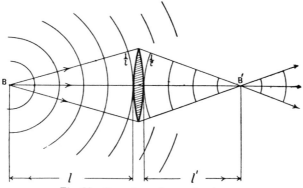

Fig. 50.—Curvature of wave fronts

In the divergent pencil the curvatures of the wave fronts are continuously decreasing as they recede from the object point and in the convergent pencil continuously increasing as they approach the focus.

The curvature of the wave front may be used to express the amount of divergence or convergence of the light at any point. In Fig. 50 the light has a divergence $\frac{1}{l}$ on reaching the lens and a convergence $\frac{1}{l'}$ immediately on emergence.

46. Sign Convention

To denote the directions in which various distances, lengths and angles are to be measured, positive and negative signs are given to the values. Thus in Fig. 50 the distances l and l' are each measured from the lens, and as these are distances in opposite directions the values will be given opposite signs. The sign convention to be adopted throughout the work is as follows (Fig. 51):

(*a*) *Distances along the lens or mirror axis.*

All distances are measured *from* the lens or mirror. Those in the same direction as the *incident* light is travelling are positive, and in the opposite direction negative.

In diagrams the incident light will ordinarily be represented as travelling from left to right.

(*b*) *Distances perpendicular to the axis.*

All distances are measured from the axis. Those above are positive and below, negative.

(*c*) *Angles.*

Anti-clockwise rotation is positive.

Slope of a ray.—This is defined as the acute angle described by a ruler in being moved from the direction of the ray to the direction of the axis.

Angles of incidence, reflection and refraction.—Angles are to be measured *from* the normal *to* the ray.

In Fig. 58 all the distances and angles are positive; it may be referred to as the "*all positive diagram.*"

Fig. 51.—Sign convention

Applying the sign convention to Fig. 50, it will be seen that l and therefore $\frac{1}{l}$ are negative, and l' and $\frac{1}{l'}$ positive. The radius and curvature of the first surface of the lens are positive and of the second surface, negative.

An arrow head shown on a line fixing the limits of a given distance is to indicate the direction in which that distance is measured.

The question of signs is of fundamental importance in optical work, and the student should make himself thoroughly conversant with the above convention.

47. Refraction at a Plane Surface (Change in Curvature of the Wave Front)

In § 32 it was shown that when a narrow pencil of light, diverging from an object point, meets a plane surface bounding two different media it will, after refraction, be diverging from an image point in some other position depending on the refractive indices of the media. The same result may be obtained by considering the change that takes place in the curvature of the wave front on passing from one medium to another.

In Fig. 52 DAE is a plane surface separating two media having refractive indices n and n' respectively, n being the greater. Light is diverging from the object point B, and DME represents the form of a wave front if the light had continued to travel in the first medium. As the light on passing into the second medium has its velocity increased, the light, which would have been at M, now reaches M' in the same time and light, that would have reached any other point on the front DME, is advanced a corresponding amount. The curve of the incident wave front DME is therefore changed on refraction to a curve DM'E.

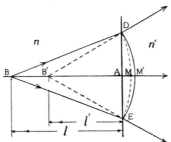

Fig. 52.—Refraction at a plane surface, $n > n'$

As the light travels a distance AM' in the medium n' in the same time as it would have travelled AM in the medium n, from the definition of refractive index (§ 12)

$$n' \ AM' = n \ AM.$$

As the sags AM and AM' are small when the pencil DBE is a narrow one, they are proportional to the curvatures $\dfrac{1}{l}$ and $\dfrac{1}{l'}$ of the wave fronts DME and DM′E.

$$\text{Hence } \frac{n'}{l'} = \frac{n}{l}$$

$$\text{or } \quad l' = \frac{n'}{n}l$$

48. Reduced Distance and Vergence

When the second medium is air, $n' = 1$ and $l' = \dfrac{l}{n}$. This is called the *equivalent air distance* or the **reduced distance** corresponding to the real or *absolute* distance l existing in the medium of index n. Similarly $\dfrac{1}{l'} = \dfrac{n}{l}$ is the *equivalent air curvature or vergence* or the **reduced vergence**. The idea of reduced distances and vergences is very useful in calculations on lenses and lens systems.

A reduced vergence will be represented by the capital letter corresponding to the small letter used for the corresponding absolute distance, thus

$$L = \frac{n}{l}, \quad L' = \frac{n'}{l'}.$$

It should be specially noticed that when an image is virtual, its distance l' is measured back from the surface, but the emergent light is actually travelling in the second medium and n' is the refractive index involved in the reduced distance.

49. Refraction at a Curved Surface

The normal to the surface at the point of incidence will be a radius of the curve passing through that point. The line joining the **centre of curvature**, C, to the **vertex** or **pole**, A, of the surface, that is, the centre of the aperture, is known as the **optical axis**. In most optical systems the various surfaces have a common axis, and such systems are said to be *coaxial* or *centred*. If the curves are spherical, the centred system is symmetrical about its optical axis.

When light falls on a curved surface between two different media, some of it is reflected (see Chapter VII), and the remainder is refracted on passing into the second medium. The direction of any part of the refracted light may be calculated from the law of refraction or may be found by a graphical construction, such as that given in § 31.

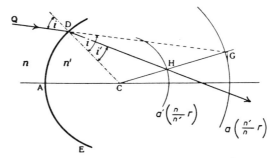

Fig. 53.–Young's construction. Positive surface, $\frac{n'}{n} = 1.5$

50. Young's Construction

A simple and more convenient construction for the refraction at a curved surface is that due to Thomas Young (1773–1829).*

In Fig. 53 QD is the direction of a ray incident on a spherical refracting surface DAE, having its centre at C. With C as centre and with radii equal to $\frac{n'}{n}$. r and $\frac{n}{n'}$. r describe the arcs a and a'. The incident ray QD intersects the arc a at G, and the straight line joining G and C intersects the arc a' at H; then DH is the direction of the refracted ray. The construction for a ray incident on a concave surface is shown in Fig. 54.

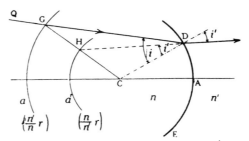

Fig. 54.–Young's construction. Negative surface, $\frac{n'}{n} = 1.5$

Proof.

As $$CG = \frac{n'}{n}.r, \quad CH = \frac{n}{n'}.r, \quad \text{and } CD = r$$

$$CG:CD = CD:CH = n':n$$

*THOMAS YOUNG, *Lectures on Natural Philosophy*, Vol. 2 Art. 425, 1807.

and therefore the triangles CDG and CHD are similar.

Hence $\qquad \widehat{CDG} = \widehat{CHD}$

and $\qquad \widehat{CDH} = \widehat{CGD}.$

In the triangle CDG

$$\sin CDG : \sin CGD = CG : CD = n' : n.$$

As $\widehat{CDG} = i$ and $\sin i : \sin i' = n' : n$ then $\widehat{CGD} = \widehat{CDH} = i'$, and the line DH is the path of the refracted ray.

The Ray Plotter.—A simple and convenient method of applying Young's construction is by means of the **ray plotter.** From the above proof it is clear that so long as $CG = \dfrac{n'}{n}\,CD$ and $CH = \dfrac{n}{n'}\,CD$, C may be any point on the normal to the surface and not necessarily the centre of curvature. In this way it is possible to construct a simple instrument to give the direction of the refracted ray at any surface between two given media.

Fig. 55.—The ray plotter

One form of ray plotter devised by F. G. Smith* consists of a piece of celluloid in the form shown in Fig. 55, with the points D, C, H and G marked on a straight line, small holes being drilled through these points. DC is taken of any convenient length, say 5 cm., and CH is made equal to $\dfrac{n}{n'}\,DC$ and CG equal to $\dfrac{n'}{n}\,DC$, where n and n' are the two refractive indices for which the instrument is intended to be used.

The instrument is placed on the diagram with D over the point of incidence and the line DG over the normal, and a needle is put through the hole at C to act as a pivot. The instrument is turned until the point G lies on the incident ray, produced if necessary, when the refracted ray will pass through the point H, the position of which can be pricked through to the paper.

*F. G. SMITH, *A Ray Plotter.* Trans. Optical Society. Vol. xxi, No. 3 (1919–20).

The line through D perpendicular to DG allows of the instrument being used by placing this line tangentially to the surface, D being over the point of incidence. The construction is then carried out as before, pivoting the instrument at C.

51. Spherical aberration. Caustic curve

In Fig. 56 light is shown diverging from an object point B on the axis of a spherical refracting surface. The direction of any part of the refracted light may be found as described in the previous paragraph,

Fig. 56.—Refraction at spherical surface. Spherical aberration, caustic curve

and a number of refracted rays are shown in the figure. It will be seen that any pair of rays at an equal distance from the axis intersect at some point on the axis. As the surface is symmetrical about the axis, all light from an axial point meeting the surface at a given distance from the axis, that is, in the same zone of the surface, passes through the axis at the same point. The greater the distance of a particular zone of the surface from the axis, the closer to the surface will the light from that zone meet the axis, so that light which started from an object point will not be brought to a point again after refraction. This effect is known as **spherical aberration.** A similar effect is produced when the light is divergent after refraction, the light from each zone of the surface diverging from a different point on the axis.

It will be noticed in the figure that consecutive pairs of rays intersect on a double curve GB′H, which is known as a **caustic curve.** There will consequently be a concentration of light over a surface of the form generated by the revolution of the caustic curve about the optical axis. This surface is a **caustic surface.** The point or *cusp* of the curve will lie on the axis at the point at which rays near the axis intersect.

52. Paraxial Rays.

Rays which lie very close to the axis of a spherical surface and which therefore meet the surface near the vertex will, after refraction, pass approximately through a point focus, as will be seen in Fig. 56, or will diverge as though from a point. Such rays are known as **paraxial rays,** and the region very close to the axis as the **paraxial region.**

For light in the paraxial region the angles of incidence and refraction will be very small, and may be considered as proportional to their sines in applying the law of refraction. Also the apertures of the wave fronts being very small, their curvatures are proportional to the sags. Whenever therefore the aperture of an *axial* pencil is quite small we may assume that all the refracted light will be converging to or diverging from a *point* on the axis, the refracted or reflected pencil thus being homocentric.

Much of the future work will apply strictly only to the paraxial light with surfaces of quite small aperture and object points very close to the axis. It is possible, however, by a suitable choice of curvatures, separations of surfaces, and refractive indices to produce lens systems that will give approximately a point image of a point object for quite large apertures and for object points at considerable distances from the axis. Many of the results obtained for the paraxial region may be applied to such *corrected systems*.

53. Refraction at a Spherical Surface (Change in Vergence)

In Fig. 57 light converging towards a focus at B is intercepted by a spherical surface DAE, separating the media of refractive indices n and n'. The incident wave front SAP meets the vertex of the surface at A

$$n \text{ SD} = n' \text{ AM}$$
$$n \text{ (TD} - \text{TS)} = n' \text{ (AN} - \text{MN)}$$
$$n' \text{ MN} - n \text{ TS} = n' \text{ AN} - n \text{ TD}$$

For small portions of the curves, that is, in the paraxial region, TD = AN, and the sags are proportional to the curvatures hence:

$$\frac{n'}{l'} - \frac{n}{l} = \frac{n' - n}{r} \qquad \ldots \qquad \ldots \qquad \ldots \quad (5.02)$$

and, if n' is greater than n, the centre of the wave front is retarded with respect to the *rim* portion on passing into the second medium. DME represents the form of the wave front immediately after refraction. If the light is confined to the paraxial region (the aperture is shown greatly exaggerated in the figure), the refracted wave front is spherical and the light is now converging to a focus at B'. B' is therefore the image of the *virtual object* B. As the light which travels along the axis meets the surface normally, it is unchanged in direction and therefore, if the object point is on the axis, the image must also be formed on the axis.

The centre of the refracted wave front travels from A to M in medium n' in the same time as the rim portion travels from S to D in medium n.

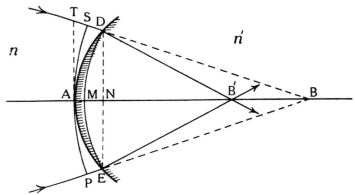

Fig. 57.—Refraction at a spherical surface $(n' > n)$. Paraxial region

where l and l', the distances of object and image from the refracting surface, are the radii of curvature of the incident and refracted wave fronts respectively, and r is the radius of curvature of the surface. As $\dfrac{n}{l}$ and $\dfrac{n'}{l'}$ are the reduced vergences of the incident and refracted light respectively, equation (5.02) may be written

$$L' - L = \frac{n' - n}{r} = (n' - n)\,R \quad \dots \quad \dots \quad \dots (5.02A)$$

$L' - L$ is the change produced by the surface in the reduced vergence of the light, and is known as the **focal power** of the surface. It will be denoted by F.

Hence $\qquad\qquad\qquad L' - L = F \dots \quad \dots \quad \dots \quad \dots$ (5.03)

and $\qquad\qquad\qquad F = (n' - n)\,R \quad \dots \quad \dots \quad \dots$ (5.04)

The focal power is thus a constant for a particular surface, that is, a given surface will always produce the same total change in the vergence of the light. Focal power is usually expressed in dioptres and will be positive for a surface that tends to produce convergence (Fig. 57), and negative for one that tends to produce divergence.

The equations (5.02) or (5.02A) will be called the **fundamental paraxial equation** for a surface.

When the first medium is air $(n = 1)$ and the second, glass of refractive index n, equation (5.04) becomes

$$F = (n - 1)\,R$$

This is the quantity given by the opticians' lens measure (§ 44).

54. Refraction at a Spherical Surface (Change in Ray Path)

The expression (5.02) may also be derived by considering the change produced by the surface in the path of a single ray, as follows: Fig. 58. From the law of refraction

$$n \sin i = n' \sin i'$$

and when i and i' are small

$$n i = n' i'.$$

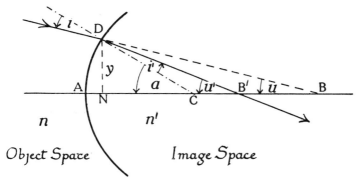

Fig. 58.—Refraction at a spherical surface. Change in ray path. All positive diagram

When a, u and u' are small, the arc AD may be considered as a short line y perpendicular to the axis; then

$$u = \frac{y}{AB}; \quad u' = \frac{y}{AB'}; \quad a = \frac{y}{AC}.$$

From the figure

$$i = a - u; \quad i' = a - u'$$

therefore

$$n (a - u) = n' (a - u')$$
$$n' u' - n u = (n' - n) a.$$

Substituting the above values for u, u' and a, and dividing through by y

$$\frac{n'}{AB'} - \frac{n}{AB} = \frac{n' - n}{AC}$$

$$AB' = l'; \quad AB = l; \quad AC = r.$$

79

Hence $$\frac{n'}{l'} - \frac{n}{l} = \frac{n'-n}{r} \qquad \dots \qquad \dots \qquad \dots \quad (5.02)$$

55. Conjugate Foci. Principal Foci. Focal Lengths

It follows from the above that for any object point B there will be a corresponding image point B', the position of which may be found from expressions (5.02) or (5.02A), giving the appropriate signs to the various distances. As the path of the light is reversible, the positions of object and image are interchangeable. These pairs of object and image points are therefore known as **conjugate points** or **foci** and the distances of object and image from the surface as the **conjugate distances.**

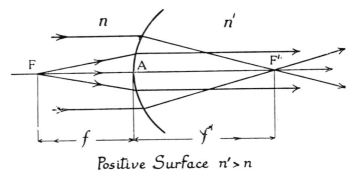

Fig. 59A.—Refracting surface. Principal foci and focal lengths

Two of these pairs of conjugate points are of special importance. When the object is at an infinite distance and the incident light therefore parallel (Figs. 59A and B), the refracted light will be converging to a real image in the case of a surface of positive power, or diverging as though from a virtual image in the case of a surface of negative power. This image, the conjugate of the infinitely distant object, is termed the **second principal focus,** F', of the surface, and its distance *from* the surface is the **second focal length,** f'. There must also be an object position, such that the image is at an infinite distance and the refracted light therefore parallel. This will be a real object position from which the incident light is diverging in the case of a positive surface, and a virtual object position towards which the incident light is converging in the case of a negative surface. This object position is termed the **first principal focus,** F, and its distance *from* the surface is the **first focal length,** f.

It will be seen that both principal foci are real with a positive surface and virtual with a negative surface. *The two focal lengths are of opposite sign; the sign of the second is that of the focal power.*

The planes perpendicular to the axis passing through the principal foci are known as the **focal planes.**

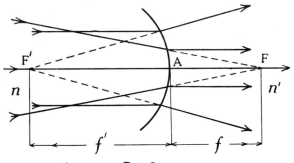

Negative Surface $n' > n$

Fig. 59B.—Refracting surface. Principal foci and focal lengths

Any incident ray parallel to the axis will, after refraction at a spherical surface, pass through or diverge from the second principal focus, and any incident ray passing through or travelling towards the first principal focus will, after refraction, travel parallel to the axis. These facts are useful in graphical construction (§ 73).

Focal Lengths.

When l' is infinity, $L' = 0$ and

$$-L = -\frac{n}{l} = F.$$

Similarly when l is infinity,

$$L' = \frac{n'}{l'} = F.$$

These special values of l and l' are the first and second focal lengths respectively, hence

$$-\frac{n}{f} = \frac{n'}{f'} = F = (n' - n)R \quad \dots \qquad \dots \qquad \dots \quad (5.05)$$

and

$$\frac{f'}{f} = -\frac{n'}{n}$$

The power of a surface is equal to the reciprocal of the "reduced" second focal length.

Also we have $$f' + f = r$$

and $$CF' = -f; \quad CF = -f'$$

Example.—An *aphakic* eye, that is, one from which the crystalline lens has been removed, may be regarded as a single spherical surface, the cornea, of $+ 7·8$ mm. radius of curvature between air, $n = 1$, and a medium of refractive index, $n' = 1·336$. Find the focal power and the positions of the principal foci of such an eye. What will be the position of the object of which an image is formed on the retina 24 mm. from the cornea?

$$r = +7·8 \text{ mm.} = 0·0078 \text{ } m.$$

$$F = \frac{0·336}{0·0078} = +43·1 \text{ D.}$$

$$f = -\frac{1000}{43·1} = -23·2 \text{ mm.}$$

$$f' = \frac{1336}{43·1} = +31·0 \text{ mm.}$$

Therefore the first principal focus is 23·2 mm. in front of the cornea, and the second principal focus 31·0 mm. behind the cornea.

$$l' = +24·0 \text{ mm.} = +0·024 \text{ m}$$

$$L' = \frac{n'}{l'} = \frac{1·336}{0·024} = +55·66 \text{ D}$$

$$L = L' - F = 55·66 - 43·1 = +12·56 \text{ D}$$

$$l = \frac{1000}{12·56} = +79·6 \text{ mm.}$$

The incident light must therefore be converging towards a point 79·6 mm. behind the cornea in order that, after refraction, it may focus on the retina.

56. Images of Object Points not on the Axis

As, in practice, any object will be of finite size, only one point of the object can lie on the optical axis. It will be necessary therefore to consider the light coming from points off the axis—**extra-axial points.** A good image of such a point will only be formed by a single surface when the light is in the paraxial region, that is, when the light makes a small

$$f = -\frac{n \times 1000}{F}$$

$$f' = \frac{n' \times 1000}{F}$$

angle with the axis, and the aperture is small. For the present we shall limit the work to such cases.

Of the light from any object point Q (Fig. 60), one ray will be incident on the surface normally, and will therefore be unchanged in direction on passing into the second medium. This ray is called the **chief ray** or **principal ray** from the particular point and will, of course, be the ray that passes through the centre of curvature C. C is frequently called, on this account, the **nodal point** of the surface. Since, in the paraxial region, all rays from an object point meet at the corresponding image, this image must be on the principal ray from the object point. If the direction of a second ray is known, such as an incident ray parallel to the axis, which after refraction must pass through the second principal focus, the position of the image is fixed by the intersection of these two rays.

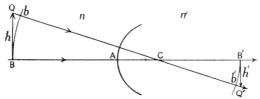

Fig. 60.—Spherical refracting surface. Image of extra-axial point

If Bb is an arc struck with C as centre, the image of b is b′ on an arc B′b′. Clearly then, the image of the perpendicular BQ is a curved line more curved than B′b′, as Q is further from the surface than b, and its image therefore closer to the surface than b′. When BQ is very small, however, we can take B′Q′ as the image, and in the paraxial region we shall assume that for a plane object perpendicular to the axis the image is also in a plane perpendicular to the axis. Object and image are said to be in **conjugate planes.**

In Fig. 61 BQ is an object in front of a spherical refracting surface DAE. The position of the axial image point B′ can be found from the fundamental paraxial equation. The position of any extra-axial point such as Q can be found graphically by considering any two rays, the refracted directions of which are known. Thus, the chief ray QC will be undeviated, a ray QD parallel to the axis will be refracted to pass through the second principal focus F′, and a ray QE passing through the first principal focus F travels parallel to the axis after refraction. The intersection of any two of these rays gives the position of the image Q′. As the chief ray QC is undeviated,

$$\frac{h'}{h} = \frac{CB'}{CB}$$

where h and h' are the lengths of object and image respectively. The ratio of the length of the image to the length of the object is termed the **lateral magnification** of the image, and will be denoted by m.

Then
$$m = \frac{h'}{h} = \frac{l'-r}{l-r}$$

As the angles w and w' are very small

$$\frac{\tan w'}{\tan w} = \frac{\sin w'}{\sin w} = \frac{n}{n'}$$

then
$$\frac{h'}{h} = \frac{l' \tan w'}{l \tan w} = \frac{nl'}{n'l} = \frac{L}{L'} \qquad \ldots \qquad \ldots \qquad (5.06)$$

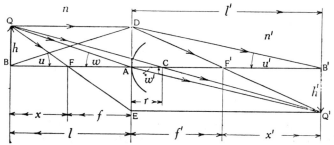

Fig. 61.—Spherical refracting surface. Formation of image. The construction applies only to the paraxial region, and the position of the surface is therefore represented by the straight line DE. (See § 73).

The magnification will be positive when the image is erect, and negative when the image is inverted. Images will be erect when they lie on the same side of the surface as the object, and inverted when on the opposite side.

Example:

Seen through the cornea the pupil of an eye appears to be situated 3 mm. from the cornea and of 4·5 mm. diameter. Treating the cornea as a single spherical surface of 7·8 mm. radius of curvature between air and a medium of refractive index $\frac{4}{3}$, find the actual position and size of the pupil.

In solving problems the student will always be helped by making a diagram showing all the given data.

In this example the actual pupil is the object and light from it is refracted at the cornea on leaving the eye and diverges as though from the virtual image, the apparent pupil, situated -3 mm. from the cornea and of $4\cdot5$ mm. diameter. Thus $n = \frac{4}{3}$, $n' = 1$, $l' = -3$ mm. and $r = -7\cdot8$ mm. (Surface is concave towards the incident light).

$$\frac{n'}{l'} - \frac{n}{l} = \frac{n'-n}{r}$$

$$-\frac{1}{3} - \frac{4}{3l} = \frac{1-\frac{4}{3}}{-7\cdot8}$$

$$\frac{1}{l} = -0\cdot282$$

$l = -3\cdot55$ mm. (distance of pupil from cornea)

$$m = \frac{h'}{h} = \frac{nl'}{n'l}$$

$$\frac{4\cdot5}{h} = \frac{\frac{4}{3}\times3}{3\cdot55}$$

$h = 3\cdot99$ mm. (diameter of pupil)

From Fig. 61 it will be seen that the lateral magnification may also be expressed in the following forms:

$$m = \frac{B'Q'}{AD} = -\frac{x'}{f'} \left.\vphantom{\frac{B'Q'}{AD}}\right\}$$

and $$m = \frac{AE}{BQ} = -\frac{f}{x} \qquad \qquad \dots \quad \dots \quad \dots \quad (5.07)$$

where x and x' are the distances of object and image from the first and second principal foci respectively. These distances are called the **extrafocal distances.** From expression (5.07) we have.

$$ff' = xx' \qquad \dots \quad \dots \quad \dots \quad (5.08)$$

This is known as **Newton's Relation,** an expression that is frequently useful.

57. Smith-Helmholtz or Lagrange's Law. Convergence Ratio

$$\frac{h'}{h} = \frac{nl'}{n'l} \text{ we have}$$

$$n'h'l = n\,h\,l'.$$

In the paraxial region, as the angles u and u' (Fig. 61) are very small.

$$\frac{AD}{l} = u, \text{ and } \frac{AD}{l'} = u'.$$

Hence $\qquad n'\,h'\,u' = n\,h\,u \qquad \ldots \qquad \ldots \qquad \ldots \quad (5.09)$

This optical invariant has important applications, and has been credited in various forms to Robert Smith (1689–1768), Helmholtz (1821–1894) and Lagrange (1736–1813). It is usually known as either the **Smith-Helmholtz** or the **Lagrange's Relation.**

From (5.09) we have

$$\frac{u'}{u} = \frac{n\,h}{n'\,h'} = \frac{n}{n'} \cdot \frac{1}{m}$$

This is called the convergence ratio.

58. Image of a Distant Object

All rays from any one point Q on a distant object are parallel on reaching the surface, and, after refraction, reunite in a point Q' in the second focal plane of the surface (Fig. 62). The position of Q' is de-

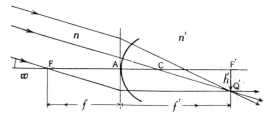

Fig. 62.–Spherical refracting surface. Image of distant object

termined by the chief ray which passes undeviated through the centre of curvature C.

If w denotes the angle subtended by the object standing on the axis or its **apparent size,** then

$$F'Q' = h' = f \tan w = f w, \text{ when } w \text{ is small } \ldots \qquad \ldots \qquad \ldots \quad (5.10)$$

This gives another definition of the first focal length which is often useful:—

$$f = \frac{h'}{w} = \frac{\text{size of image in } F' \text{ plane}}{\text{apparent size of object}} \qquad \ldots \qquad \ldots \quad (5.11)$$

1. What is meant by the curvature of a surface? Show that it is equal to the reciprocal of the radius of curvature.

2. In measuring the curvature of a surface with an Aldis spherometer having a distance of 25 mm. between the centres of the centre and outer spherical feet, the readings are -0.16 mm. on a plane and $+2.54$ mm. on the curved surface. The diameter of the spherical feet is 5 mm. Find the curvature of the surface. What percentage error is introduced by using the approximate formula?

3. The radius of the circle containing the three fixed legs of a spherometer is $1\frac{1}{2}$ inches. The instrument is placed on a convex spherical surface of curvature 3 dioptres. By how much will the central leg have to be raised above the outer legs to be just in contact with the surface?

4. A plano-concave lens has a diameter of 27 mm., radius of curvature 66 mm. and central thickness 1·5 mm.; what is its edge thickness?

5. An optician's lens measure has its outer points $\frac{3}{4}$ inch apart, and a depression of the centre point of 1/32 inch gives a reading of $+9.25$ D on its scale. For what refractive index was it scaled? (S.M.C.)

6. A lens measure applied to a brass tube of 9 inches external diameter gave a reading of 4·75 dioptres. For what refractive index was it adjusted? (S.M.C.)

7. A lens made of glass of refractive index 1·523 and having surface powers of $+6D$ and $-11D$ is to have a diameter of 40 mm. and 2 mm. central thickness. What will be the edge thickness of this lens and the least thickness of a plane glass plate from which it can be ground?

8. What must be the distance between the two outer points of a lens measure if a power of $+1D$ is to give a depression of $\frac{1}{4}$ mm. to the centre point for a refractive index of 1·53?

9. A pencil of light is made to converge by means of a lens to a point 20 cm. from the lens. Find the curvature of a wave front (a) on leaving the lens, (b) at 1 cm. from the lens, (c) at 19·9 cm. from the lens, (d) at 21 cm. from the lens.

10. A small source of light is placed 75 cm. from a lens; what is the curvature of the wave front at the lens? If the effect of the lens is to add 4D convergence to the light, find the position at which the light will focus.

11. Where must a small source of light be placed in order that the wave front of the divergent pencil shall have a curvature of 7D on reaching a lens. If the lens brings the light to a focus at a point 30 cm. beyond the lens, what will be the change in curvature of the wave front produced by the lens?

12. Show that when a pencil of convergent light is intercepted by a plane parallel plate of glass of thickness t and refractive index n, the position of the focus is moved outwards a distance $t - \dfrac{t}{n}$.

13. A plane wave front is incident in air on a curved glass surface of $+5$ cm. radius of curvature, refractive index of the glass 1·5, the line bisecting the wave front at right angles passing through the centre of curvature of the surface. Find by Huygens' construction the form of the refracted wave front. What do you notice about its form?

14. Derive the fundamental paraxial equation:—

$$\frac{n'}{l'} - \frac{n}{l} = \frac{n' - n}{r}$$

for refraction at a spherical surface.
Indicate clearly in your proof that the expression only applies to the paraxial region. Explain the convention adopted as to signs.

15. An object 1 cm. high is placed 15 cm. from a convex spherical refracting surface of 20 cm. radius of curvature, which separates air from glass ($n = 1\cdot52$). Find the position and size of the image. What are the positions of the principal foci of this surface?

16. A spherical refracting surface of radius $+8$ cm. separates air on the left from glass of refractive index $1\cdot5$ on the right. A parallel pencil is incident on it, the chief ray of the pencil passing through the centre of curvature. Find by graphical construction the refracted rays corresponding to incident rays at 3, 4, 5, 6 and 7 cm. above the chief ray. Note where they cut the chief ray.

17. What is the difference in the apparent thickness of a biconvex lens ($n = 1\cdot5$) having radii or curvature of 8 cm. and 20 cm. and central thickness 2 cm., the lens being examined first from one side and then from the other?

18. A plano-convex lens is viewed from above (a) with its plane surface uppermost, (b) with its convex surface uppermost. In (a) the greatest apparent thickness is $13\cdot3$ mm. and in (b) $14\cdot6$ mm. If the lens is 20 mm. thick, what is the refractive index of the material, and what is the radius of curvature of the convex surface? (L.U.)

19. A glass sphere has a diameter of 10 cm., and refractive index $1\cdot53$. Two small bubbles in the glass appear to be, one, exactly at the centre of the sphere and the other midway between the centre and the front surface; find their actual positions.

20. A glass paper-weight ($n = 1\cdot53$) has a picture fastened to its base. Its top surface is spherical and has a radius of curvature of $2\frac{1}{2}$ inches and its central thickness is $1\frac{1}{2}$ inches. Where will the picture appear to be when viewed through the top surface and what will be its magnification?

21. For the purpose of calculation the eye may be considered as a single refracting surface of $+60$D power. having air on one side and a medium of refractive index $\frac{4}{3}$ on the other. What must be the distance of the retina—the receiving screen—from this surface, if the eye be correctly focused on a distant object? What will be the distance of the object that is focused by an eye in which the distance between refracting surface and retina is 24 mm.?

22. Treating the cornea as a single spherical refracting surface of radius $+7\cdot8$ mm. separating air from the aqueous of refractive index $\frac{4}{3}$, find the position and size of the image of the eye's pupil, the latter being $3\cdot6$ mm. behind the cornea and 4 mm. in diameter.
What effects would be produced on the retinal image in an eye as the pupil is reduced in diameter? (S.M.C.)

23. A narrow beam of parallel light is incident normally on the flat surface of a glass hemisphere of 10 cm. radius. Find where the rays of light are brought to a focus (a) when the hemisphere is in air, (b) when it is in water. The refractive indices of glass and water may be taken to be $\frac{3}{2}$ and $\frac{4}{3}$ respectively.

24. A glass rod 2 cm. long and of refractive index $1\cdot60$ is plane at one end and has a convex spherical surface of radius of curvature $1\cdot875$ cm. worked on the other end. It is set up with the curved surface facing a distant point source. Find the curvature

of the wave front emerging from the plane surface and the position of the point where the light comes to a focus. (S.M.C.)

25. Considering the eye as a single refracting surface of $+5 \cdot 56$ mm. radius of curvature, having air on one side and a medium of refractive index $\frac{4}{3}$ on the other, what will be the size of the image formed of a distant object that subtends an angle of $1°$ at the eye?

26. Explain and prove Young's construction for the path of a ray of light refracted at a spherical surface of radius r separating two media of refractive index n and n'.

VI - The Thin Lens

59. Introduction

The lens may be defined as a portion of a transparent substance bounded by two polished surfaces, both of which may be curved, or one only may be curved and the other plane. The curved surfaces may be of a number of different forms, spherical, cylindrical, toroidal, or paraboloidal, but the commonest form of lens is that with spherical surfaces, and the work of this chapter will be limited to this form. Such lenses are usually called **spherical lenses.**

The term *lens*—in French *lentille*—is derived from the Latin word for the lentil, from the similarity of the convex lens to the form of the lentil seed. The use of lenses appears to have been known from very early times, lenses having been found that date back to about 1600 b.c. These lenses, which are of crystal, were no doubt used as magnifiers.

60. Forms of Lenses

Thin lenses may be divided into two classes according to the curvature of their surfaces and therefore the effect that they produce on the light. Lenses of the forms shown in Fig. 63 have either both surfaces or the steeper of the two surfaces convex.

Such lenses, if *thin* and of higher refractive index than the surrounding medium, tend to produce convergence of the light—a positive effect. They will therefore be termed **convex, convergent** or **positive.** Lenses having both surfaces or the steeper of the two surfaces concave, as shown in Fig. 64, will, under the same conditions, tend to make the light divergent. These will therefore be termed **concave, divergent** or **negative.**

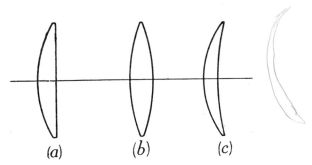

Fig. 63.—Convex, convergent or positive thin lenses

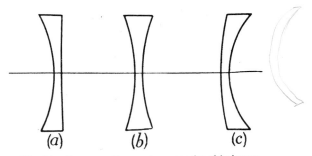

Fig. 64.—Concave, divergent or negative thin lenses

The following terms are applied to the various forms of lenses.

Positive.	Negative.
1. Plano-convex (Fig. 63a).	Plano-concave (Fig. 64a).
2. Bi-convex (Fig. 63b).	Bi-concave (Fig. 64b).
3. Equi-convex.	Equi-concave.
4. Convex meniscus (Fig. 63c).	Concave meniscus (Fig. 64c).
5. Convex periscopic.	Concave periscopic.

Periscopic (5) is a term applied to an ophthalmic lens of meniscus form having either one of the surfaces of comparatively small curvature (surface power = 1·25 D), the term "meniscus" being used in ophthalmic work for the more "bent" form.

The surface towards the incident light will be called the first surface of the lens, and positions in the space occupied by the incident light will be said to be in front of the lens.

Fig. 65 represents a section through a biconvex spherical lens, the spheres of which its surfaces are portions having centres at C_1 and C_2 respectively. The straight line C_1C_2 joining the centres of curvature of the two surfaces is called the **optical axis** of the lens. If one surface is plane, it may be considered as a portion of a sphere of infinite radius,

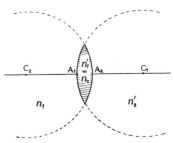

Fig. 65.—Optical axis of lens

and the optical axis will be the line perpendicular to the plane surface passing through the centre of curvature of the other surface. The points A_1 and A_2, where the axis meets the surfaces, are termed the **front** and **back vertex** or **pole** of the lens respectively, and A_1A_2 is the lens thickness. When the lens is so thin that the thickness may be neglected, the common vertex A is the **optical centre**. Lenses are usually "edged" so that the optical centre coincides with the geometrical centre of the lens aperture. This is not always the case with ophthalmic lenses, such lenses sometimes being "decentred". At the optical centre the two surfaces of the lens are parallel and any rays passing through this point are therefore unchanged in direction. The ray passing through the optical centre is the chief ray from any object point.

61. Refraction by a Lens

The change in direction of any ray on passing through a lens may be found by considering the refraction at each surface in turn. This may either be calculated from the law of refraction or may be found graphically by one of the methods previously described. It will be noticed that pencils of light on emerging from the lens are no longer homocentric. As will be expected from the result of refractions at a single curved surface, there will be spherical aberration, the rays through the more peripheral portions of the lens crossing the optical axis closer to the lens than the rays from the more central portions of the lens aperture.

Rays close to the optical axis, that is, in the paraxial region, meet the axis after refraction approximately in one point, and we shall assume that for the paraxial region point images are formed of point objects.

62. Focal Power

The function of the lens is to change the vergence of the light or the curvature of the wave fronts, thus producing an image—real or virtual—of the object from which the light is coming. This change in reduced vergence that the lens produces is termed its **focal power**. As this depends, in the case of a thin lens, only on the curvature of the surfaces and the refractive indices, the focal power of any one lens is a constant.

The expressions obtained in Chapter V for the focal power of a surface may be applied to each surface of the lens in turn, and we have for the paraxial region

First surface.

$$\frac{n'_1}{l'_1} - \frac{n_1}{l_1} = \frac{n'_1 - n_1}{r_1}$$

Second surface.

$$\frac{n'_2}{l'_2} - \frac{n_2}{l_2} = \frac{n'_2 - n_2}{r_2}$$

$$n_2 = n'_1.$$

If the thickness of the lens is neglected, the vergence $\dfrac{1}{l'_1}$ of the refracted light at the first surface is the vergence $\dfrac{1}{l_2}$ of the incident light at the second surface, and $\dfrac{n_2}{l_2} = \dfrac{n'_1}{l'_1}$. Then from the expressions above for the separate surfaces

$$\frac{n'_2}{l'_2} - \frac{n_1}{l'_1} = \frac{n'_1 - n_1}{r_1} + \frac{n'_2 - n_2}{r_2}$$

As $\dfrac{n'_2}{l'_2}$ and $\dfrac{n_1}{l_1}$ are the reduced vergences of the incident and emergent light respectively, the expression gives the focal power of a *thin* lens with media of any refractive index in the object and image spaces.

Thin lens in air

If the lens is in air, then $n_1 = n'_2 = 1$ and

putting $\qquad\qquad n'_1 = n_2 = n \qquad\qquad$ we have

$$\frac{1}{l'_2} - \frac{1}{l_1} = \frac{n-1}{r_1} + \frac{1-n}{r_2} = F_1 + F_2$$

where F_1 and F_2 are the surface powers.

$$\frac{1}{l'_2} - \frac{1}{l_1} = (n-1)\left(\frac{1}{r_1} - \frac{1}{r_2}\right)$$

or, dropping the suffixes of l'_2 and l_1

$$\frac{1}{l'} - \frac{1}{l} = (n-1)\left(\frac{1}{r_1} - \frac{1}{r_2}\right) = L' - L \quad \ldots \quad \ldots \quad (6.01)$$

$L' - L$ is the focal power, F, of the lens and we have

$$F = (n-1)\left(\frac{1}{r_1} - \frac{1}{r_2}\right) = (n-1)(R_1 - R_2) \quad \ldots \quad (6.02)$$

or

$$\left. \begin{array}{c} F = L' - L \\ L' = L + F \end{array} \right\} \quad \ldots \quad \ldots \quad \ldots \quad (6.03)$$

For a lens in air L and L' are the vergences at the lens of the incident and emergent light respectively, and the expression (6.03) may be stated as follows:

For a lens in air, the final vergence L' of the light is equal to the incident vergence L plus the impressed vergence F.

63. Conjugate Foci and Principal Foci

Applying expression (6.03) to the case of a positive lens, we see that with an object at an infinite distance, the light emerging from the lens is convergent, and a real image is formed. The position of this image is the **second principal focus,** F'. As the object moves up towards the lens, the incident light becomes more and more divergent, and therefore the emergent light becomes less convergent, and the image moves away from the lens until the object is in such a position that the image is at infinity, that is, the emergent light is parallel. The object is now at the **first principal focus,** F. If the object is brought still closer to the lens, the power of the lens is insufficient to overcome all the divergence of the incident light, and the emergent light diverges from a virtual image on the same side of the lens as the object, but farther from the lens than the object is.

With a negative lens, as its effect is to add divergence to the light, a virtual image will always be formed of a real object, and this image will be closer to the lens than the object. The position of the virtual image when the object is at an infinite distance will be the second principal focus, F'.

When the incident light is convergent, the point to which it is converging may be considered as a *virtual object*. The positive lens will always form a real image of this virtual object closer to the lens, since it adds convergence to the convergent incident light. The negative lens may, however, in this case, form either a real or a virtual image according to whether the convergence of the incident light at the lens is greater or less than the lens power. If the convergence of the incident light at the lens is equal to the power of the negative lens, the emergent light will be parallel, and the image will be at infinity. The position of the virtual object is then the first principal focus, F. With the virtual object closer to the negative lens than the first principal focus—"inside" the

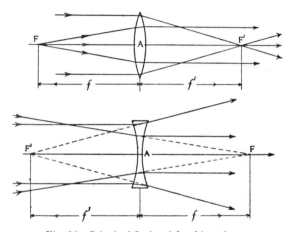

Fig. 66.—Principal foci and focal lengths

first principal focus—a real image is formed, and with the virtual object "outside" the first principal focus the image is virtual. The positions of the principal foci of a positive and a negative lens are shown in Fig. 66.

As with a single surface, the conjugate foci are interchangeable, and therefore, for any one pair of real object and real image positions, except when the lens is situated exactly midway between these positions, there will be two possible positions of the lens, the object distance in one position being equal numerically to the image distance in the other and vice versa.

64. Focal Lengths

The distance AF (Fig. 66), *from* the lens to the first principal focus, is the **first** (or anterior) **focal length,** f, and the distance AF' *from* the lens to the second principal focus is the **second** (or posterior) **focal length,** f'.

When the object is at infinity, $L = 0$, and for a thin lens in air we have

$$L' = F$$

$$l' = f' = \frac{1}{F} \qquad \dots \qquad \dots \qquad \dots \qquad (6.04)$$

Also, when the image is at infinity, $L' = 0$, and therefore

$$-L = F$$

$$l = f = -\frac{1}{F} \qquad \dots \qquad \dots \qquad \dots \qquad (6.05)$$

Thus, for a lens in air, or in any case where the first and last media are of equal refractive index, the two focal lengths are numerically equal but of opposite sign, the second focal length having the same sign as the power. Except in the case of ophthalmic lenses, which are expressed in focal power, most lenses are specified in terms of their second focal lengths.

Expression (6·03) may also be written

$$\frac{1}{l'} = \frac{1}{l} + \frac{1}{f'} \qquad \dots \qquad \dots \qquad \dots \qquad \dots (6.03\text{A})$$

A table of reciprocals will be found of great convenience in solving problems on conjugate foci.

65. Chromatic Aberration

As the power of a lens depends upon the refractive index and this varies with the colour of the light, it is evident that the power of any lens is different for light of different colours.

An axial object point emitting white light will therefore be imaged as a short spectrum, lying along the axis, the violet being focused closest to the lens and the red farthest away, as in Fig. 67. The effect is known as **chromatic aberration.** In order that a lens shall produce good images it is necessary that light of different colours shall, as far as possible, be brought to a focus at the same place. This aberration can be largely corrected by combining a positive and a negative lens of suitable powers and made of different types of glass. Such a combination is known as an **achromatic lens.** Chromatic aberration and the principles of its correction are more fully considered in Chapter XIII. By suitably choosing the curves and glasses of a combination of lenses, both spherical and chromatic aberration can be largely corrected; the expressions obtained for the paraxial region will then be true for large apertures.

Fig. 67.—Chromatic aberration

66. Images of Extra-Axial Points

As with a single refracting surface, the pencil of light from an object point not on the lens axis will, after refraction, no longer be homocentric, even when the lens aperture is small. The light will be affected by various aberrations that are described in Chapter XVIII. The paths of any rays from an extra-axial object point can be found by calculation from the law of refraction or by one of the graphical constructions.

When the distance of the object from the axis is small as compared with its distance from the lens and when the aperture of the lens is also small, all light reaching the lens from the object will be in the paraxial region, and in this case a reasonably good image of the object will be formed by a single lens. By suitably combining a number of lenses it is possible to correct the oblique aberrations to a great extent, and in the case, for example, of photographic objectives and certain eyepieces, good images are produced of objects at considerable angles from the axis and with lenses of large aperture. When only a small aperture of the lens is to be used, as in the spectacle lens, considerable reduction of the oblique aberrations may be effected by a suitable choice of curves for the two surfaces of a single lens. The expressions obtained for the positions of the images of extra-axial points lying in the paraxial region may also be used in dealing with objects at considerable distances from the axis in the case of corrected systems.

If Q (Fig. 68) is an extra-axial object point of which the lens forms a good image, all the light from the object which passes through the lens will converge to a real image point or diverge, from a virtual image point, and the position of this image Q' is fixed if the intersection of two refracted rays can be found. From the properties of the lens that have already been considered the directions of certain rays after refraction

97

are known, and any two of these may be used to find the image position. These known ray directions are as follows:

(a) Rays passing through the optical centre of the lens are undeviated —chief rays.

(b) Rays passing through the first principal focus leave the lens parallel to the axis.

(c) Rays incident parallel to the axis pass, after refraction, through the second principal focus.

In Figs. 68 and 74 three rays QA, QD and QE from a point Q are shown converging after refraction to a real image Q′ in the case of the positive lens and diverging as from a virtual image Q′ in the case of the negative lens.

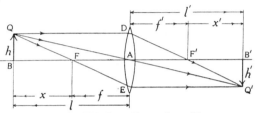

Fig. 68.–Thin lens. Formation of image

67. Object of Finite Size. Lateral Magnification

The position and size of the image of an object of finite size can be determined by finding the image positions for a number of points on the object by the above method. It will be found that, within the paraxial region, for an object perpendicular to the axis an image is formed also perpendicular to the axis, as in Fig. 68. The position of the image is determined from the conjugate foci relation (6.03), and its size as follows:

$$\frac{h'}{h} = \text{lateral magnification of the image} = m$$

From Fig. 68

$$\frac{B'Q'}{BQ} = \frac{AB'}{AB}$$

or
$$\frac{h'}{h} = m = \frac{l'}{l} = \frac{L}{L'} \qquad \dots \qquad \dots \qquad \dots \quad (6.06)$$

The magnification will be positive when the image is erect, that is, when Q′ is on the same side of the axis as Q, and negative when the image is inverted.

Example:
When an object is a certain distance from a thin positive lens a real inverted image one-half the size of the object is formed. The object is moved 10 cm. nearer the lens and the real image formed is now the same size as the object. What is the focal length of the lens?

$$m_1 = -\frac{1}{2} = \frac{l'_1}{l_1}; \quad l'_1 = -\frac{l_1}{2}$$

$$m_2 = -1 = \frac{l'_2}{l_2}; \quad l'_2 = -l_2$$

$$\frac{1}{f'} = \frac{1}{l'_1} - \frac{1}{l_1} = -\frac{2}{l_1} - \frac{1}{l_1} = -\frac{3}{l_1}$$

$$\frac{1}{f'} = \frac{1}{l'_2} - \frac{1}{l_2} = -\frac{1}{l_2} - \frac{1}{l_2} = -\frac{2}{l_2}$$

$$\frac{3}{l_1} = \frac{2}{l_2} = \frac{2}{l_1 + 10}$$

$$l_1 = -30 , l'_1 = +15$$

$$\frac{1}{f'} = \frac{1}{15} + \frac{1}{30} = \frac{3}{30}$$

$$f' = 10 \ cm.$$

Further expressions for magnification are easily derived from (6.03) and (6.06), thus

$$m = \frac{L}{L'} = \frac{L'-F}{L'}$$

$$= 1 - l'F = 1 - \frac{l'}{f'} \quad \ldots \quad \ldots \quad \ldots \quad (6.07)$$

This is an expression useful in problems on the projection lantern. Also

$$\frac{1}{m} = \frac{L'}{L} = \frac{L+F}{L}$$

$$= 1 + l F = 1 + \frac{l}{f'} \quad \ldots \quad \ldots \quad \ldots \quad (6.08)$$

an expression useful in problems on the camera and magnifier.

Note.—In expressions (6.07) and (6.08), *l* and *l'* must be expressed in metres when used with *F*, which is in dioptres.

With any lens there must be a position of object and image such that the image is of the same size as the object but inverted, then $m = -1$, and $l' = -l$. The object and image are then said to be in the **symmetrical planes** of the lens, and the distance between these planes for a thin lens, $l' - l = 4f'$. This is the minimum distance between a real object and real image in the case of a positive lens or between a virtual object and virtual image in the case of a negative lens.

Fig. 69.—Thin lens. Image of distant object

The positions of object and image such that $m = +1$, that is, where object and image are of the same size and on the same side of the axis, are known as the **principal planes** of the lens, and in a thin lens these coincide at the lens. These planes are of great importance in the case of thick lenses and lens systems and are dealt with in Chapter IX and XIX.

68. Newton's Relation

In the same way as for a single refracting surface, the conjugate foci and magnification expressions for a lens may be expressed in terms of the extra-focal distances, *x* and *x'*. Thus, from Fig. 68 we have

$$m = -\frac{f}{x} = \frac{f'}{x} \qquad \dots \qquad \dots \qquad \dots \quad (6.09)$$

and
$$m = -\frac{x'}{f'} \qquad \dots \qquad \dots \qquad \dots \quad (6.10)$$

Hence
$$-xx' = f'^2 \qquad \dots \qquad \dots \qquad \dots \quad (6.11)$$

Newton's relation, expression (6.11), is the basis of various methods of measuring the focal lengths of lens systems and thick lenses (§ 102).

69. Image of a Distant Object

When the object is very distant the rays from any one point are parallel to one another on reaching the lens. The position of the image of a distant object point not on the axis may be determined by finding the point at which two parallel incident rays intersect after refraction. The two most convenient rays to consider will be the chief ray, and the ray incident through the first principal focus and therefore refracted parallel to the axis (Fig. 69). It will be found that the image is formed in the second focal plane, and its position in this plane is therefore determined by the chief ray.

If w is the angle subtended at the lens by a distant object standing on the axis, or the apparent size of the object, then the size of the image.

$$\left. \begin{array}{l} h' = -f' \tan w = f \tan w \\ \quad = fw, \text{ when } w \text{ is small} \end{array} \right\} \qquad \dots \qquad \dots \quad (6.12)$$

When the object is situated symmetrically to the axis, as is usually the case,

$$h' = 2f \tan \frac{w}{2} \qquad \dots \qquad \dots \qquad \dots (6.12\text{A})$$

where w is the angle subtended by the whole object. When w is small

$$2 \tan \frac{w}{2} = \tan w = w$$

Problems involving the determination of the size of the image of a distant object frequently occur in dealing with the eye, the telescope and the camera.

70. Magnification of Images in the Direction of the Axis

It is sometimes necessary to consider the size of images of objects lying along the lens axis, or the separation of the images of objects at different distances from the lens. Let $B_1 B_2$ (Fig. 70) be the two ends of an object lying along the axis and $B'_1 B'_2$ the corresponding image; then

$$\frac{1}{l'_1} - \frac{1}{l_1} = \frac{1}{l'_2} - \frac{1}{l_2} = \frac{1}{f'}$$

$$\frac{l'_2 - l'_1}{l'_1 l'_2} = \frac{l_2 - l_1}{l_1 l_2}$$

$$\frac{l'_2 - l'_1}{l_2 - l_1} = \frac{l'_1 l'_2}{l_1 l_2}$$

$l'_2 - l'_1$ is the length of image corresponding to the object of length $l_2 - l_1$, and $(l'_2 - l'_1)/(l_2 - l_1)$ is termed the longitudinal or axial magnification.

When the lateral magnification, $m = \dfrac{l'}{l}$, is very small, as with a comparatively distant object, $\dfrac{l'_1}{l_1}$ and $\dfrac{l'}{l_2}$ are nearly equal, and the longitudinal magnification.

$$\frac{l'_2 - l'_1}{l_2 - l_1} = m^2 \text{ (approx.)} \qquad \dots \qquad \dots \quad (6.13)$$

Thus, if an image in a photograph, for example, is one-hundredth the size of the object, the "depth" of the image along the axis is only

Fig. 70.—Longitudinal or axial magnification

one ten-thousandth the depth of the object. This fact is of importance in considering depth of focus and allows of reasonably good images of objects at different distances being produced on a plane surface, such as the photographic plate.

71. Prismatic Effect of a Lens

Neglecting aberration, all rays from an axial point are deviated at the lens in such a way that they pass through a single image point on the axis, and each ray therefore suffers an amount of deviation depending on the distance from the axis at which it meets the lens. We may therefore consider the lens as consisting of a series of innumerable prisms, of which the refracting angles, and hence the deviations produced, increase steadily from the centre to the periphery. The refracting angles of these prisms are towards the periphery in the positive lens and towards the centre in the negative lens, while at the optical centre, that is on the axis, the surfaces are parallel, and there is no deviation. The deviation produced on a ray passing through any point on a lens will be termed the **prismatic effect** of the lens at that point, and this increases from the optical centre towards the edge of the lens.

This prismatic effect may be easily seen on looking through a lens at an object. When the eye is looking through the optical centre of the lens, that is, when the line joining the object to the eye passes through the optical centre, the object will appear unchanged in position laterally.

On moving the lens at right angles to the axis the object will appear displaced, in the opposite direction to the lens in the case of a positive lens, and in the same direction as the lens in the case of a negative lens, the rays from the object entering the eye having been deviated by the prismatic portion of the lens through which they have passed. This apparent displacement of an object seen through a lens, as it is moved laterally, forms the basis of the neutralisation method commonly used for determining the power of ophthalmic lenses. Spectacle lenses are often "decentred" before the eye in order to give a prismatic effect in addition to the focusing effect of the lens.

The amount of prismatic effect produced at any point on the lens must depend on the power of the lens and the distance from the axis and may be found as follows:

Any ray BD (Fig. 71a) parallel to the axis, meeting the lens at a distance $AD = c$ from the optical centre, is deviated by the lens (neglecting aberrations) towards the second principal focus F'. Then if d is the angle through which the ray is deviated,

$$\tan d = \frac{c}{f'} = c\,F$$

$\tan d \times 100$ = prismatic effect P in prism dioptres (§ 39) hence P in prism dioptres = c in cm. $\times F$ in dioptres ... ... (6.14)
c is known as the **decentration** of the particular ray.

Fig. 71.—Prismatic effect of lenses

In Fig. 71b a ray incident towards B is deviated by the lens and passes through B'.
Then

$$d = u' - u$$

As the angles are small—paraxial region—

$$u = \frac{c}{l}; \ u' = \frac{c}{l'} \text{ and } d = \frac{c}{f'}$$

hence

$$d = c\left(\frac{1}{l'} - \frac{1}{l}\right) = \frac{c}{f'} = CF.$$

Therefore for all paraxial rays meeting a lens at the same height above the axis, the deviation produced by a given thin lens is a constant, and is equal to $\frac{c}{f'}$ radian $= cF$ prism dioptres, when c is in centimetres and F in dioptres of power.

72. Systems of two or more Lenses—Elementary Treatment

In practice, single thin lenses are very seldom used and most optical instruments consist of a system of two or more lenses. The most frequent use of a single and comparatively thin lens occurs in the ophthalmic lens, but this is used in conjunction with the eye, which is itself a compound system. Systems of lenses and thick lenses are considered in Chapter IX and XIX, but the position and size of the image formed by a system of thin lenses may be found by applying the results obtained in the present chapter to each of the lenses of the system in turn.

Thin Lenses in contact.

When a number of thin lenses are placed in contact, their axes being coincident, the vergence of the light on leaving any one lens is the vergence of the light on entering the next. It may be shown that the power of the system is then the sum of the powers of the separate lenses.

$$L_1' = L_1 + F_1 \text{ and } L_1' = L_2$$

$$L_2' = L_2 + F_2 \text{ and } L_2' = L_3$$

$$L_3' = L_3 + F_3 \text{ etc.}$$

$$L_2' = L_2 + F_2 = L_1' + F_2 = L_1 + F_1 + F_2$$

$$L_3' = L_3 + F_3 = L_2' + F_3 = L_1 + F_1 + F_2 + F_3$$

$$L_3' - L_1 = F_1 + F_2 + F_3$$

$L_3' - L_1$ is the power of the complete system, F

Hence $F = F_1 + F_2 + F_3$ ... ... (6.15)

In practice, usually not more than two or three lenses in contact can still be considered as a thin lens.

Separated Thin Lenses.

When the lenses of a system are separated, the vergence of the light leaving one lens is changed before reaching the next, and expression (6.15) no longer applies. The position and size of the final image formed by a system of coaxial thin lenses may, however, be found by considering each lens in turn, and treating the image formed by one lens as the object for the next. Thus we have

$$L_1' = L_1 + F_1 \text{ and } l_2 = l_1' - d$$

where d is the distance between the lenses.

$$L_2' = L_2 + F_2$$

Hence l_2' the distance of the image from the second lens.
Also

$$m_1 = \frac{h'_1}{h_1} = \frac{l'_1}{l_1} \text{ and } h'_1 = h_2$$

$$m_2 = \frac{h'_2}{h_2} = \frac{h'_2}{m_1 h_1} = \frac{l'_2}{l_2}$$

hence

$$m = \frac{h'_2}{h_1} = m_1 \times m_2 = \frac{l'_1}{l_1} \times \frac{l'_2}{l_2}$$

The above method is cumbersome where much work is to be done with complicated systems, and the methods of Chapter IX are preferable.

73. Graphical Constructions

Many problems involving the determination of the position and size of the image formed by mirrors, lenses and lens systems may be quickly solved by graphical methods, and the student should make himself quite familiar with the constructions used, and practised in the production of accurate drawings.

Choice of scale will be an important factor in obtaining accurate results. Dimensions perpendicular to the axis are usually very small as compared with those along the axis, and these vertical dimensions, such as object and image sizes, may be drawn to a greatly magnified scale, while dimensions along the axis can be drawn in many cases full size or smaller. Angles will therefore be shown magnified, and some that are really equal will appear unequal; but all distances along and perpendicular to the axis will be shown in their relative scales.

As the principles upon which the graphical constructions are based apply only to rays in the paraxial region, that is, to surfaces and lenses of very small aperture, the surface or lens may be represented graphically as a straight line perpendicular to the axis. The diagrams so obtained will not, of course, represent the actual path of the light, but this may easily be put in when the positions of object and image have been found. The constructions are based on the following facts:

(*a*) Rays passing through the optical centre of a lens, or the centre of curvature of a surface, are unchanged in direction.

(*b*) Rays passing through the first principal focus of a lens or surface leave the lens or surface parallel to the axis.

(*c*) Rays incident parallel to the optical axis pass, after refraction by a lens or surface, through the second principal focus.

(*d*) All rays that are parallel to one another in the object space intersect after refraction in the second focal plane.

(*e*) All rays proceeding from a point in the first focal plane emerge from the lens or surface, after refraction, parallel to one another.

The following typical constructions should, with a few exceptions, be self-explanatory:

Positive Lens. Real Object. Real Image (Inverted). **Fig. 68.**
Positive Lens. Real Object. Virtual Image (Erect). **Fig. 72.**

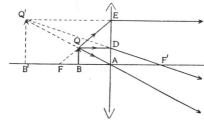

Fig. 72.--Positive lens. Real object. Virtual image (erect).

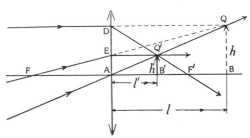

Fig. 73.—Positive lens. Virtual object. Real image (erect).

Positive Lens. Virtual Object. Real Image (Erect). **Fig. 73.**
Positive Lens. Distant Object. Real Image (Inverted). **Fig. 69.**
Negative Lens. Real Object. Virtual Image (Erect). **Fig. 74.**

Positive Lens. Image of Axial Object Point, and Construction of the Refracted Ray for any Incident Ray. Fig. 75.

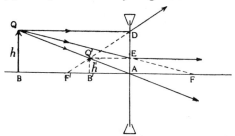

Fig. 74.—Negative lens. Real object. Virtual image (erect).

A ray from the given object point B meets the lens at D and the first focal plane at E. Join EA; the emergent ray DB′ is parallel to EA (rule *e* above), hence B′. This construction also enables the complete path of any portion ST of an incident ray to be found; it is extremely useful and will be frequently used. An alternative construction is to draw

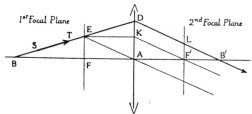

Fig. 75.—Positive lens. Image of axial object, and construction for the refracted ray of any incident ray

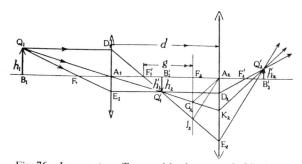

Fig. 76.—Lens system. Two positive lenses, real object

a line from E parallel to the axis meeting the lens at K; join KF′. Then the emergent ray DB′ is parallel to KF′.

Negative Lens. Image of Axial Object Point.
The construction is as in Fig. 75.

Lens System. Two Positive Lenses. Real Object. Fig. 76.
The commencement of the construction is as in Fig. 68. The rays continuing beyond Q'_1 meet the second lens at D_2, K_2 and E_2, the last two cutting the F_2 plane in G_2 and J_2. Draw the emerging rays $K_2Q'_2$ and $E_2Q'_2$ parallel to G_2A_2 and J_2A_2 respectively. The ray E_1D_2 is parallel to the axis and thus passes through F'_2 intersecting the other two emerging rays in the final image Q'_2.

Lens System. Distant Object. Equivalent Lens. Fig. 77.
The image $F'_1Q'_1$ is formed in the second focal plane of the first lens. It is a virtual object for the second lens. Rays $A_2Q'_1$ and $E_2F'_2$ fix the position of Q'_2. This final image is in the second focal plane of the

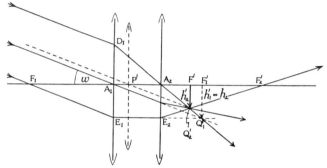

Fig. 77.—Lens system. Distant object. Equivalent lens

system. Q'_2P' drawn parallel to the original incident beam fixes P', the position of a thin lens producing the same effect. This is called the **equivalent thin lens** (see Chapter IX).

74. Effective Power of a Lens

The vergence of the emergent light at A (Fig. 78) immediately on leaving the lens is $\dfrac{1}{l'} = L'$. At some place X at a distance d from the lens, the vergence has changed to

$$\frac{1}{l'-d} = \frac{1}{l'_x} = L'_x.$$

$$L'_x = \frac{1}{l'-d} = \frac{L'}{1-d\,L'}$$

If the incident light is parallel, $L' = F$, and substituting F_x for L'_x

in this case, we have

$$F_x = \frac{F}{1 - dF} \qquad \dots \qquad \dots \qquad \dots \quad (6.16)$$

d, of course, being in metres when F is in dioptres. This is termed the **effective power** of the lens at the position X. The idea of effective power is very useful in dealing with systems of lenses, e.g., the spectacle lens and the eye. It will be seen that the effective power of a positive lens increases, while that of a negative lens decreases, as the position X moves away from the lens.

Fig. 78.–Effective power of a lens

75. The Measurement of Focal Length and Power

The effect of a lens is specified by its focal power or focal length, and it is necessary to be able to measure these quantities, often with considerable accuracy. A large number of methods are available, and a few of the most useful and typical will be described here.*

When the refractive index of the glass is known, the power of a thin lens may be found from a spherometer measurement of its curvatures. This is done in the case of the optician's lens measure. (§ 44).

Neutralisation. The method commonly used by the ophthalmic optician is that known as **neutralisation.** The power of the lens is determined by finding a lens that, when placed in contact with the lens to be measured, gives a combination having zero power, as judged by observing that a lateral movement of the combination before the eye produces no apparent movement of an object seen through it. If the two lenses are considered as thin, the lens tested is of equal but opposite power to the known lens. The test is carried out with the optician's trial case, which consists of a set of positive and negative lenses of known power increasing by small steps.

* Fuller descriptions of methods of measuring focal length will be found in B. K. JOHNSON, *Practical Optics*, A. F. WAGNER, *Experimental Optics*, R. S. CLAY, *A Treatise on Practical Light*.

76. The Optical Bench

Most methods involve the use of an **optical bench**. This consists of a horizontal bar, usually of either one or two metres length, graduated in centimetres and millimetres. A number of stands to carry lenses, screens, objects, etc., slide along the bar, their positions being read by an index against the scale. As an object, cross-lines stretched across a frame, or a series of small holes, in the form of a cross, drilled in a metal plate, may be placed in front of a frosted electric lamp. To receive a real image a ground glass screen is used, the image being viewed *through* the glass. Adjustable holders may be provided to carry lenses and systems of various diameters, and it is very convenient to have a lens holder which will carry the standard trial case lenses of 38 mm. diameter, a large selection of lenses thus being easily available for experimental purposes. Other bench fittings include a collimator and telescope (see below), and a measuring rod to enable distances between lenses, screens, etc., to be determined when these are not exactly over the index mark.

77. Optical Bench Methods. Positive Lenses

Collimator Method.

When a suitable distant object is available the focal length of a thin positive lens may be found directly by receiving the image of the object on a screen, its distance from the lens being the second focal length. In the laboratory, however, a convenient distant object is not usually available, but the same effect may be obtained by means of a piece of apparatus known as a **collimator.** This consists of a positive achromatic lens mounted at one end of a tube with an illuminated object of some form at the other; the object may be a small circular aperture, a narrow slit, cross-lines or a scale, according to the purpose for which the collimator is required. The object is placed at the first principal focus of the lens, so that light from any point on the object is parallel on emerging from the lens, as though coming from a distant object. The collimator is an important piece of apparatus in many optical experiments.

To use the collimator in determining the focal length it is mounted with its axis in line with the bench and the image formed by the lens under test is received on the screen as with an actual distant object.

Telescope Method.

The position of the first principal focus and hence the focal length of a thin positive lens can be found if we can determine when the light emerging from the lens is parallel. This can be done by means of a telescope. The telescope can be of the astronomical form (see Chapter X) with cross-lines in the focal plane of the eyepiece. To adjust the telescope the eyepiece is first sharply focused on the cross-lines; the telescope is then directed to a very distant object and focused so that the image of the distant object falls exactly in the plane of the cross-lines. This is checked by noting that there is no parallax between cross-lines and image, that is, there is no apparent movement of one against the other as the eye is moved slightly from side to side. An object, the lens to be tested and the adjusted telescope are now set up in line on the bench, and the distance between lens and object varied until a clear image of the object is seen in the telescope. Since the telescope has been focused for parallel light, the object must now be at the principal focus of the lens, and the focal length can be read on the bench.

Mirror Method.

A simple method of determining when the light leaving a lens is parallel is by means of a plane mirror placed behind the lens to reflect the emergent light back through the lens. When the object is at the first principal focus of the lens, the parallel light reflected by the mirror must focus, after passing again through the lens, in the plane of the object. If the object is inside the first principal focus, the divergent emergent light will be still divergent after reflection and will therefore focus, after again passing through the lens, outside the principal focus. Similarly if the object is outside the first principal focus, the reflected image will be inside.

Object, lens and plane mirror are set up on the bench, and the reflected image received on a small screen placed in the plane of the object; or a small object, such as a pin, may be used, this being adjusted until it and the reflected aerial image coincide as judged by parallax.

This method of accurately placing an object at the principal focus of a lens by obtaining coincidence of the object and its reflected image is known as **auto-collimation.**

Conjugate Foci Method.

When the focal length of the lens to be measured is sufficiently short for the object and its real image to come within the length of the bench, the focal length or the focal power may be determined by forming a real image of the object on a screen and measuring the distances l and l' of the object and image from the lens. The focal length or focal power is then found from the conjugate foci expression (6.03) or (6.03A). It should be remembered that in order that a real image of the object shall be received on the bench, the length of the bench must be at least four times the focal length of the lens.

Fig. 79.—Measurement of focal length of negative lens.
Conjugate foci method

78. Optical Bench Methods. Negative Lenses

As the negative lens will not form a real image of a real object, methods of measuring negative lenses are a little more complicated than those described above. In most cases an additional positive lens is used to give convergent light incident on the negative lens, the image formed by the positive lens acting as a virtual object for the negative lens. The following are two simple methods of measuring the focal length or focal power of negative lenses.

Conjugate Foci Method.

A positive lens A_1 (Fig. 79) is arranged to give a real image of an object B on a screen at B'_1. The negative lens to be measured, A_2, is placed between the positive lens and the screen, and the distance $A_2 B'_1$ is then the object distance l_2 for the negative lens. If this distance l_2 is less than the focal length of the negative lens, the light on leaving this lens will be still convergent, and the screen may be moved back to the new position B'_2, where a sharp image is again formed. The distance $A_2 B'_2$ is then the image distance l'_2 for the negative lens, and the focal power or focal length can be found from expressions (6.03) or 6.03A). Both object and image distances are positive in this case.

This method is also useful for measuring the focal length and power of weak positive lenses. The lens to be measured is placed between the lens A_1 and the first image, as with the negative lens, and the screen must then be moved towards the lens to focus the new image.

Telescope Method.

Using the same arrangement as in the previous experiment, but replacing the screen in its second position by a telescope focused on a distant object, the negative lens may be adjusted until the light emerging from it is parallel (Fig. 80), when a sharp image of the object will be seen in the telescope. The image formed by the positive lens is now at the

Fig. 80.—Measurement of focal length of negative lens
Telescope method.

principal focus of the negative lens, and the focal length can be read off directly on the bench.

Other methods that are suitable also for measuring the focal lengths of thick lenses and systems, are described in Chapter IX.

CHAPTER VI   *EXERCISES*

1. The radius of curvature of one surface of a biconvex lens made of glass having a refractive index of 1·53 is six times that of the other surface. If the power of the lens is 5D, what are its radii of curvature?

2. Distinguish between a real and a virtual image, illustrating your answer by describing the formation of (a) a real image, (b) a virtual image, by a convex lens.

3. Explain what is meant by the focal power and the focal length of a thin lens, illustrating your answer by the case of both positive and negative lenses. A positive lens is to be made of glass having a refractive index of 1·55 and is required to have a focal length of 15 cm.; what will be the radii of curvature of its surfaces if the lens is made (a) equi-convex, (b) plano-convex?

4. A thin plano-convex lens is to be made from glass of refractive index 1·62, so that its power will be +5 dioptres. Find the radius of the tool with which the convex surface must be ground.

5. A very thin watch glass, the radius of curvature of which is 4 inches, is placed on a table; water is poured into it to a depth of ⅛ inch. Find the aperture of the water lens thus formed and its focal power and focal length.

6. Draw, full size, a section through an equi bi-convex lens of 80 mm. aperture and surface radii 50 mm. From a point on the axis 40 mm. from the front surface draw a ray inclined to the axis at 30°. Trace its path through the lens and find the corresponding image point on the axis. Assume the refractive index to be 1·5.

7. Establish a formula which gives the focal lengths of a thin lens in terms of the positions of an axial object and its image.
A rod AB 15 cm. in length lies along the axis of a thin converging lens of focal length 20 cm. If the nearer end of the rod is 25 cm. from the lens, find the length of the image and show its position on a diagram. (L.U.)

8. Where must an object be placed in front of a convex lens of 8 cm. focal length to give an image:
        (a) 25 cm. in front of the lens,
        (b) 25 cm. behind the lens.
State in each case whether the image is real or virtual.

113

9. An object is placed 33·3 cm. from an equi-convex lens and a virtual image is formed 2 metres from the lens. Calculate the radii of curvature of the lens if its refractive index is 1·523.

10. A small object is moved along the optical axis of a positive thin lens of focal length 25 cm. Find the position of the image and the vergence of the emergent pencil (or curvature of emergent wave front) for each of the following object distances: (a) $-\infty$; (b) -200 cm.; (c) -100 cm.; (d) -50 cm.; (e) -25 cm.; (f) -10 cm.

11. An object 5 mm. high is placed 60 cm. in front of a negative lens of focal length 25 cm. Find (a) by calculation, (b) by graphical construction to scale, the position, size and attitude (erect or inverted) of the image.

12. A positive lens forms a real image of a luminous object. If the lens is gradually covered over, what happens to the image? Explain clearly with diagrams.

13. Give a graphical construction for the image formed by a convergent lens. A small object is placed upright on the axis of a convergent lens at a distance of 10 cm. from the lens, and it is observed that a virtual erect image of twice the linear dimensions of the object is formed. Determine the focal length and power of the lens. (B.O.A.)

14. A lamp and screen are 3 feet apart and a + 7D lens is mounted between them. Where must the lens be placed in order to give a sharp image, and what will be the magnification? (S.M.C.)

15. How does the transverse magnification produced by a lens of focal length f depend on the distance of the object from the lens? Calculate the focal length of the lens required to throw an image of a given object on a screen with a linear magnification of 3·5, the object being situated 20 cm. from the lens. (L.U.)

16. Prove for (a) a positive lens, (b) a negative lens, that the magnification of an image is changed by unity for a motion in either direction of the image equal to the focal length of the lens.

17. What is meant by the magnification of an image? An object is placed 30 mm. from a positive lens and a virtual image four times the size of the object is required. Find graphically and by calculation the necessary focal length of the lens.

18. An object is to be photographed three-quarters full size using a lens of 6 inches focal length. What must be the distance of the object from the lens?

19. In photographing a person 6 feet high and standing 12 feet from the camera it is required to obtain an image 4 inches high. What must be the focal length of the lens?

20. An object 1 cm. long is placed perpendicular to the optical axis of a positive lens, the lower end of the object being 5 cm. above a point on the axis 20 cm. from the lens. The lens has a focal length of 8 cm. and an aperture of 3 cm. Find graphically the size and position of the image.

21. When a real image is formed 9 inches from a convex lens the magnification is unity. Where will a real image be formed when the magnification is fifty? What is the focal length of the lens?

22. It is required to form a real image 80 cm. from an object and magnified four times. Find the position and focal length of the required lens.

23. The distance between an object and its real image formed by a lens is 32 inches. If the image be three times the size of the object, find the position and focal power of the lens.

24. The difference in the positions of the image when the object is first a great

distance and then at 5 metres from the anterior principal focus of a convex lens is 3 mm. Calculate the focal length of the lens.

25. Apply carefully a graphical construction to find the position and size of the image of an object 2 mm. high formed by a positive lens of focal length 25 cm. The object distance is (a) − 50 cm., (b) − 80 cm. Explain the reasons for the various steps in the construction.

26. With a projection lantern it is required to project an image 20 feet in diameter on a screen 50 feet from the lens. What focal length is required, the diameter of the object being 3 inches?

27. Find the size and position of the image formed by a convex lens of 10 inches focal length, of the following object, viz., an arrow 1 foot 8 inches long lying along the axis of the lens so that its middle point is 2 feet 1 inch from the lens. What would be the size and position of the image if the arrow were turned through 90° about its middle point?

28. When an object is 10·5 inches in front of a lens the virtual image formed is 4·3 inches from the lens. What will be the position of the image when the object is 43 inches in front of the lens?

29. Calculate, and show with a diagram, the position of the image when an object is placed 16 inches in front of a convex lens, the focal length of which is 20 inches. What will be the magnification of the image?

30. A ray of light is incident in a direction inclined at 30° to the axis of a positive thin lens of 2 cm. focal length, cutting the axis at a point 5 cm. in front of the lens. A point moves along this ray from infinity on one side of the lens to infinity on the other. Find graphically the conjugate ray and trace the movement of the conjugate image point.

31. Light is made to converge to a point 30 cm. from a convex lens. When a concave lens is placed 10 cm. from the convex lens and in the converging pencil of light, the point to which the pencil converges is moved to 10 cm. farther from the convex lens. Find the focal length of the concave lens.

32. A real inverted image of an object is formed on a screen by a thin lens, and the image and object are equal in size. When a second thin lens is placed in contact with the first, the screen has to be moved 2 cm. nearer the lenses in order to obtain a sharp image, and the size of this image is three-quarters that of the first image. Find the focal length of the two lenses. (S.M.C.)

33. An image formed on a screen by a convex lens is 5 cm. long. Without moving either the object or screen, which are 150 cm. apart, a second image can be produced which is 20 cm. long. Show how this is possible and find the size of the object and the focal power of the lens.

34. A lens + 6·55D is 10 cm. in diameter and is fixed horizontally 24 cm. above a point source of light. Calculate the diameter of the circle of light projected on a ceiling situated 8·2 metres above the lens. (B.O.A.)

35. A distant object subtends an angle of 30°. Give a careful diagram showing the formation of an image of this object by a lens of + 5D power, and from the diagram find the size of the image.

36. A distant object subtends an angle of 10° at the centre of a positive lens of 15 cm. focal length and 5 cm. aperture, one end of the object being on the lens axis. Show graphically the path of a pencil of light from each end of the object and filling

the aperture of the lens. Find from the diagram and by calculation the size of the image formed.

37. A distant object subtends an angle of 5°; what focal length lens is necessary to form a real image 8 mm. long?

38. An image of a distant object is formed by a + 2D lens. Find the position of a second lens of + 5D power in order that a real image may be formed half the size of the original image.

39. A distant object subtends an angle of 10°; find the size of the image formed by a + 4D lens. Where must a − 8D lens be placed to form a real image four times the size of the original image?

40. Find the focal length of a thin lens which will form an image of the same size as that formed by the system in question 39, using the same object.

41. A small object stands upright on the axis of a convergent lens A. Prove by means of a diagram, that if another lens is placed in the anterior focal plane of the lens A the size of the image remains unaltered while its position is changed. (B.O.A.)

42. A converging lens forms an image of a distant object at a plane A and when a diverging lens is placed 1 inch in front of A (between the converging lens and A) a real image is formed 5 inches beyond A. Find the focal length and focal power of the diverging lens.

43. What is meant by chromatic aberration of a lens? The light from a distant small source of light is brought to a focus by means of a thin equi-convex lens with curves of 20 cm. radius of curvature. If the refractive indices for red and blue light be 1·515 and 1·524 respectively, find the chromatic aberration of the lens.

44. The refractive index of a certain kind of crown glass is: for yellow light 1·5186, red 1·5161, blue 1·5247. The radii of curvature of a lens made of this glass are + 55 mm. and + 87 mm. Find the power and focal length of the lens for each colour.
What appearances would you expect at the place where the lens focuses a pencil of white light?

45. Parallel rays of white light fall axially on a converging lens of glass. Draw a diagram showing the convergence of the resultant red and blue rays to separate foci, and show that a white screen can be placed in such positions that the image is (1) a white spot fringed with red or (2) a white spot fringed with blue. (B.O.A.)

46. A thin converging lens of focal length f is used to form an image of a luminous object on a screen. Determine the least possible distance between object and screen.
A lens of 10 cm. focal length forms a sharp image of a fixed object on a fixed screen for two positions of the lens which are 15 cm. apart. If the distance from object to screen be doubled, what will be the distance apart of the two positions of the lens for which there will again be a sharp image? (L.U.)

47. A ray of light on passing through a + 5D lens is deviated through the same angle as it would have been when refracted by a prism of 10° refracting angle, $n = 1·62$. At what distance from the optical centre is the ray meeting the lens?

48. When a positive lens is placed close to the eye and moved laterally objects seen through the lens are seen to move in the opposite direction, while with a negative lens the apparent movement of the object is in the same direction as the lens. Explain why this is and show how the effect may be used to find the power of a lens.

49. Two thin watch glasses each of radius of curvature 30 cm. are cemented together in the form of a double convex lens. The lens is placed in water, the refractive index of which is $\frac{4}{3}$. Calculate the focal length and prove any formula you use. (B.O.A.)

50. A thin bi-convex lens having curvatures of 3D and 5D and $n = 1 \cdot 62$ is immersed in a large tank of water. An object is placed in the water one metre from the lens; what will be the position of the image if that is also formed in the water?

51. An object 33 mm. high is placed at 444 mm. from a lens, which brings the light to a focus at an equal distance beyond the lens, but of this distance 124 mm. is travelled in glass of refractive index $1 \cdot 55$. Find the power of the lens and the size of the image.

52. A plane mirror is set up behind a convex lens and a pin is placed in front. By adjusting the distances of the pin and mirror from the lens it is found that in two positions the image of the pin coincides with the pin, the image in one case being erect, in the other inverted. Draw diagrams to illustrate the formation of the image in each case and show how the focal length may be determined. (L.U.)

53. Describe a method of measuring the focal length of a negative lens.

54. Explain exactly how to determine the refractive index of a given concave lens by using an optical bench and its usual fittings.

55. In measuring the focal length of a lens by the telescope method, the telescope was focused on an object 8 feet away instead of on a distant object. With the telescope placed close to the lens a sharp image was obtained when the object was 6 inches away from the lens. What was the focal length of the lens?

56. Describe carefully how you would determine the focal length of a weak positive lens, say of 0·5D power. Illustrate the method by giving a typical set of bench readings, and show how the focal length could be found from these.

57. An illuminated object A is placed at the first principal focus of a 20D positive lens L on the other side of which, at some distance, is a telescope which is adjusted until the object is seen distinctly. When a lens of unknown power is placed at the second principal focus of L, between L and the telescope, it is found necessary to move the object 2 cm. further away from L in order to see it distinctly in the telescope. Calculate the power of the unknown lens. (S.M.C.)

VII – Reflection at Curved Surfaces

79. Introduction

It has been shown in Chapter III that the effect of a plane reflecting surface is to change the direction of the light without altering its vergence, and an image of an object is therefore formed at the same distance from the surface as the object, but on the opposite side of it. If, however, the surface is curved, the reflected light will be changed both in direction and vergence.

The curved reflecting surface may be either convex or concave to the incident light, and may be of various forms, usually spherical, paraboloidal, or cylindrical. For the present, only the spherical form will be considered.

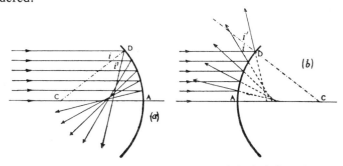

Fig. 81.–Reflection at spherical mirror. Spherical aberration

In Fig. 81a and 81b DA represents the principal section of a concave and of a convex reflecting surface; C is the **centre of curvature,** and A, the centre of the aperture, is the **pole** or **vertex** of the surface; the line

AC is the **principal** or **optical axis.** Clearly any directions, such as AC and DC passing through the centre of curvature will be normal to the portions of the surface through which they pass. The direction of any part of the light reflected at a curved surface may therefore be found from the law of reflection, the angle of incidence for any ray being the angle which the ray makes with the radius of curvature through the point of incidence. Figs. 81a and 81b show the light from a distant object point on the axis reflected at a concave and a convex surface respectively. It will be seen that the light which comes from the object point is no longer travelling exactly from or towards a point image after reflection; the rays meeting the surface close to the axis will cross the axis farther from the surface than those more remote from the axis. Thus there is spherical aberration, and the form of the reflected beam is very similar to that of the beam after refraction at a spherical surface (§ 49). Only when the object point is at the centre of curvature of a spherical mirror will the reflected pencil be homocentric, as in this case all the rays meet the surface normally and are reflected along their original paths; object and image are therefore coincident.

80. Focal power

If the light is limited to the paraxial region of the mirror, the light from a point object will, after reflection, be converging to or diverging

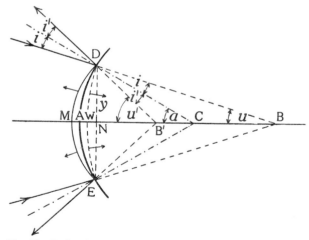

Fig. 82.—Reflection at spherical surface. Change in vergence

from an approximate point, and an image of the object will be formed either before or behind the mirror. The effect produced by a curved mirror, and hence the position of the image, may be found, as in the case of the refracting surface and the lens, either by considering the change in vergence or curvature of the wave fronts, or the change in the direction of the rays.

Change in Vergence.—Fig. 82 shows light, converging towards a virtual object point B, intercepted by a convex mirror at A, and diverging after reflection from the virtual image B′

DAE represents the mirror, of curvature $\frac{1}{r} = R$.

DWE represents the incident wave front, of curvature $\frac{1}{l}$.

By applying Huygens' construction as for a plane reflecting surface (§ 18), the exact form of the reflected wave front can be found. It will be seen that when the mirror aperture is large, the section of the reflected wave front is not circular, that is, the wave front is not spherical, and the reflected light does not therefore diverge from a point. If, however, we consider only a very small portion of the mirror near the axis, the reflected wave front will be approximately spherical with its centre at B′.

In the time that the centre of the wave front would have travelled a distance AW, the reflected wave front will have travelled back an equal distance AM, and similarly for all other points on the incident wave front. DME is therefore the reflected wave front of curvature $\frac{1}{l'}$, corresponding to the incident wave front DWE. Then

$$MA = AW = AN - WN$$

$$MN = MA + AN = AW + AN = AN - WN + AN$$

$$= -WN + 2AN$$

For small portions of the curves, the sags are proportional to the curvatures; hence

$$\frac{1}{l'} = -\frac{1}{l} + \frac{2}{r} \qquad \ldots \qquad \ldots \qquad \ldots \qquad (7.01)$$

As the reflected light is travelling in the opposite direction to the incident light, we shall regard

$$L' = -\frac{1}{l'}$$

when the light is in air. That is, the reflected light is divergent, negative vergence, L' negative, when l' is positive, and convergent, positive vergence, L' positive, when l' is negative.

Expression (7.01) may then be written

$$L' = L - 2R$$

$L' - L$, the change in vergence produced by the mirror, is the **focal power**, F, of the mirror and, for a mirror in air,

$$\left.\begin{array}{c} L' - L = F = -2R \\ \text{or} \\ L' = L + F \end{array}\right\} \quad \dots \quad (7.02)$$

Change in Ray Path.—From the law of reflection, angle i = angle i'. When the angles a, u and u' are small, the arc AD may be considered as a short line of length y perpendicular to the axis and

$$a = \frac{y}{AC}, \quad u = \frac{y}{AB}, \quad u' = \frac{y}{AB'}$$

From the figure,

$$a = i + u, \text{ and } u' = i' + a = i + a$$

Hence

$$a = u' - a + u$$
$$2a = u' + u$$

Substituting the above values for a, u and u'

$$\frac{2}{AC} = \frac{1}{AB'} + \frac{1}{AB}$$

or

$$\frac{1}{l'} = -\frac{1}{l} + \frac{2}{r} \text{ as before} \dots \quad \dots \quad \dots \quad (7.01)$$

81. Reflection Regarded as a Special Case of Refraction

If we apply the sign convention for angles to the case of reflection, we see that the angles of incidence and reflection are always of opposite signs. The law of reflection may therefore be considered as a special case of the law of refraction where $n' = -n$, and $n' \sin i' = n \sin i$ then becomes $i' = -i$, i' being the angle of reflection.

By means of this device any expression derived for refraction at a spherical surface may be applied to reflection at a spherical surface, $-n$ being substituted for n'. Thus the fundamental paraxial expression (5.02)

$$\frac{n'}{l'} - \frac{n}{l} = \frac{n' - n}{r}$$

becomes for reflection

$$-\frac{n}{l'} - \frac{n}{l} = -\frac{2n}{r}$$

or

$$\frac{1}{l'} = -\frac{1}{l} + \frac{2}{r} \qquad \cdots \qquad \cdots \qquad \cdots \quad (7.01)$$

Similarly the expressions (5.05)

$$\frac{n'}{f'} = -\frac{n}{f} = F = \frac{n'-n}{r}$$

become for reflection in air

$$-\frac{1}{f'} = -\frac{1}{f} = F = -\frac{2}{r} \qquad \cdots \quad (cf.\ 7.04)$$

For a mirror in any medium of refractive index n, its focal power

$$F = -\frac{n}{l'} - \frac{n}{l} = L' - L = -\frac{2n}{r} = -2nR \qquad \cdots \quad (7.03)$$

82. Conjugate Foci. Principal Focus. Focal Length

The positions of object and image for any mirror can be found from expressions (7.01) or (7.02), the same sign convention being used as for the refracting surface or lens. It will be seen that the general effect of a concave mirror is similar to that of a positive lens, and the effect of the convex mirror is similar to that of a negative lens, with the addition, however, in each case that the general direction of the light is changed on reflection.

Thus, parallel light from a distant object will, after reflection at a concave mirror, be converging to form a real image in front of the mirror (Fig. 83a); the position of the image is then the **second principal focus, F'**, of the mirror, and its distance AF' from the mirror, the **second focal length.** As the object moves up towards the mirror, the image moves away, and object and image coincide at the centre of curvature. When the object reaches the **first principal focus,** which in the case of the mirror coincides with the second principal focus, the image is at infinity, and the reflected light is parallel. With the object inside the principal focus, the reflected light will be diverging from a virtual image behind the mirror, and at a greater distance from the mirror than the object, until on reaching the mirror surface object and image again coincide.

With a convex mirror, as it makes the light more divergent, a virtual image is always formed with a real object. When the object is very distant, this virtual image is at the second principal focus F', its distance AF' from the mirror being the second focal length, f' (Fig. 83b).

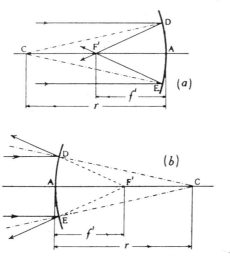

Fig. 83.—Principal focus and focal length of spherical mirror

When the incident light is converging towards a virtual object, in the case of a concave mirror a real image will always be formed between the principal focus and the mirror. In the case of the convex mirror, however, as with the negative lens, the image may be real or virtual according to the position of the virtual object. When the incident light is converging towards the first principal focus, which coincides with the second principal focus, the reflected light is parallel, and the image therefore at infinity. If the virtual object is between the mirror and the principal focus, a real image is formed in front of the mirror, while if the virtual object is outside the principal focus, the reflected light will be diverging from a virtual image. Object and image coincide at the centre of curvature and at the surface.

The image positions for a number of different object positions, should be found graphically and by calculation.

When $l = \infty$, $l' = f'$ and when $l' = \infty$, $l = f$. Therefore from (7.01) and (7.02)

$$f' = -\frac{1}{F} = \frac{r}{2} \qquad \dots \qquad \dots \qquad \dots \quad (7.04)$$

The principal focus of a mirror is therefore situated midway between the mirror and its centre of curvature.

123

83. Extra-axial Points. Graphical Construction. Magnification

When an object point is situated some distance from the axis of the mirror, the reflected light does not pass through a point, even if the aperture of the mirror is quite small, as may be seen by finding graphically from the law of reflection the reflected directions of a number of rays from such a point. If, however, the object is close to the axis and the mirror aperture small, so that the incident light is in the paraxial region, a reasonably good image will be formed. This image must lie somewhere on the ray from the object point that passes through the centre of curvature, for this ray is reflected back along its original direction, and the position of the image can be determined by finding where this ray is intersected by some other ray. A convenient second ray to consider will be one incident parallel to the axis, that after reflection will pass through the principal focus. The image, formed by a mirror, of an object of finite size, may be determined by finding the images of a number of points on the object. The image of a small plane object perpendicular to the mirror axis will be a plane also perpendicular to the axis.

As with the refracting surface and lens, the position and size of the image formed by a curved mirror may conveniently be found graphically. The constructions being true only for the paraxial region, the mirror may be represented by a straight line passing through the vertex perpendicular to the axis. Suitable scales may be chosen, as explained in § 73. The constructions are based on the following facts:

All rays parallel to the axis will, after reflection, pass through the principal focus and vice versa.

All rays incident through the centre of curvature are reflected back along their original paths.

All rays incident at the vertex are reflected at an equal angle on the opposite side of the axis.

Fig. 84.—Concave mirror (convergent). Real inverted image

Three examples of these constructions are shown, viz.:

Concave mirror, real inverted image. Fig. 84.

Concave mirror, object inside principal focus, virtual, erect and enlarged image. Fig. 85. This illustrates the formation of the image in the ordinary concave shaving glass.

Fig. 85.–Concave mirror (convergent). Virtual erect image

Fig. 86.–Convex mirror (divergent). Virtual, erect and diminished image

Convex mirror, virtual, erect and diminished image. Fig. 86. In the convex driving mirror of the motor car, the diminished image permits of a greater extent of image being seen than would be seen with a plane mirror of the same size.

In Figs. 84, 85 and 86

$$\frac{h'}{h} = m = -\frac{AB'}{AB} = -\frac{l'}{l} = \frac{L}{L'} \qquad \dots \qquad \dots \quad (7.05)$$

The magnification may also be expressed in terms including f' as for a thin lens.

$$\left.\begin{array}{l} m = \dfrac{L}{L'} = \dfrac{L'-F}{L'} = 1 - \dfrac{l'}{f'} \\[2mm] \dfrac{1}{m} = \dfrac{L'}{L} = \dfrac{L+F}{L} \quad 1 - \dfrac{l}{f'} \end{array}\right\} \qquad \dots \qquad \dots \quad (7.06)$$

125

Other expressions, in terms of object and image distances measured from the centre of curvature or from the principal focus (Newton's relation) are easily derived.

84. The Lens Mirror

Some light is reflected at each of the surfaces of a lens, and that reflected at the second surface again passes through and is refracted by the first. Mirrors are sometimes produced by silvering the second surface of a lens; examples occur in the Mangin mirror (§ 87) and the large double convex lens silvered on the second surface that is used in place of the concave shaving glass. Such a system is known as a **lens mirror** and its general properties are considered in Chapter XIX (§ 250). When, however, the lens may be considered as thin the focal power of the lens mirror may be found simply as follows:

The light passing through the first surface of power F_1 is reflected at the second surface, which is a mirror having a power $-2nR_2$, and again passes through the first surface. The total change produced in vergence, that is, the power of the lens mirror, is therefore:

$$F_M = 2F_1 - 2nR_2 .. \qquad .. \qquad .. \qquad (7.07)$$

Example:

A thin biconvex lens of $+5D$ power is set up on an optical bench. It is found that the image formed by reflection from the second surface coincides with the object at a distance of 10 cm. On reversing the lens, object and reflected image coincide at 12·5 cm. from the lens. Find the curvatures of the surfaces and the refractive index of the glass.

Treating the lens-mirror as a single concave mirror, since object and image coincide the radius of curvature of the equivalent mirror would be -10 cm. in the first place and $-12·5$ cm. when the lens is reversed. Therefore the powers of the lens-mirror are $+20D$, and $+16D$ when reversed.

The power of a lens-mirror (from 7.07) is

$F_M = 2F_1 - 2nR_2 = 2(n-1)R_1 - 2nR_2$

$20 = 2(n-1)R_1 - 2nR_2$ or $10 = nR_1 - R_1 - nR_2$ (1)

$16 = -2(n-1)R_2 + 2nR_1$ or $8 = -nR_2 + R_2 + nR_1$ ...   (2)

(In (2) the signs of R_1 and R_2 are changed as the lens is reversed)

For the lens

$$F = (n-1)(R_1 - R_2)$$
$$5 = (n-1)(R_1 - R_2) \qquad ... \qquad ... \qquad ... (3)$$

Solving the simultaneous equations 1, 2 and 3 we obtain $R_1 = +3D$, $R_2 = -5D$, $n = 1·625$.

85. Measurement of Focal Power and Focal Length

As the power of a mirror depends only upon its curvature, the power may be determined from a spherometer measurement of curvature. Optical methods are, however, often preferable, especially in the case of surfaces of very short or very long radius, and such methods avoid the possibility of damage to the surface, as may occur when a spherometer is used on a finely polished surface. The optical methods involve the use of an optical bench and fittings, as described in § 76.

Concave Mirrors—Radius of Curvature—Parallax Method.

The mirror to be measured is mounted vertically on the optical bench, and a well illuminated small object, such as a pin, is placed in front of it. On moving the object along the bench, a position will be found where the real inverted aerial image formed by the mirror coincides in position with the object. This coincidence is judged by observing that object and image do not appear to separate as the head is moved across the bench, that is, there is no parallax. Object and image must then be at the centre of curvature of the mirror, and the radius of curvature can be measured.

Focal Length. Collimator Method.

When light from a distant object is available, the focal length of the concave mirror can be obtained directly. In the laboratory a collimator will generally be used. The collimator and the mirror are set up on the bench with their axes almost coinciding—it may be necessary to tilt the mirror slightly in order to separate the incident and reflected beams. A small screen covering not more than half the aperture of the mirror, is mounted between them, and adjusted until a well defined image of the collimator object is received on it from the mirror. The screen is then at the principal focus of the mirror, and the focal length can be read off on the bench.

Focal Length. Telescope Method.

A telescope, previously focused on a distant object, is mounted on the bench, and directed towards the mirror. A small object, such as a pin, well illuminated, is mounted vertically between the mirror and telescope, the head of the pin being as near as possible on the axis of the mirror. On moving the object along the bench, a position will be found where a sharply focused image of the object is seen in the telescope. The light reflected from the mirror must now be parallel, and the object is therefore at the principal focus.

Focal Power or Focal Length. Conjugate Foci Method.

The focal power or focal length of the concave mirror may be determined by the method of conjugate foci, as described for a positive lens. A real image of an object is received on a small screen, and the distances *l* and *l'* measured; the power is then obtained from expression (7.02).

Fig. 87.—Convex mirror. Measurement of radius of curvature

Convex Mirrors—Radius of Curvature.

When a convergent beam is intercepted by a convex mirror in such a position that the beam is converging towards the centre of curvature of the mirror, the reflected light will return along its original path. This fact is the basis of a useful method of measuring the radius of curvature of a convex mirror. An illuminated object, positive lens and the mirror to be measured, are set up on the optical bench, and adjusted until a sharp image of the object is formed on the mirror surface, the distance from the lens to this position of the mirror being made greater than the radius of curvature of the mirror. The mirror is then moved towards the lens until an image is formed in the plane of the object; this position is most accurately found by parallax. The distance A_2A_1 between the two positions of the mirror is then the radius of curvature of the mirror.

The same arrangement can be used to determine the radius of curvature of a concave mirror, the mirror in this case being moved away from the lens to its second position.

Measurement of Surfaces of Short Radius—Drysdale's Method.*

A modification of the last method, suggested by Dr. C. V. Drysdale, permits of the measurement of surfaces of very short radius of curvature, such as the surfaces of microscope lenses.

A microscope (Fig. 88), with scale and vernier to read the position of the body, is fitted with a **vertical illuminator.** This consists of a short length of tube between the end of the microscope body and the objective with an aperture on one side and a small clear glass reflector in the tube at 45° to the axis to reflect light from a source down through the

* C. V. DRYSDALE. *On a Simple Direct Method of Measuring the Curvature of Small Lenses.* Trans. Opt. Soc. 2 (1900–01).

objective. The source B, a small electric lamp, is attached to the microscope tube in line with the aperture and at approximately the same

Fig. 88.—Measurement of surfaces of small radius.
Drysdale's method

Fig. 89.—Radius of curvature of convex surface of short radius.
Ophthalmometer method

distance from the reflector as is the focal plane of the eyepiece, B″. The surface to be measured is placed on the stage of the microscope, and the microscope is focused on it; some fine dust may be placed on the surface for this purpose. The source is then adjusted until a sharp image of it is focused in the microscope, that is, the image B′ lies on the surface. On racking the microscope down in the case of a convex

129

surface and up in the case of a concave surface, a second position is found where a sharp image of the source is again focused; the image B' now lies at the centre of curvature of the surface, and the distance the microscope has moved between the two positions is the radius of curvature of the surface. This method is the principle of a number of instruments now made for the measurement of the curvature of contact lenses.

Convex Surfaces of Short Radius—Ophthalmometer Method.

If the size of the image, formed by a mirror, of an object of known size at a known distance can be found, the radius or focal length of the mirror can be calculated. Thus in Fig. 89 BQ is an object in front of a convex mirror at A, and B'Q' is the virtual image formed. Then

$$m = \frac{B'Q'}{BQ} = \frac{h'}{h} = \frac{f'}{-x} = \frac{r}{-2x}$$

$$r = \frac{-2xh'}{h}$$

When the object distance is large as compared with the radius of curvature, the image is formed very close to the principal focus, and x is practically equal to d, the distance between image and object; thus

$$r = \frac{-2dh'}{h} \qquad \dots \qquad \dots \qquad \dots \qquad (7.08)$$

This is the principle of the **ophthalmometer** or **keratometer** an instrument that is used to measure the curvature of the cornea of the eye, or more usually, the difference in its curvature in different meridians. A large illuminated object, or two separated small objects are attached to a long focus microscope having a scale in the focal plane of the eyepiece. The microscope is directed towards the surface to be measured, and focused on the virtual image of the object; as the object moves with the microscope in focusing, the distance d between object and image remains constant. The size of the image h' is then found by measuring its image on the scale in the eyepiece, and h being the size of the object or the separation of the two small objects, r can be found from expression (7.08).

In measurements on the living eye it is impossible, because of movement, to measure the size of the image on a scale, and in the actual ophthalmometer the size of the image is determined by means of a device in the microscope for doubling the image. Various forms of the instrument are made; in some the measurement of the image is made by varying the amount of doubling, while in others the amount of doubling is constant and the object length h is varied to give an image of constant size.

86. Measurement of Curvature of the Surfaces of a Lens

The curvature of the surfaces of a lens may be measured by the above methods, treating the surface as a mirror. The following method based on the principle of the lens mirror may sometimes be useful, particularly in the case of a positive lens with shallow curves.

The lens is placed in front of an illuminated aperture in a white screen, the surface to be measured being turned away from the aperture. The distance between lens and screen is adjusted until an image of the aperture is formed on the screen by means of the light internally reflected from the second surface; then if l is the distance of the screen from the lens, f' the focal length of the lens and r_2 the radius of curvature of the second surface,

$$r_2 = \frac{lf'}{l+f'}$$

The proof of this is left as an exercise.

87. Mirrors of Large Aperture

Two of the most frequent applications of the curved reflecting surface are the use of the concave mirror for producing a parallel beam, and its use in bringing the light from a distant object to a focus. In the searchlight and the motor headlight a very intense beam of light is required, and it is therefore necessary that the beam shall be approximately parallel; this is obtained by placing a small intense source at the principal focus of a concave mirror. The concave mirror of the reflecting telescope replaces the objective in producing an image of a star or other distant object; thus the mirror in this case must bring parallel incident light to a focus. In each of these cases the aperture of the mirror is large —in the headlight the aperture is often more than twice the focal length—and, as there must be little or no spherical aberration, it is obvious that a spherical mirror cannot be used.

From Fig. 81 it will be seen that the effect of the spherical mirror on a parallel incident pencil is too great towards the periphery; and in order that a parallel incident pencil shall be brought to a good focus, or that the light from a small source at the focus shall be reflected as a parallel pencil, the curvature of the mirror must get less and less towards the periphery. In Fig. 90 B is a point source at the focus of a concave mirror of large aperture and, if the reflected beam is to be parallel, the reflected wave fronts must be plane; therefore, the various paths by

131

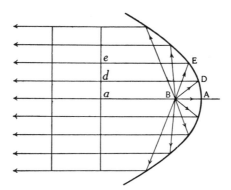

Fig. 90.–Paraboloidal mirror

which the light travels, BAa, BDd, BEe, etc., must be equal. The curve having this property is the parabola; and mirrors of large aperture are made of paraboloidal form when either the incident or reflected light is parallel.

A similar result is often obtained by using a concave spherical reflector, and placing in front of it a negative lens, the curves being

Fig. 91.–Mangin mirror

chosen so that the increasing thickness of glass towards the periphery corrects the spherical aberration. In practice, the reflector consists of a suitably designed negative meniscus lens silvered on the convex surface (Fig. 91). This form of reflector, which is frequently used in headlights, is known as the **Mangin mirror.**

In the headlight and searchlight fitted with a paraboloidal or a Mangin mirror only the reflected light from a point on the axis will be truly parallel, but if, as is usually the case, the source is small the light from any point on the source will be reflected as a practically parallel pencil, the various pencils diverging slightly from the axis. The area illuminated by the reflected light therefore increases as the distance from the lamp is increased, but the light, that would have diverged in all directions, is concentrated in a comparatively small area, and hence a greatly increased illumination of distant objects is obtained. When a mirror is to be used to focus light to a point at some finite distance from the mirror, other than at the centre of curvature, an ellipsoidal mirror must be used. The conjugate points are then situated at the two foci of the ellipsoid. Mirrors of this form are frequently used in the illuminating system of the cinematograph projector.

CHAPTER VII   *EXERCISES*

1. Prove that parallel rays of light are reflected through a single point by a concave spherical mirror provided that only a small portion of the whole sphere of which the mirror is part is employed.

What is the effect if a large part of the spherical surface is employed?

2. A small source of light is placed at the principal focus of a concave spherical mirror, the radius of curvature of which is 10 cm. and aperture 16 cm. Find graphically the paths of the reflected rays corresponding to incident rays that strike the mirror at points 2, 3, 4, 6 and 8 cm. from the axis.

In what manner would the shape of the mirror need to be altered to make the reflected beam parallel?

3. A pencil of light has a convergence of 5·5D on reaching a concave mirror whose focal length is 22 cm. Find the position of the image.

4. The curvature of a concave mirror is 5D. The focal power of another concave mirror is 5D. Explain the difference between the mirrors. In each case find the position of the image of an object 2 metres from the mirror.

5. What is meant by the focal power and focal length of a concave reflecting surface? How are these values related to the curvature of the surface?

6. A parallel pencil of light of 3 cm. diameter is incident on a concave mirror of 25 cm. radius of curvature. What is the area of the cross section of the reflected pencil 40 cm. from the mirror?

How far from the mirror must a second concave mirror of 3D curvature be placed in order that the light may be reflected back along its original path?

7. Deduce the relation between object distance, image distance and focal length for a spherical mirror, directly from the laws of reflection. Describe how the focal length of a convex spherical mirror may be determined experimentally. (L.U.)

8. Prove, by consideration of the light fronts, that for an object in any position the change in vergence produced by a convex mirror is equal numerically to twice the mirror's curvature. What limitations are there to the truth of this statement?

9. How far from a concave spherical mirror must a real object be placed in order that the image shall be (*a*) real and four times the size of the object and (*b*) virtual and four times the size of the object? (S.M.C.)

10. An object 3 inches long is placed 15 inches from the principal focus of a concave mirror, and a real image $1\frac{1}{4}$ inches long is formed. Find the curvature of the mirror and the distance of object and image from the mirror.

11. When an object is 31 cm. in front of a concave mirror the image is 19 cm. in front of the mirror. Where will the image be if the object is a virtual one and 31 cm. behind the mirror?

12. Find graphically and by calculation the positions and sizes of the images in the following cases, the object in each case being 3 cm. long.
   (a) Object 10 cm. from convex mirror of 10 cm. radius.
   (b) Object 5 cm. from concave mirror of 15 cm. radius.

13. Draw figures showing the change which takes place in the size and position of the image when an object on the axis of a concave mirror is moved from a great distance until it is in contact with the mirror.

14. The focal length of a concave mirror is 17 inches and a virtual image is required 5 inches from the mirror. Where must the object be placed?

15. The radius of curvature of a convex mirror is 25 cm. Standing on its axis and 60 cm. in front of the mirror is an object 5 mm. high. Find the position and size of the image (a) graphically, (b) by calculation. Is the image real or virtual?

16. Find the position of the image of an object which is 9 inches from a concave mirror whose radius of curvature is 12 inches.

17. A concave mirror forms a real image of an object at 20 cm. in front of the mirror; if the object is 133·3 cm. from the mirror, find (1) the focal power, (2) the curvature, (3) the radius of curvature of the mirror.

18. Explain the formation of an image by reflection of light from a convex spherical mirror.
   A small object is placed on the axis of a convex spherical mirror at a distance of 20 cm. from it, and the virtual image formed is observed to have half the linear dimensions of the object. Determine the focal length and the radius of curvature of the mirror. (B.O.A.)

19. A concave mirror has a radius of curvature of 2 feet; show graphically and by calculation where an observer must be situated in order to see a virtual image of his face magnified $2\frac{1}{2}$ times.

20. A convex mirror having a radius of curvature of 25 cm. forms, at 20 cm. from the mirror, a virtual image of an object 2 cm. in height. Calculate the magnification and give a diagram showing the path of the light.

21. When a real inverted image is formed 14 inches from, and in front of a mirror, the magnification produced is two. What will be the position of the object when the magnification is four and the image is erect?

22. A convex mirror for a motor car has a radius of curvature of 18 inches. What will be the size and position of the image of a car 6 feet broad and 5 feet high at a distance of 10 feet from the mirror? (S.M.C.)

23. An object is placed on the axis of a concave mirror and perpendicular to it. Obtain an expression, in terms of the object distance and the focal length of the mirror, for the ratio of the linear dimensions of the image to that of the object. Represent the result graphically, taking the focal length of the mirror as 10 cm.

24. An object is placed 75 cm. from a mirror and a real image is formed 120 cm. from the mirror. What will be the focal length of the mirror? If the back surface of

the mirror be plane, its diameter 5 cm. and its central thickness 3 mm., find its edge thickness.

25. A convex motor driving mirror has a radius of curvature of 15 inches; it is rectangular in shape, its dimensions being 5 inches horizontally, and 3 inches vertically. What extent of object 20 yards from the mirror could be seen by reflection by a driver whose eye is 3 feet from the mirror?

26. The front surface of the cornea of the eye has a radius of curvature of 8 mm.; what will be the size and position of the image formed by the reflection of an object 20 cm. in front of the cornea and 6 cm. long?

27. An object and its image are one metre apart; if the object is four times the size of the image, find the radius of the concave mirror forming the image.

28. A gas flame is 4 feet from a wall, and it is required to throw on the wall a real image of the flame which shall be magnified three times. Find the position and focal length of a spherical mirror to give the required image; give a diagram.

Describe very briefly a method for finding the radius of curvature of a steeply curved mirror, such as a ball-bearing, for example, without touching it. (S.M.C.)

29. A point source of light is placed 15 cm. from a concave mirror of 45 cm. radius of curvature and 8 cm. diameter. What will be the diameter of the cross section of the reflected pencil at 10, 35 and 100 cm. from the mirror?

30. Explain (a) the use of convex mirrors as back view reflectors on motor cars, (b) the optical arrangement of a motor car headlight. (L.U.)

31. Deduce expressions for the magnification of the image formed by a curved reflecting surface in terms of the distances of object and image from (a) the centre of curvature, (b) the principal focus.

32. If the sun's disc subtends an angle of 32 minutes at the earth's surface, find the focal length of the concave mirror which would produce a real image of the sun upon a screen, the diameter of the image being 5·5 cm.

33. A convex mirror and a plane mirror are placed facing each other and 28 cm. apart. Midway between them and on the principal axis of the convex mirror a small luminous object is situated. On looking into the plane mirror the first two images of the luminous object are seen. Draw a diagram showing how the two images are formed and calculate the radius of curvature of the convex mirror if the image formed by two reflections is situated at a distance of 38 cm. behind the plane mirror.

34. A concave and a convex mirror, each of 20 cm. radius of curvature, are placed opposite to each other and 40 cm. apart on the same axis. An object 3 cm. high is placed midway between them. Find the position and size of the image formed when the light is reflected first at the convex and then at the concave mirror. Give a diagram showing the path of the light from a point on the object to a corresponding point on the image.

35. A concave mirror is immersed in water. The image of an object 30 cm. from the mirror is formed at 9 cm. in front of the mirror. If both object and image are in the water, what is the radius of the mirror?

36. An object and its image formed by a concave mirror are in front of the mirror and on opposite sides of its centre of curvature. If the object and image are 30 cm. and 7·5 cm. respectively from the centre of curvature and the image is nearer to the mirror than the object, find the focal power of the mirror.

37. Two concave mirrors of focal length 20 and 40 cm. are turned towards each

other, the distance between their vertices being one metre. An object 1 cm. high is placed between the mirrors at a distance of 10 cm. from the more steeply curved mirror. Find the position and size of the image produced by light which is reflected first from the nearer mirror and then from the farther mirror. Give a diagram showing the path of the light.

38. Explain carefully with diagrams the advantages of a convex over a plane surface for the driving mirror of a motor car.

39. Two small illuminated objects are placed 50 cm. apart and 40 cm. in front of an eye, the line joining the objects being perpendicular to the visual axis. Find the position and distance between the images reflected from the cornea, the radius of whichis 8 mm.

40. An object 5 cm. high is perpendicular to and has one end on the axis of a + 5D lens, the object being 20 cm. from the lens. At 30 cm. on the other side of the lens is placed a concave mirror of 30 cm. radius of curvature, its axis coincident with that of the lens. Find graphically and by calculation the position and size of the final image. Show in a diagram the path of the light from each end of the object.

41. Describe two methods of measuring the curvature of a polished concave surface.

42. When an object is held at a distance of 8 cm. from one face of a thin lens of glass $n = 1.5$, the image of the object formed by reflection in this face is found to lie in the same plane as the object. If the object is placed at a distance of 30 cm. from the lens, the image produced by the lens is inverted and of the same size as the object. Find the radii of the surfaces.

43. A lens made of glass of refractive index 1.5 is set up on the optical bench. With the object at 40 cm. it is found that the image formed by the lens coincides with the image produced by reflection at the first surface of the lens. When the object is placed 80 cm. from the lens the lens forms a real image which is $2\frac{1}{3}$ times the size of the object. Find the radii of curvature of the lens surfaces.

44. A thin convergent lens gives an inverted image the same size as the object when the latter is 20 cm. from the lens. An image formed by reflection from the second surface of the lens is found to be coincident with the object at 6 cm. from the lens. Find the radius of curvature of the second surface of the lens and, assuming $n = 1.5$, the radius of curvature of the first surface.

45. A concave mirror of focal length 11.5 cm. is placed on a bench and liquid is poured into it until the greatest depth is 0.5 cm. It is found that, on placing a luminous point on the vertical axis of the mirror at a height of 15.9 cm. from the liquid surface, the luminous point and its image coincide. Determine the refractive index of the liquid and give the theory underlying your calculation. (B.O.A.)

46. A thin converging lens and a convex spherical mirror are placed coaxially at a distance of 30 cm. apart, and a luminous object, placed 40 cm. in front of the lens, is found to give rise to an image coinciding with itself. If the focal length of the lens is 25 cm., determine the focal length of the mirror. Explain fully each step in your calculation. (B.O.A.)

47. A parallel beam of light falls normally on the plane surface of a thin plano-convex lens and is brought to a focus at a point 48 cm. behind the lens. If the curved surface be silvered, the beam converges to a point 8 cm. in front of the lens. What is the refractive index of the glass, and what is the radius of curvature of the convex surface? (B.O.A.)

48. Derive Newton's expression for the focal length of (a) a convex mirror, (b) a concave mirror.

49. A concave mirror is formed of a block of glass ($n = 1.62$) central thickness 5 cm. The front surface is plane and the back surface, which is silvered, has a radius of curvature of 20 cm. Find the position of the principal focus and the position of the image of an object 20 cm. in front of the plane surface.

50. A $+ 1.25$D equi-convex lens of refractive index 1.53 has one of its surfaces silvered, and parallel light falls axially on the other surface. What will be the effect? (S.M.C.)

51. A thick glass rod of refractive index n has a convex spherical surface of radius r worked on one end. Of a very distant bright object lying on the axis of the rod an image is formed by reflection and a second image by refraction, at the surface. Show that the distance separating the images is given by $\dfrac{r}{2}\left(\dfrac{n + 1}{n - 1}\right)$ and that the refracted image is numerically $\dfrac{2}{(n - 1)}$ times as large as the reflected image.

52. A convex lens of focal length f is placed at a distance of $4f$ from a concave mirror of radius of curvature f, and an object is placed midway between the two. Compare the sizes of the images formed by the light refracted through the lens (a) directly, and (b) after one reflection at the mirror. (S.M.C.)

53. Two small lamps A and B are set up 20 cm. apart facing symmetrically a polished steel ball of diameter 2 cm. The perpendicular distance from the midpoint of AB to the nearer surface of the ball is 25 cm. If this ball is replaced by a second ball of 2·2 cm. diameter, its nearer surface occupying the same position as that of the first ball, by how much must the separation of A and B be changed in order that the separation of the reflected images shall remain unaltered? (S.M.C.)

54. A ray of light travels parallel to, and at a distance y from, the axis of a concave spherical mirror of radius r. Find the distance from the centre of curvature of the point at which the ray intersects the axis of the mirror after reflection. Show that the spherical aberration of this ray is given by $\dfrac{r}{2}\left(\dfrac{r}{\sqrt{r^2 - y^2}} - 1\right)$.

55. A ray of light is incident at an angle θ on a concave spherical mirror parallel to a radius CA where C is the centre of curvature and A a point on the mirror surface. Show that the reflected ray intersects CA at a distance $r(1 - \frac{1}{2}\sec\theta)$ from A; r is the radius of curvature of the mirror.

If this reflected ray strikes the mirror again and the second reflected ray passes through the end B of the diameter BCA of the spherical surface of which the mirror forms part, what must be the angle θ? (S.M.C.)

VIII – Cylindrical and Sphero – cylindrical Lenses

88. Introduction

The commonest form of surface used for optical purposes is the spherical. Such surfaces can be produced comparatively easily to a high degree of accuracy by mechanical means, and the lenses of almost all optical instruments have spherical or plane surfaces. It has been shown in previous chapters that good images will not be formed by refraction at a single surface or a spherical lens when the aperture is at all large, but it is possible to obtain good images with a suitable combination of plane and spherical surfaces. Such a combination of surfaces is more easily and accurately manufactured than the aspherical surfaces required to obtain the same result with fewer surfaces. The increasing use of plastics in place of glass has made possible the manufacture of more complicated forms of surface, and such surfaces may be more widely used in future in producing simpler optical systems.

There are, however, a number of cases where the use of a lens is required for other purposes than the direct production of a good image of an object. The most frequent example of this is the use of the lens to correct the astigmatic eye, which has different powers in different meridians and so does not form a sharp image on the retina. The correcting lens must therefore be one having corresponding different powers and hence different curvatures in various meridians. Another example is the "astigmatiser" of the rangefinder, a lens that draws out the images of distant point sources into lines, thus making coincidence settings easier.

89. The Cylindrical Surface

The simplest form of "non-spherical" surface is the cylindrical. In practice, usually only one surface of a lens will have a cylindrical curve, the other being plane or spherical.

The cylindrical surface (Fig. 92) has zero curvature in the section DOO'D', the plane of which is parallel to the axis MN of the cylinder. In the section EAE'G the surface has its maximum curvature R, and the curve is circular. These two directions DD' and EE' are the **principal meridians** of the surface. Along any intermediate oblique meridian, such as HH', the curve will be elliptical, and the curvature will have a value somewhere between zero and the maximum R.

If only a small portion of the surface is considered, the curve in an

Fig. 92.—Cylindrical surface

oblique meridian may be regarded as circular, and the curvature may be found as follows: For a given sag AG the section in the principal meridian of maximum curvature has an arc EE', while the section in an oblique meridian for the same sag has an arc HH' and

$$HH' = \frac{EE'}{\sin \theta}$$

where θ is the angle that the oblique section makes with the axis. For small portions of the curve the curvatures of two curves having the same sag will be inversely proportional to the square of their chords (§ 44); therefore if R_θ is the curvature in a meridian at an angle θ with the axis and R the maximum curvature,

$$\frac{R_\theta}{R} = \left(\frac{EE'}{HH'}\right)^2$$

hence $\qquad\qquad R_\theta = R \sin^2 \theta \qquad \ldots \qquad \ldots \qquad \ldots \qquad (8.01)$

Similarly, the curvature at A along a meridian at right angles to HAH', and therefore making an angle (θ + 90°) with the axis, is given by

$$R_{(\theta+90')} = R \sin^2 (\theta+90°)$$
$$= R \cos^2 \theta \qquad \ldots \qquad \ldots \qquad \ldots \qquad (8.02)$$

Hence $\qquad R_\theta + R_{(\theta+90°)} = R (\sin^2 \theta + \cos^2 \theta) = R$

that is, *the sum of the curvatures along any two directions of a cylindrical surface intersecting each other at right angles is equal to the maximum curvature.* These theorems are a special case of a more general theorem concerning curved surfaces due to the mathematician Euler (1707–1783).

90. Refraction by Cylindrical Lens. Line Focus

Light from a luminous point B (Fig. 93) in front of a thin cylindrical lens continues unchanged in divergence in the axis meridian after leaving the lens. In any other meridian the light is refracted by an amount depending on the power in that meridian, and in the figure the light in

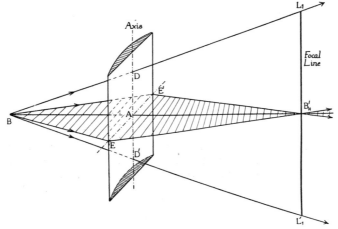

Fig. 93.–Refraction by cylindrical lens. Line focus

the meridian at right angles to the axis is shown converging to B'_H. The refraction in the oblique meridians is such that all the light from the object point on leaving the lens will pass through an **astigmatic** or **line**

focus at B'_H. The position of this line focus can be found from the usual conjugate foci expression taking the maximum power of the lens; and its direction, as will be seen from the figure, is parallel to the cylinder axis.

Between the lens and B'_H the light is also travelling as though it had come from a virtual line focus at B perpendicular to the cylinder axis, and an eye placed close behind the lens sees the luminous point B drawn out into a line perpendicular to the axis. This is the principle of the **Maddox Rod,** an important device used in sight-testing for disassociating the images seen by the two eyes in order to test for muscle imbalance. The Maddox Rod is a glass cylinder of about ⅛ inch diameter mounted in an opaque disc. The disc is placed before one eye with the rod in line with the pupil, and the eye then sees a distant small source as a line perpendicular to the axis of the rod. In practice, a glass disc ground with a series of parallel cylindrical corrugations generally takes the place of a single rod.

91. Sphero-Cylindrical Lenses

An eye that has *regular astigmatism* requires for its correction a lens which the power varies from a minimum in one meridian to a maximum in the meridian at right angles. In a great many cases, however, the minimum power required will not be zero, as would be given with a cylindrical lens. The required lens could be made up in the following different forms:

1. Each surface cylindrical with the axes of the cylinders at right angles—*crossed cylinders.*
2. One surface spherical and the other cylindrical—*sphero-cylinder.*
3. One surface of a form having different curvatures in different meridians—toroidal or toric curve, the other surface usually being spherical—*sphero-toric.*

Forms 2 and 3 are those generally used in practice.

Thus, if a thin lens is to have a power of, say + 5D in the vertical meridian, and + 3D in the horizontal, it would have in the crossed-cylinder form, one cylindrical surface of + 5D power with its axis horizontal and another cylindrical surface of power + 3D with its axis vertical. In the sphero-cylindrical form the lens could be made in two ways. One surface could be spherical with a power of + 3D and the other surface cylindrical of + 2D power with axis horizontal to give the additional power in the vertical meridian; or the spherical surface could have a power of + 5D, and the cylindrical surface a power of − 2D with its axis vertical. The toric form of this lens will be considered in § 95. The changing of the form of a lens of given power in this way is known as transposition.*

92. Refraction by Sphero-Cylindrical Lens. Astigmatic Beam

In considering the refraction by a lens of the form described in the previous paragraph, we may regard the sphero-cylindrical lens as typical of any lens having different powers in the principal meridians, and the general effect of such a lens will be the same in whichever form it is made. The principal meridians of the sphero-cylindrical lens will be the meridian containing the axis of the cylinder and the meridian perpendicular to the axis.

Fig. 94 shows the form of the pencil from a luminous point B on the optical axis after passing through a sphero-cylindrical lens having positive power in each of its principal meridians, the axis of the positive cylindrical surface in this case being vertical. Light passing through the vertical section DD' is affected only by the power of the spherical surface, and converges to B'_v; in sections parallel to DD' the light converges to points on a horizontal line $L_2L'_2$ passing through B'_v. In the horizontal section EE' the power is the sum of the powers of the spherical and cylindrical surfaces, and the light in this section converges to B'_H, sections parallel to EE' converging to points on a vertical line $L_1L'_1$ through B'_H. A pencil of this form, where the light originating at a point passes through two perpendicular lines, is called an **astigmatic pencil.**

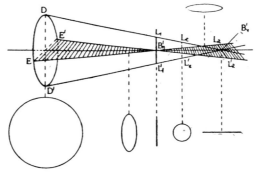

Fig. 94.—Astigmatic pencil

Cross-sections of such a pencil are shown in Fig. 94. If the aperture of the lens is circular, the cross-section of the emergent pencil is at first

* Fuller particulars of transposition will be found in EMSLEY and SWAINE *Ophthalmic Lenses.*

an ellipse with its major axis parallel to the cylinder axis, narrowing down to a line parallel to the cylinder axis at B'_H. The cross-section then becomes circular at $L_cL'_c$, after which it becomes elliptical with the major axis perpendicular to the cylinder axis and degenerates into a line again at B'_v. The circular cross-section at $L_cL'_c$ where the pencil will have its smallest cross-sectional dimension, is known as the **circle of least confusion.**

The properties of the astigmatic pencil were investigated by the mathematician Sturm (1838), and the pencil of the form shown in Fig. 94 is known as Sturm's Conoid. The distance between the focal lines is known as the **interval of Sturm.**

93. Positions and Lengths of Line Foci, etc

The positions of the line foci are found by applying the conjugate foci expression $(L' = L + F)$ to each meridian in turn.

Lengths of Line Foci.

To find the lengths of the line foci and the position and diameter of the circle of least confusion, it is convenient to represent the sections of the pencil in the two principal meridians in one plane, as in Fig. 95, where the section in a horizontal meridian is shown in full lines and in a vertical meridian in broken lines.

From Fig. 95

$$\left. \begin{aligned} \frac{L_1L'_1}{DD'} &= \frac{B'_HB'_V}{AB_V} = \frac{l'_2 - l'_1}{l'_2} \\[2mm] \text{or} \quad L_1L'_1 &= DD'\left(\frac{L_1 - L_2}{L'_1}\right) \end{aligned} \right\} \qquad \dots \quad \dots \quad (8.03)$$

$$\left. \begin{aligned} \frac{L_2L'_2}{DD'} &= \frac{B'_HB'_V}{AB'_H} = \frac{l'_2 - l'_1}{l'_1} \\[2mm] \text{or} \quad L_2L'_2 &= DD'\left(\frac{L_1 - L_2}{L'_2}\right) \end{aligned} \right\} \qquad \dots \quad \dots (8.03\,\text{A})$$

Circle of Least Confusion.

$$\frac{L_cL'_c}{DD'} = \frac{l'_c - l'_1}{l'_1} = \frac{l'_2 - l'_c}{l'_2} \qquad \dots \quad \dots \quad (8.04)$$

$$l'_2l'_c - l'_1l'_2 = l'_1l'_2 - l'_1l'_c$$

$$l'_c = \frac{2\,l'_1l'_2}{l'_1 + l'_2}$$

or $$L'_c = \tfrac{1}{2}(L'_1 + L'_2) \quad \dots \quad \dots \quad \dots \quad (8.05)$$

giving the position of the confusion circle.

Equation (8.04) may also be written

$$z = L_c L'_c = DD' \left(\frac{L'_1 - L'_2}{L'_1 + L'_2}\right) \quad \dots \quad \dots \quad (8.06)$$

where z is the diameter of the confusion circle.

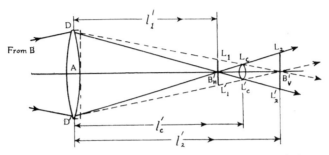

Fig. 95.—Astigmatic pencil. Positions and lengths of line foci

94. Image of Object of Finite Size

The optical axis of a sphero-cylindrical lens is a line passing through the centre of curvature of the spherical surface and perpendicular to the cylinder axis. In the case of a thin lens the point on the lens through which the optical axis passes is the optical centre.

As with a spherical lens, any ray passing through the optical centre will be undeviated and will be a chief ray from the object point from which the ray started. Since one ray from every point of the object passes undeviated through the sphero-cylindrical lens, the image formed when the aperture is quite small, will correspond in its general shape to the object, and its size may be found in the same way as with the image formed by a spherical lens. Each object point is, however, imaged as a line, parallel to one or other of the principal meridians, the length of the line being dependent upon the aperture of the lens and the distance between the two focal lines. The form of the image will therefore only agree with that of the object when the length of each line focus is quite small as compared with the size of the image.

The nature of the image formed by a sphero-cylindrical lens with its principal meridians horizontal and vertical is shown in Fig. 96. It should be noticed that lines sharply defined in the image are perpen-

144

| Object. | Image at vertical line focus. | Image at horizontal line focus |

Fig. 96.—Image formed by sphero-cylindrical lens

dicular to the meridian focusing in the image plane. The form of the image produced by astigmatic beams is of considerable importance in the case of the astigmatic eye.

95. Toric Lenses

When the astigmatic lens is required in meniscus form, as is often the case with ophthalmic lenses, it will be necessary for one surface of the lens to be of a form having different curvatures in its principal meridians. Such a surface is the **toroidal** or **toric** surface, which takes its name from the *torus*, the architectural term for the moulding at the base of an Ionic column.

A toric surface, two forms of which are illustrated in Fig. 97, is generated by the revolution of the arc of a circle about an axis OO′ which lies in the plane of the circle but does not pass through its centre. Common examples of a toric surface occur in the motor tyre, anchor ring and capstan.

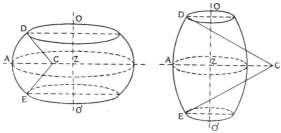

Fig. 97.—Toric surfaces

A straight line AZ, which passes through the centre of the circle perpendicular to the axis of revolution will be the optical axis of the surface. One principal meridian of the surface will be a section containing the axis of revolution, and the other, a section through the optical axis

145

and perpendicular to the axis of revolution; in the figure these meridians are vertical and horizontal respectively. As with the cylindrical surface, the principal meridians are the directions of maximum and minimum curvature. The principal meridian of the toric curve along which the curvature, and hence the power, is numerically the smaller is known as the base curve.

By the use of a toric surface, any given astigmatic lens can be made up in an infinite number of forms. The toric surface will have the required difference in power in the principal meridians, and the other surface will be spherical and of the necessary power to give, in conjunction with the toric surface, the required power of the lens. For example, the lens given in § 91 could be transposed into toric form with a toric surface of + 6D power in the horizontal meridian and + 8D power in the vertical, the second surface then being spherical of − 3D power.

CHAPTER VIII   EXERCISES

1. An optician's lens measure divided for a refractive index of 1·523 is placed on a cylindrical surface of 10 cm. radius of curvature. What will be the reading when the line containing the three legs of the measure is (a) perpendicular, (b) parallel, (c) at 30° to the cylinder axis?

2. A sphero-cylindrical lens has surface powers of + 3D.S. and + 2D.C. ($n = 1.523$). It is edged circular 45 mm. diameter and its minimum edge thickness is 1 mm. Find its maximum edge thickness and its central thickness.

3. A thin lens has cylindrical surfaces of equal curvature, the axes of the cylinders being at right angles; show that the effect of the lens is the same as that of a spherical lens with a curvature equal to either of the cylinders.

4. Show that the sum of the curvatures along any mutually perpendicular directions of a cylindrical surface is equal to the maximum curvature.

Find the sphero-cylinder equivalent to the combination
+ 4·0 D.C. ax 80° / − 2·0 D.C. ax 170°.

5. A sphero-cylindrical lens has surface powers of + 4·50D. sphere and − 2·75D cylinder, and is made of glass of refractive index 1·622. Find the radii of curvature of the tools with which the surfaces are ground.

6. A point source is placed 50 cm. from a cylindrical lens of + 5D power with axis vertical and 40 mm. diameter. Find the position, length and direction of the line focus.

If a + 3D spherical lens be placed in contact with the cylindrical lens, what will be the positions and lengths of the line foci and the position and diameter of the circle of least confusion?

7. A small source of light is placed 75 cm. from a sphero-cylindrical lens + 6D.S./ − 3·5D.C. ax. V., diameter 50 mm. Find the position and lengths of the focal lines.

8. A point source of light is situated 70 cm. from the following lens + 6D.S./ − 2·25D.C. ax. V 40 mm. diameter. Find the positions, lengths and directions of the line foci and the position and diameter of the circle of least confusion. What will be the appearance of the image in the plane of each of the focal lines when the point source is replaced by an illuminated circular object 5 mm. diameter?

9. The lens $+ 10\cdot00$ D.S./$- 4\cdot00$ D.C., the aperture of which is circular and of 6 cm. diameter, is set up with a distant luminous point on its optical axis. The focal lines are found to be respectively horizontal and vertical, the horizontal one being the nearer to the lens. What is the shape and size of the patch of light formed on a vertical screen fixed $12\cdot5$ cm. behind the lens? What lens placed in contact with the above lens will convert the combination into a sphere the principal focus of which is formed on the screen? (S.M.C.)

10. Two thin cylindrical lenses, one of power $+ 5$D axis vertical and the other $+ 4$D axis horizontal and 45 mm. diameter, are placed in contact 50 cm. from a point source of light. Find the dimensions of the patch of light on a screen parallel to the plane of the lenses and (a) 20 cm., (b) 40 cm., (c) 60 cm. from the lenses.

11. A narrow illuminated slit 15 mm. long is placed $16\cdot67$ cm. from a $+ 10$D cylindrical lens of 9 mm. diameter. Find the nature and size of the "image" when the slit is (a) parallel, (b) perpendicular to the cylinder axis.

12. A sphero-cylindrical lens has powers $+ 5\cdot0$D.S./ $-2\cdot0$D.C. ax. V. It is required to make a lens equivalent to this with one surface plane; what powers must be worked on the other surface? What type of surface is this? Give a diagram.

13. A lens made of glass of refractive index $1\cdot523$ has a toric surface the powers of which are $+ 10$D in the vertical meridian and $+ 6$D in the horizontal; the other surface is plane. Find the radii of curvature of the surfaces when a lens of the same power is made up in the crossed cylinder form.

14. A polished metal cylinder 2 inches in diameter stands on a table with its axis vertical and at a distance of 10 feet from a window 6 feet broad. What would be the breadth of the image seen by reflection from the cylinder? (S.M.C.)

15. A small crossline chart (the limbs of the cross being horizontal and vertical) and a white screen are set up vertically on an optical bench one metre apart. The lens $+ 5\cdot0$ D sph./$- 1\cdot0$ D cyl. axis V. is moved along the bench between them. How far from the chart must the lens be placed in order that (a) the horizontal line (b) the vertical line of the chart shall be in focus on the screen? (S.M.C.)

16. The lens $\dfrac{- 4\cdot0 \text{ D sph.}}{+ 9\cdot0 \text{ D cyl. ax. } 30^\circ / + 12\cdot0 \text{ D cyl. ax. } 120^\circ}$ is set up facing a small source of light. The astigmatic difference, or interval of Sturm, is found to be $16\frac{2}{3}$ cm. How far from the lens is the source situated? For what real position of the source is the astigmatic difference a minimum? (S.M.C.)

17. An object B and a screen B′ are placed $62\cdot5$ cm. apart. Between them is a combination consisting of the following two thin lenses, of aperture 20 mm. each, placed in contact:

Lens (1) $\dfrac{+ 16\cdot00 \text{ S}}{- 5\cdot75 \text{ C. ax. H}/ - 6\cdot25 \text{ C. ax. V}}$

Lens (2) $\dfrac{- 5\cdot75 \text{ S}}{+ 5\cdot50 \text{ C. ax. H}/ + 6\cdot00 \text{ C. ax. V}}$.

Find a position for the combination such that a sharp and diminished image of B will be formed on the screen.

A point source is placed at B and the lens (2) is rotated through 90°. What will now be the shape and size of the appearance on the screen? (S.M.C.)

IX – Lens Systems and Thick Lenses – Elementary

96. Introduction

Most optical instruments consist of a number of lenses separated by air spaces, or, as in the eye, of a number of media of different refractive indices separated by curved surfaces. In most cases the systems are coaxial or centred, the centres of curvature of the surfaces being on a common axis.

The work of this chapter will be limited to coaxial systems of thin lenses in air, and to single thick lenses having air on each side.

It has been shown in Chapter VI, § 72, that the position and size of the image formed by a system of thin lenses may be determined by calculating through the system, treating the image formed by one lens as the object for the next. Such a method, while simple in principle, becomes tedious where a number of image positions have to be found for the one system, and efforts were made, notably by Euler, Lagrange, Möbius and Gauss, to simplify the treatment of the lens system. As a result of the work of these investigators, particularly that of Gauss (1777–1855), it is possible to express the effect of any lens system in terms of an equivalent thin lens or surface. Once the power and positions of this equivalent lens, relative to the system have been found, any further work is greatly simplified, as the simple formulæ already found for the thin lens and surface can be applied to the system.

97. Equivalent Focal Length and Power. Principal Points

In Fig. 77 (Chapter VI) it is seen that a thin lens of a certain focal length placed at P′ would produce an image of a distant object of the same size and in the same position as that produced by the system. The focal length P′ F′ of this thin lens, which will theoretically replace the system, is termed the **equivalent focal length** of the system, and is usually the most important quantity in the specification of optical systems, such as photographic and microscope objectives and eyepieces. The reciprocal of this distance, usually in metres, is the **equivalent power.** The equivalent focal length of a system may therefore be defined as *the focal length of a thin lens, which will produce an image of a distant object of the same size as is produced by the system.*

We must now consider the position of this equivalent lens with respect to the system. Fig. 98a shows the path of a ray B_1D_1 incident parallel to the axis through the thin lenses at A_1 and A_2. The effect is the same as if the incident ray had been refracted at H' by the equivalent lens placed at P'. F' is the second principal focus of this lens, or of the combination of A_1 and A_2. In the lower half of the figure, a ray FE_1 is shown incident from a point F, such that after refraction by the system, it emerges parallel to the axis. The effect is the same as if the incident ray had been refracted at H by the equivalent lens placed at P. F is the first principal focus of this lens or of the combination. It will be found that $FP = P'F'$ or $f' = -f$ where f and f' are the first and second equivalent focal lengths respectively of the system.

NOTE.—The path of the ray from F may be found graphically by starting from B_2 and proceeding from right to left.

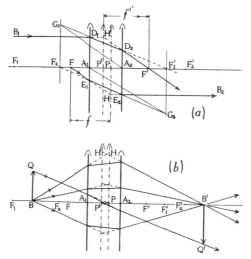

Fig. 98.—Lens system. Equivalent lens and principal planes

In Fig. 98b, the paths of a number of rays from a near object BQ have been traced through the same system (the construction lines are not shown in the figure), and again the effect of the system is seen to be the same as though the incident light had entered the equivalent lens when it was in the position PH, and had left when the lens was in the position P' H'. The effect of the system is therefore that of a single thin lens with focal lengths PF and P' F' situated at two different positions, at P with respect to the incident light, and at P' with respect to the emergent light. These two positions are called the **principal planes** of the system, and the points P and P', where these planes intersect the axis, are the **principal points.**

PH and P'H' (Fig. 99) represent the principal planes of a system with its principal foci at F and F'. A ray QH' from an infinitely distant axial point may be considered as refracted once at H', and the refracted ray passes through F'; similarly a ray RH may be considered as refracted at H and passing through F. Considering QH and FH as two rays intersecting in an object point at H then, as H'F' and H'R are the two conjugate rays, H' is the image of the object point H formed by the system. Hence *the principal planes are conjugate planes in which the magnification is* + 1. The principal planes are sometimes known as **unit planes.**

Fig. 99.—Principal or unit planes

Fig. 100.—Lense system. Principal and nodal points. Graphical construction of image

Two further points on the axis may be found for any system which have the property that any incident ray directed towards the first leaves the system as though from the second, and with its direction unchanged. These are the **nodal points,** N and N' (Moser 1844; Listing 1845), and are useful in considering the size of the image. When the object and image spaces have the same refractive index, the principal and nodal points coincide as in Fig. 100, and are then sometimes known as the **equivalent points.**

There are therefore six points on the axis of a lens system or thick lens that together are known as the **cardinal points** of the system; these are the two principal foci, F and F′, the two principal points, P and P′, and the two nodal points, N and N′. The significance of the first four points as a means of simplifying the theoretical treatment of optical systems was first suggested by Gauss (1840), and these points are sometimes named after him.

98. Graphical Construction of Image formed by a Lens System

When the positions of principal foci and principal points are known, the position and size of an image formed by a centred lens system, however complicated, may be found easily by the graphical methods described for a thin lens (Chapter VI, § 73). Thus in Fig. 100 an object BQ is situated in front of a lens system, the principal foci and principal points of which are at F, F′, P and P′ respectively. Two rays from the object point Q are sufficient to determine the position and size of the image. These rays are one leaving Q parallel to the axis, and the other ray passing through the first principal focus F. The first of these meets the first principal plane at H and, from the properties of the principal planes, must, whatever refractions it undergoes in the system, travel in the image space as though from a point H′ on the second principal plane, H′ being at the same distance from the axis as H. Since the incident ray is parallel to the axis, the emergent ray must pass through the second principal focus F′.

The incident ray passing through the first principal focus F must emerge parallel to the axis, as though from a point E′ on the second principal plane. The distance of E′ from the axis will be the same as that of E, the point at which the ray met the first principal plane. The intersection of these two emergent rays determines the position of the image Q′.

99. Expressions for Equivalent Power, Back Vertex Power, etc. Two Thin Lenses in Air

In addition to the equivalent focal length, the distances of the first and second principal foci from the front and back vertices of the system respectively will often be required; these distances are called the **front** and **back vertex focal lengths,** f_v and f'_v. The reciprocals of these distances, with the sign changed in the case of the front vertex focal length, are known as the **front** and **back vertex powers,** F_v and F'_v or F.V.P. and B.V.P. These vertex powers are of considerable importance in dealing with spectacle lenses. In Fig. 98

$$\text{F.V.P. or } F_V = -\frac{1}{f_V} = -\frac{1}{A_1 F}$$

$$\text{B.V.P. or } F'_V = \frac{1}{f'_V} = \frac{1}{A_2 F'}$$

In Fig. 101 parallel light is incident on the first lens of a system of two thin lenses in air. The effective power at A_2 of the first lens A_1

$$= \frac{F_1}{1 - dF_1} \text{ from (6.16)}$$

This is the L value for the second lens (L_2).

Then the vergence of the light on emergence from the second lens A_2

$$= L_2 + F_2 = \frac{F_1}{1 - dF_1} + F_2$$

$$= \frac{F_1 + F_2 - dF_1 F_2}{1 - dF_1} = F'_V = \frac{1}{f'_V} \quad \ldots \quad \ldots \quad (9.01)$$

The equivalent lens of focal length $P'F'$ placed at P' will produce the same deviation of the ray as is produced by the system. To find its power, let $A_1 D_1 = 1$ cm., then the prism effect at $D_1 = F_1$ prism dioptres (from 6.14).

Fig. 101.—Lens system, equivalent and back vertex focal lengths

The deviated ray from D_1 meets the second lens at D_2 and

$$A_2 D_2 = \frac{f'_1 - d}{f'_1} = 1 - dF_1$$

Hence, the prism effect at D_2

$$= F_2 (1 - dF_1)$$

and the total prism effect

$$= F_1 + F_2 (1 - dF_1) = F_1 + F_2 - dF_1F_2.$$

This would be the prism effect produced by the equivalent lens at H′, P′H′ being 1 cm.

Hence the equivalent power

$$F = F_1 + F_2 - d\,F_1F_2 \quad \ldots \quad \ldots \quad \ldots \quad (9.02)$$

and

$$f' = -f = \frac{1}{F}$$

From (9.01)

$$F'_V = \frac{F}{1 - dF_1} = \frac{1}{f'_V} \ldots \quad \ldots \quad \ldots \quad (9.03)$$

Similarly, by considering a ray incident through the first principal focus, we obtain

$$F_V = \frac{F}{1 - dF_2} = -\frac{1}{f_V} \ldots \quad \ldots \quad \ldots \quad (9.04)$$

The distance e of the first principal point *from* A_1 and the distance e' of the second principal point *from* A_2 may be found when f, f', f_v and f'_v are known, for

$$e = f_V - f \ldots \quad \ldots \quad \ldots \quad \ldots \quad (9.05)$$

and

$$e' = f'_V - f' \quad \ldots \quad \ldots \quad \ldots \quad (9.06)$$

Example.

Find the equivalent focal length and the positions of the principal foci and principal points of a Huygens eyepiece, consisting of two thin lenses of focal length, $f'_1 = +45$ mm. and $f'_2 = +18$ mm., the lenses being separated 31·5 mm.

$$F_1 = \frac{1000}{45} = 22 \cdot 22 \text{D}, \; F_2 = \frac{1000}{18} = 55 \cdot 56 \text{ D}$$

$$F = F_1 + F_2 - dF_1F_2$$

$$= 77 \cdot 78 - (0 \cdot 0315 \times 22 \cdot 22 \times 55 \cdot 56)$$

$$= 38 \cdot 86 \text{ D}$$

Therefore $f' = +25 \cdot 73$ mm. and $f = -25 \cdot 73$ mm.

$$F_V = -\frac{1}{f_V} = \frac{F}{1-dF_2}$$

$$= \frac{38\cdot86}{1-(0\cdot0315 \times 55\cdot56)} = -51\cdot19 \text{ D}$$

$$f_V = +19\cdot27 \text{ mm.}$$

$$F'_V = \frac{1}{f'_V} = \frac{F}{1-dF_1}$$

$$= \frac{38\cdot86}{1-(0\cdot0315 \times 22\cdot22)} = +129\cdot5 \text{ D}$$

$$f'_V = +7\cdot72 \text{ mm.}$$

The first principal point is therefore situated $f_v - j = +45\cdot0$ mm. from the first lens, and the second principal point at $f'v - f' = -18\cdot0$ mm. from the second lens. The positions of the various points are shown to scale in Fig. 102.

The following further expressions are sometimes useful in dealing with the lens system, and students should derive these for themselves as exercises.

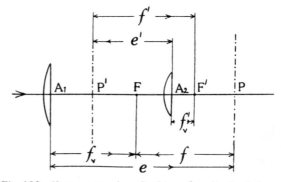

Fig. 102.—Huygens eyepiece. Position of cardinal points

$$ff' = xx' \quad \dots \quad \dots \quad \dots \quad \dots \quad (9.07)$$

Newton's relation (see Chapter VI)

$$F = -g F_1 F_2 \quad \dots \quad \dots \quad \dots \quad (9.08)$$

where g is the distance from the second principal focus of the first lens to the first principal focus of the second lens.

$$F = F_2 \left(1 - aF_1\right) \quad \ldots \quad \ldots \quad \ldots \quad (9{\cdot}09)$$

where $a = d + f_2$, the distance from the first lens to the first principal focus of the second lens.

100. Thick Lens in Air

The thick lens, with air on either side of it, may be dealt with in a similar manner to the lens system above.

The focal powers of the two surfaces will be, from (5.04)

$$F_1 = (n - 1) R_1$$
$$F_2 = (1 - n) R_2$$

Where n is the refractive index of the lens and R_1 and R_3 the curvatures of the surfaces.

A ray BD_1 (Fig. 103) parallel to the axis is refracted by the first surface towards F'_1, the second principal focus of the first surface, and

Fig. 103.—Thick lens in air

after refraction at the second surface crosses the axis at F'. F' is the second principal focus of the thick lens. $P'H'$ is the position of the equivalent lens with respect to the emergent light, that is, the second principal plane and $P'F'$ the second equivalent focal length.

As before, taking A_1D_1 as 1 cm., the prismatic effect due to the first surface $= F_1$ prism dioptres, and the prismatic effect at D_2 due to the second surface $= A_2D_2 \times F_2$.

155

$$A_2 D_2 = \frac{f'_1 - d}{f'_1} \text{ and } f'_1 = \frac{n}{F_1} \text{ (from (5.05))}$$

$$\text{or } A_2 D_2 = 1 - \frac{d}{n} F_1$$

Therefore the total prismatic effect produced

$$= F_1 + \left(1 - \frac{d}{n} F_1\right) F_2$$

$$= F_1 + F_2 - \frac{d}{n} F_1 F_2$$

This would be the prism effect produced by the equivalent lens at H', P'H' being 1 cm.

Hence the equivalent power is given by

$$F = F_1 + F_2 - \frac{d}{n} F_1 F_2 \quad \ldots \quad \ldots \quad \ldots \quad (9.10)$$

$$A_2 F' = P'F' \times \frac{A_2 D_2}{P'H'} = \frac{1}{F}\left(1 - \frac{d}{n} F_1\right)$$

Hence

$$\frac{1}{A_2 F'} = \frac{1}{f'_V} = F'_V = \frac{F}{1 - \frac{d}{n} F_1} \quad \ldots \quad \ldots \quad (9.11)$$

Similarly

$$-\frac{1}{f_V} = F_V = \frac{F}{1 - \frac{d}{n} F_2} \quad \ldots \quad \ldots \quad \ldots \quad (9.12)$$

It should be noticed that the expressions for the thick lens are of the same form as those for two thin lenses in air, the value d in the latter being a special case of the reduced separation $\frac{d}{n}$.

101. Position and Size of Image

In Fig. 104 a ray BH incident on the first principal plane PH of a lens system or thick lens is refracted along H'B' from the second principal plane P'H'. This ray has been deviated through the angle KLH' = d and from the figure

Fig. 104.–Lens system. Prismatic effect. Position of image

$$d = u' + u$$

or, as u is a negative angle, taking signs into account

$$d = u' + (-u)$$

$$u = \frac{c}{\text{PB}}, \quad u' = \frac{c'}{\text{P'B'}}$$

As the heights of intercept in the two principal (unit) planes must be equal, $c' = c$, and

$$d = \frac{c}{\text{P'B'}} - \frac{c}{\text{PB}}.$$

The total prismatic effect in prism dioptres produced on a ray at 1 cm. from the axis is equal to the equivalent focal power of the system (§ 99), or the deviation in radians $= \dfrac{c}{f'}$

Hence

$$d = \frac{c}{\text{P'B'}} - \frac{c}{\text{PB}} = \frac{c}{f'}$$

and

$$\frac{1}{\text{P'B'}} - \frac{1}{\text{PB}} = \frac{1}{f'}$$

Representing the distances PB and P'B', the object and image distances measured from the principal planes, by l and l' we have for a system or thick lens

$$\frac{1}{l'} - \frac{1}{l} = \frac{1}{f'} = F.$$

Thus, the conjugate foci expression obtained for a thin lens may be used for a lens system or thick lens, if object and image distances are measured from the first and second principal points respectively.

157

Also, since a ray from any object point passing through the first nodal point emerges from the second nodal point undeviated by the system, the various expressions for the magnification of the image formed by a thin lens are applicable to a system if object and image distances are measured respectively from the first and second nodal points.

102. The Measurement of Focal Length and Power

The measurement of the vertex focal lengths and thence the vertex powers of a system or thick lens may be made by determining the positions of the principal foci by the methods described in Chapter VI.

The determination of the equivalent focal length is, however, less simple, owing to the difficulty of finding directly the positions of the principal points. The following are a few of the methods applicable to lens systems.

Newton's Method—Positive Systems.

The system to be measured is set up about midway along an optical bench, and the positions of the principal foci found, using either a telescope or a collimator (Chapter VI, § 77). An illuminated object is then placed at some distance x outside the first principal focus, and an image is received on a screen. The distance of the image from the second principal focus is x', and the focal length is found from the expression $f' = \sqrt{-xx'}$. Most accurate results will be obtained when x and x' are nearly equal.

Magnification Methods.

Methods depending on the measurement of the magnification of the image formed by a system are often very convenient. Providing the image is sufficiently free from distortion, the focal length can be obtained quickly and accurately by such methods.

Two Magnifications.

An illuminated object of known size, such as a rectangular aperture, is set up on the axis of the lens to be tested. The image is focused on a screen, and its size measured. This is repeated with the object in a different position. From the magnification m_1 and m_2 thus found and the distance moved by the object, $l_2 - l_1$, or by the image, $l'_2 - l'_1$, the focal length may be calculated.

$$f' = \frac{l'_2 - l'_1}{m_1 - m_2} = \frac{l_1 - l_2}{\dfrac{1}{m_1} - \dfrac{1}{m_2}} \qquad \dots \qquad \dots \qquad (9.13)$$

This method has been applied by Abbe (1840–1905) to the measurement of the focal lengths of microscope objectives. The two positions of object and image are obtained by altering the tube-length of the microscope, and the size of the image is measured with a micrometer eyepiece, a stage micrometer being used as object.

Two Magnifications, Negative Systems.

The two magnifications method may also be applied to negative systems by using an auxiliary positive lens to form a real image of the object (Fig. 105). The negative system is inserted in the convergent light, and the image formed by the positive lens becomes a virtual object for

Fig. 105.—Negative system. Measurement of focal length.
Two magnifications method

the negative system. By measuring the magnification of the final image with the system being tested in two different positions, A_1 and A_2, the focal length can be found from (9.13).

Inclined Collimators.

If two distant objects subtend a known angle w at the lens, then their images formed in the second focal plane of the lens are separated by a distance

$$h' = 2f \tan \frac{w}{2} \quad (6.12A)$$

and

$$f = \frac{h'}{2 \tan \frac{w}{2}}$$

The distant objects may conveniently be replaced by two collimators fixed at an angle w between their optical axes. The method is a simple and rapid one for the measurement of photographic objectives, which have a flat field reasonably free from distortion.

The Foco-Collimator.*

This method, which is particularly useful in the measurement of short focus systems such as eyepieces, is similar in principle to the method

* T. H. BLAKESLEY, *Proc. Opt. Convention*, 1905, p. 185. F. J. CHESHIRE, *Trans. Opt. Soc.* XXII, No. 1, 19; *Journ. Roy. Micro. Soc.*, 1914, pp. 513–19.

of inclined collimators. A collimator is provided with one or more pairs of lines QS (Fig. 106) ruled on a plane glass placed at the focus of the

Fig. 106.—The foco-collimator

objective. Each pair of lines subtends an angle w at the collimator objective. Images Q'S' are formed in the focal plane of the lens under test, and their separation h' will depend on the focal length of this lens.

From Fig. 106 we have

$$\frac{1}{2 \tan \dfrac{w}{2}} = \frac{f_C}{QS} = \text{collimator constant.}$$

Then f, the focal length of the lens under test, is given by

$$f = \frac{Q'S'}{2 \tan \dfrac{w}{2}} = Q'S' \times \text{collimator constant} \qquad \dots \quad (9.14)$$

The collimator constant is first determined by measuring the angle w by means of a theodolite or spectrometer. The lens to be tested is then mounted coaxially with the collimator, and the distance Q'S' measured by receiving the image on a screen, or, more accurately, by means of a measuring microscope or eyepiece with scale at its focus.

This method can also be used for measuring negative lenses. Using a measuring microscope with objective of long focal length the virtual image of the collimator object can be focused and its size measured.

Rotation Method—Nodal Points.

The following method provides a means of determining the positions of the nodal points, and since, for a system in air, these coincide with the principal points, the equivalent focal length of the system can be measured.

In Fig. 107 light from a distant luminous point is incident on a lens system having its nodal points at N and N', and an image is formed at F'. On rotating the system about the second nodal point N', no movement takes place in the image, for from the property of the nodal points, any ray from the distant object point directed towards the first nodal point in its new position must still emerge along N'F'. If the system is rotated about any other point, N' and therefore F' move.

Fig. 107.—Focal length of lens system. Rotation method

The lens to be measured is set up on the optical bench in a holder that can be rotated about a vertical axis, the lens sliding in the holder so that any point on its axis can be brought over the axis of rotation. Parallel light from a collimator is focused on a screen, and the position of lens and screen adjusted until a rotation of the lens produces no movement of the image. The distance from the axis of rotation to the screen is then the equivalent focal length.

This method can also be used with the telescope method (Chapter VI, § 77). An object is brought up towards the lens until its image is clearly seen in a telescope focused for parallel light, and the lens is adjusted, as above, so that the image remains stationary on rotating the lens. Owing to the magnification given by the telescope, this method will usually be more accurate than that of receiving an image on a screen.

The principle of the stationary image with the lens rotating about its second nodal point has been applied to panoramic cameras.

<u>CHAPTER IX</u>   *EXERCISES*

1. The following lenses, all assumed to be thin, are placed in contact. What will be the focal power and focal length of the combination?
    (*a*) + 4D,
    (*b*) − 6 inches focal length,
    (*c*) + 4 inches focal length,
    (*d*) − 2·62D.

2. Deduce an expression for the focal length of two thin lenses in contact.

A convex lens produces a real image of an object at a distance of 40 cm. from the lens. A concave lens, the focal length of which is 15 cm. is placed 20 cm. from the convex lens on the side away from the object. Find the new position and nature of the image. (L.U.)

3. Two equal + 4D meniscus lenses of spectacle glass ($n = 1·523$) with − 6D

concave surfaces are placed together with these concave surfaces facing one another and the space between them is filled with water. Find the power of the complete glass-water lens (neglect thickness). (S.M.C.)

4. A 10D plano concave lens of refractive index 1·57 lies on a horizontal plane sheet of glass, with its concave surface uppermost and is filled up level with water, having a refractive index of 1·33. What will be the power of the combination? (S.M.C.)

5. Two thin convex lenses of focal length 5 cm. and 7 cm. respectively are separated by 16 cm. Show by graphical construction that (within the paraxial region) incident rays parallel to the axis at all incident heights intersect their conjugate refracted rays always in a plane perpendicular to the axis.

6. Two thin lenses of powers + 20D and + 25D respectively are separated by a distance of 15 cm. Find graphically the focal length and the positions of the principal points, P. and P′, of the combination. Three rays at different inclinations are incident on the combination at the axial point P. Trace them through the two component lenses, showing that they emerge from the system through P′ parallel to their original directions.

7. Show with a diagram what are meant by the principal and nodal points of a lens. Explain how these positions are used in finding the position and size of an image formed by a lens.

8. A distant object subtends an angle of 10°; find the size of the image formed by a + 6D lens. Where must a − 10D lens be placed to magnify the image four times?

9. A distant object subtends an angle of 20°. Find graphically the size and position of the image of this object formed by a lens system consisting of a + 6D followed by a + 10D lens, the lenses being coaxial and 5 cm. apart. What must be the position of an object in front of the first lens in order that its image formed by the system may be at infinity?

10. What are meant by the principal, focal and symmetrical planes of a thick lens or lens combination? Illustrate your answer by a diagram. (S.M.C.)

11. Two cylindrical lenses, one of + 4D and the other of − 6D, are set up parallel to one another, with their axes vertical and with an interval of 12 centimetres between them. If a point source of light is placed at 20 centimetres from the positive lens, where will the image be, and what will be its character? (S.M.C.)

12. Two thin positive lenses of focal lengths 4 cm. and 3 cm. respectively, each of 2 cm. aperture, are mounted coaxially 2 cm. apart. An object 2 cm. high is placed with its foot on the common axis and 8 cm. in front of the 4 cm. lens.

On a careful diagram drawn exactly full size, construct the course of the pencil of light from the extremity of the object, through the lenses, to the final image. From your diagram give the size of this image and its distance from the second lens. (S.M.C.)

13. A screen is placed 14·5 mm. behind a + 60D lens (assumed to be thin). Find the size of the blurred image formed on the screen of a distant object that subtends an angle of 20°. What must be the power of a second lens, placed 12 mm. in front of the first, so that a *sharp* image of the same object shall be focused on the screen? (S.M.C.)

14. Derive Newton's relation in the case of a system consisting of two thin separated lenses.

15. A parallel pencil of light falls axially on a + 5·5D spherical lens combined with a − 2D cylindrical lens with axis horizontal. What lens placed 8 inches behind this combination will restore the emergent light to parallelism in both meridians? (S.M.C.)

16. The image of a distant object subtending an angle of 10° is formed by a lens combination consisting of a + 4D and a + 2D lens separated 10 cm. Find the size and position of the image and the equivalent focal length of the system.

What will be the effect on the size and position of the image and on the equivalent focal length of turning the system round so that the + 2D lens is towards the object?

17. An object 2 cm. in height is placed in front of an optical system of lenses. An inverted image 12 mm. in height is formed on the opposite side of the system. If the object were placed 10 cm. nearer the system the image would be formed at infinity. Find the focal length of the system.

18. Two positive thin lenses of focal length 8 cm. and 6 cm. respectively are coaxial and are separated by 3·5 cm. Find by graphical construction the equivalent focal length of the combination and the positions of its principal and focal points. Mark these positions clearly on your drawing. Check the results by calculation.

19. Find the equivalent focal length of a lens system consisting of a + 4D and a − 4D lens separated 15 cm. What will be the position and size of the image formed by this system of an object placed 1 metre in front of the first lens?

20. A spectacle lens of power + 10D is mounted coaxially with and 5 cm. in front of a second lens of power − 8D. Find the power of the combination and the position of the principal points. Of what well-known optical system is this a model? (S.M.C.)

21. What are meant by the equivalent focal power and principal points of a lens system?

What will be the size of the image of a distant object which subtends an angle of 8° at a lens system consisting of two lenses of + 10 inches and − 5 inches focal length respectively, separated by 7 inches?

22. Find the positions of the focal and principal points of a Ramsden eyepiece consisting of two lenses of 2 in. focal length separated 4/3 in.

23. Two positive thin lenses of focal lengths 20 cm. and 50 cm. respectively are separated by a distance of 100 cm. Find the power of the resultant system and the positions of its principal and focal points. Discuss the effect of replacing the original lenses by two of the same focal lengths but of opposite sign. Give a diagram.

24. A lens system is formed of two lenses having powers of + 6D and − 10D respectively, separated by 10 cm. Find the position and size of the image of a distant object which subtends an angle of 10°.

25. A lens system consists of a + 4D lens and a + 3D lens separated by 10 cm. Find the equivalent focal length and the positions of the principal points. An object 3 cm. long is situated 50 cm. in front of the first lens. What will be the position and size of the image produced by the combination?

26. Find the equivalent focal power and the position of the principal points of a system consisting of a + 6D and a − 8D lens separated 8 cm. Use the values obtained to find the position and size of the image of an object 3 cm. long placed 50 cm. from the first lens.

27. Two thin positive lenses of focal lengths f'_1 and f'_2 are separated by a distance d. What is the value of d for the focal power of the combination to be zero? When the lenses are in this position, find a relation between the incident height of a ray parallel to the axis in the object space and the height from the axis of the conjugate emergent ray.

28. The positions of object and image such that the magnification is − 1 are known

as the **symmetrical points**. Show that the distances of these from the principal points are respectively $2f$ and $2f'$.

29. A Huygens eyepiece consists of two thin lenses having focal lengths of $2\frac{1}{2}$ inches and 1 inch respectively, separated by $1\frac{3}{4}$ inches. Find the equivalent focal length and the positions of the principal points. Show in a diagram, drawn to scale, the positions of these points and the focal length.

30. Show that the equivalent focal power F of a system of two thin lenses of focal power F_1 and F_2 is given by the expression
$$F = - gF_1F_2$$
Where g is the distance from the second principal focus of the first lens to the first principal focus of the second lens.

31. Show that in a system of two thin lenses
$$F = F_2(1 - aF_1)$$
where $a = d + f_2$, the distance from the first lens to the first principal focus of the second lens.

32. When the distance of a real image from a lens system is increased by 5 cm. the magnification is increased from 0·5 to 0·65. What is the focal length of the system?

33. Show that for a thick lens in air
$$F = (n - 1)\left(\frac{1}{r_1} - \frac{1}{r_2}\right) + \frac{(n - 1)^2}{n}\cdot\frac{d}{r_1 r_2}.$$

34. A positive meniscus lens of central thickness 7·60 mm. is made of glass of refractive index 1·520, the back surface power being $- 6D$. What power must be given to the front surface in order that the back vertex power of the lens shall be $+ 9·7D$?

35. Find the equivalent power, back vertex power and position of the second principal focus of a meniscus lens (in air) with the following constants:—

Radius of first surface	...	...	... $+$ 50 mm.
Radius of second surface	...	...	... $+$ 250 mm.
Central thickness	...	...	... 15 mm.
Refractive index	...	...	... 1·6

36. A bi-convex lens of glass $n = 1·52$ has curvatures of 5 inches and 8 inches radius respectively and a central thickness of $1\frac{1}{2}$ inches. Find the size and position of the image of an object 2 inches long placed 15 inches from the first surface of the lens.

37. A slab of glass 3 inches long, $n = 1·5$ has its end faces ground convex so that each has a radius of curvature of $\frac{1}{2}$ inch. What are the optical properties of the thick lens so made? (S.M.C.)

38. A spherical electric bulb 3 inches in diameter is immersed in water; what will be the equivalent focal length of the optical system thus produced?

39. A $+ 10D$ equiconvex lens is inserted into a thin glass box with plane parallel sides which just contains it, and the space is filled with water of refractive index 1·33. If the power of the combination is $+ 4D$, what is the refractive index of the lens and the curvature of its surfaces? (S.M.C.)

40. Find the equivalent focal length and principal points of a sphere of glass of which the refractive index is 1·54 and the diameter 2·7 inches.

41. Describe and explain the principle of one method of determining the equivalent focal length of a lens system. Give a diagram.

42. In measuring the focal length of a lens system by the foco-collimator, the collimator used had a lens of 10 inches focal length and a graticule in the focal plane

with two marks 0·35 inches apart. The size of the focused image formed by the lens being tested was a quarter of an inch. Find the focal length of the lens.

43. A scale is placed at the first principal focus of a positive lens A_1 and is focused by A_1 and a second lens A_2; show that the image is formed at the second principal focus of A_2 and that the magnification is equal to the ratio of the focal lengths of the lenses. Is the image erect or inverted?

44. Define the term "focal length" as applied to a lens of which the thickness cannot be neglected.

Describe briefly an experimental method of finding the difference between this "focal length" and the distance between the apex of the lens and its principal focal point. Prove any formula employed. (S.M.C. Hons.)

45. An optical system consists of a + 10D lens followed at a distance of 5 cm. by a − 8D lens. Find the power of the whole system, and hence the size of the image of a very distant object subtending an angle of 0·05 radians in the field of view. How would the distance of the image from the last lens, and the size of the image, change if the system were reversed so that the light enters the negative lens first? (S.M.C.Hons.)

46. With the aid of a diagram explain why the planes passing normally to the optical axis through the principal points of a lens system are called the *unit planes*.

The powers of the first and second surfaces of a lens of total power F, in air, are F_1 and F_2; the lens has central thickness d metres and refractive index n.

Show that the principal points P_1 and P_2 of the lens are separated by the distance

$$P_1 P_2 = d \left(1 - \frac{F_1 + F_2}{nF} \right)$$

(S.M.C.)

47. A compound lens of power + 32D consists of two thin positive components of equal power separated by an air space; it forms a real image of a real object, the image being 1·5 times larger than the object. The distance from object to image is found to be 125·21 mm.

Find the separation of the principal points and show their positions on a diagram.

Also find *either* (a) the power of the individual components

        or (b) the separation of these components. (S.M.C.)

48. Two coaxial thin positive spherical lenses are separated by 20 mm. The first lens has power + 5D and the back vertex power of the combination is + 50D.

Keeping the separation and the power of the second lens unchanged, by how much must the power of the first lens be altered in order to increase the back vertex power of the combination by one-third of a dioptre? (S.M.C.)

49. The surface powers of a bi-convex lens of refractive index $n = 1·50$ and central thickness $d = 18·75$ mm. are respectively $F_1 = 5·0$ and $F_2 = 8·0$ dioptres. Calculate the positions of its principal points, P and P¹.

An expression for the separation of the principal points is $PP^1 = \dfrac{(n-1) \, ad}{na - d}$

where a is the distance C_1C_2 separating the centres of curvature of the lens surfaces. Show that this expression gives the correct numerical result for the lens in question. (S.M.C.)

X - The Principles of
Optical Instruments

The more important optical instruments may be divided into two classes: (a) those which are used directly with the eye as aids to vision, as, for example, the microscope and telescope, and (b) projection instruments, such as the camera and the projection lantern.

103. The Eye

Fig. 108 represents a horizontal section of the human right eye. The greater part of the globe of the eye is almost spherical, of about one inch diameter, and is formed by the tough opaque outer coat, known as the **sclera,** S. The front portion of the globe is the more steeply curved and transparent **cornea,** C, which is the chief refracting surface. The cornea is about ½ mm. thick, and its surfaces, in the normal eye, are approximately spherical. The average radii of curvature of the cornea are 7·7 mm. for the front surface and 6·8 mm. for the back; its refractive index is 1·38. The space immediately behind the cornea, known as the anterior chamber, is filled with a fluid, the **aqueous humour,** of almost the same refractive index as water. The cornea and aqueous humour together therefore form a positive lens of about 43D power.

Behind the anterior chamber is the double convex **crystalline lens,** L, composed of layers of transparent fibres. The curvature, refractive index and hardness of the layers increase towards the centre of the lens, and the whole is enclosed in the thin lens capsule. Under the action of the **ciliary muscle,** C.M., the lens changes in form, and the power of the system is adjusted to focus images of objects at different distances on the receiving screen. The radii of curvature of the front and back surfaces of the crystalline lens when the eye is focused on a distant object

Fig. 108.—Horizontal section of the right eye
(× 2·5 natural size)

Fig. 109.—Cardinal points of the eye

are about 10 mm. and 6 mm. respectively; its thickness is about 3·6 mm. and its refractive index varies from about 1·41 at the centre to 1·38 for the outer layers. The space behind the crystalline lens is filled with a transparent jelly-like medium, the **vitreous humour,** having the same refractive index as the aqueous.

Directly in front of the crystalline lens is the **iris,** I, the diaphragm, which controls the amount of light entering the eye. This has a circular aperture, P, varying in diameter from about 2 to 8 mm., known as the **pupil.**

167

Although the centres of the various surfaces and the pupil do not lie exactly on a common axis, we may, for most purposes consider the eye as a centred system, and the approximate positions of the focal, principal and nodal points are shown in Fig. 109.

The innermost coat of the globe in contact with the vitreous humour is the **retina,** R. This is the receiving screen upon which sharp images of external objects must be formed, and from the closely packed nerve endings of which the stimulus is conveyed along the **optic nerve,** ON, to the brain. A little to the temporal side of the end of the optic axis is a small area of the retina known as the **macula lutea,** M′, or yellow spot, having a small depression at its centre, known as the **fovea centralis.** It is here that vision is most distinct, and the image of the object on which the attention is fixed falls on this portion of the retina. The portion of the retina at which the optic nerve enters is insensitive to light and forms the "blind spot". Between the retina and the sclera is the **choroid,** a coat which contains the blood vessels which nourish the eye and a layer of pigment cells which absorbs any light passing through the retina, and prevents any light that may pass through the sclera from reaching the retina.

104. Accommodation

As the eye can see distinctly, in turn, objects both far and near, it must be provided with some adjustment whereby images of objects at different distances are brought to focus on the retina. This focusing adjustment, which is known as **accommodation,** is effected by a change in the form of the crystalline lens produced by the action of the ciliary muscle. When the eye accommodates to see a near object, the ciliary muscle contracts and relaxes the suspensory ligament attached to the lens capsule, allowing the front surface of the lens to bulge forward to a steeper curvature and therefore of increased power. At the same time, as can easily be seen by watching an eye, the iris contracts, reducing the size of the pupil. The amount of change in power produced—the **amplitude of accommodation**—is greatest in the case of young persons, being at the age of 10 about 14D, which means that a normal eye at this age can see objects distinctly as close as 7 cm. This power of accommodation gradually becomes less with advancing years, owing probably to a hardening of the crystalline lens, until it is no longer possible to see objects distinctly at the distance required for reading and other close work without the aid of lenses. This condition is known as **presbyopia.**

105. Emmetropia and Ametropia. Far and Near Points

When the optical system and the axial length of the eye are such that light from a distant object is focused exactly on the retina when accommodation is completely relaxed, the eye is said to be **emmetropic** (Fig. 110a) and the condition is that of emmetropia. When this is not the case the eye is **ametropic.** The three types of ametropia are **myopia, hypermetropia** or hyperopia and **astigmatism,** the first two usually being due to abnormalities in the axial length of the globe and the last to want of sphericity in one or more of the refracting surfaces, usually the cornea.

In myopia, or short sight, the refracting power or the axial length is too great and light from a distant object focuses in front of the retina (Fig. 110b). Such an eye will therefore not be able to see distant objects clearly without the aid of a correcting lens. Objects at some closer distance will, however, be clearly seen with accommodation relaxed. In hypermetropia, or long sight, the refracting power or the axial length is too small, and light from a distant object focuses behind the retina (Fig. 110c). The uncorrected hypermetropic eye with accommodation relaxed, will not see any object clearly, but, if the amount of error be

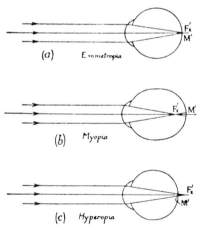

Fig. 110.—The eye. Emmetropia and ametropia

not too great, the additional power required to focus the light on the retina may be provided by accommodation. In this way the hypermetropic eye may have quite good vision for distant objects. In astigmatism the refracted beam from a point object is astigmatic and focuses through two lines (Chapter VIII); these line foci may occupy any positions with respect to the retina, but in no case will an object be seen distinctly.

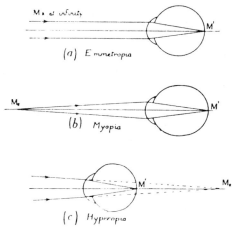

Fig. 111.—The far point of the eye

The object point conjugate to the retina, when accommodation is completely relaxed, is known as the **far point**, *punctum remotum*, M_R, of the eye. This will be (*a*) at infinity in the case of an emmetropic eye (Fig. 111*a*), (*b*) a real object point in myopia (Fig. 111*b*) and (*c*) a virtual object point behind the eye in hypermetropia (Fig. 111*c*). The vergence the incident light must have at the cornea in order that the light may focus on the retina, when accommodation is relaxed, is termed the **static refraction** of the eye.

The conjugate point to the retina, when accommodation is exerted to its fullest extent, that is, the nearest position at which an object can be seen distinctly, is the **near point**, *punctum proximum*, M_P, of the eye. The position of this will depend on the amplitude of accommodation available and the type and amount of ametropia; thus for a given amplitude the near point will be closer in myopia and more remote in hypermetropia than in emmetropia.

106. The Correction of Ametropia

Ametropia may be corrected by placing before the eye a lens that will give the light incident from a distant object the necessary vergence for it to be focused on the retina. As will be seen from Fig. 112, the necessary condition for correction will be that the power and position of the lens shall be such that the second principal focus of the lens shall coincide with the far point of the eye. Negative lenses will therefore be used to correct myopia and positive lenses hypermetropia, while in astigmatism, cylindrical or sphero-cylindrical lenses are required. In practice, the

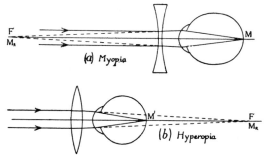

Fig. 112.—Correction of ametropia

amount of ametropia is usually expressed as the power of the lens required for correction when placed at the ordinary spectacle distance, about 12 mm. from the cornea.

107. Visual Angle and Apparent Size

The apparent size of any object or image seen by the eye depends upon the size of the retinal image, which in turn depends upon the angle subtended by the object or image at the nodal point of the eye. Objects which subtend the same **visual angle** will have the same apparent size, although their actual sizes may be very different if they are at different distances from the eye. Thus a disc $\frac{1}{4}$ inch diameter held at arm's length will appear to cover the moon, that is, the disc at that distance and the moon have the same apparent size. The principle of magnifying instruments, such as the magnifier, microscope and telescope, is to produce an image that shall subtend a larger visual angle than that subtended by the object seen directly.

108. Apparent Magnification or Magnifying Power. The Magnifier or Simple Microscope

An object situated at the near point will subtend the largest visual angle and will therefore have the largest apparent size with which it can be clearly seen by the unaided eye. If it be required to see the object apparently still larger recourse must be had to optical aid. If a positive lens of focal length shorter than the near point distance is placed close before the eye, the object can be brought up to, or a little within, the first principal focus of the lens, and the eye will be able to focus the

171

Fig. 113.—Image seen through a lens. Apparent magnification or magnifying power

parallel or divergent light leaving the lens. In Fig. 113 the light leaving the lens, as though from the virtual image B′Q′, enters the eye and, providing the image is not closer to the eye than the near point, it will be seen distinctly. The size of the retinal image and therefore the apparent size of the object seen through the lens will depend upon the angle w'.

As the distance of the near point differs in different eyes it is necessary in considering the magnification of instruments to adopt some standard value for this distance. This conventional distance is called the **least distance of distinct vision**, q, and is taken as -10 inches or -250 mm. measured *from* the eye.

The **apparent magnification** or **magnifying power**, M, produced by a lens may be expressed as the ratio of the angle, w', subtended at the nodal point of the eye by the image to the angle, w, subtended at the nodal point by the object when at the least distance of distinct vision from the unaided eye. Usually the angles are expressed in terms of their tangents. It follows that the magnifying power is also the ratio of the size of the retinal image obtained with the aid of the lens to the size of the retinal image of the object when at the least distance of distinct vision from the unaided eye.

In Fig. 114 the object to be magnified is placed at the first principal focus of a short focus positive lens, and the emmetropic eye with accommodation relaxed sees the erect image at infinity subtending an

Fig. 114.—Magnifier. Simple microscope or loupe

angle w'. Since the image is at infinity, the distance of the lens from the eye has no effect upon magnification, but in order to obtain the largest field of view the lens should be as close to the eye as possible. This represents the best conditions of working in practice, as although the magnification is slightly less than if the object were brought inside the principal focus, any strain through continual use of the accommodation is avoided.

$$M = \frac{\tan w'}{\tan w} = \frac{w'}{w} \text{ when the angles are small} \qquad \dots \text{(10.01)}$$

$$\tan w' = \frac{h}{f''}, \tan w = \frac{h}{-q}$$

Therefore

$$M = \frac{-q}{f'} = \frac{250}{f' \text{ (mm.)}} = \frac{10}{f' \text{ (in.)}} = \frac{F \text{ (D)}}{4} \qquad \dots \text{(10.02)}$$

| Achromatic | Coddington | Stanhope |

Fig. 115.–Types of Magnifiers

When the magnifier is used with the image at any finite distance the magnifying power may be calculated by finding the angles w and w' taking into account the position of the eye.

While the simple microscope possesses the distinct advantages of simplicity and erect image, the necessary short working distance and small diameter of the lens in the case of high powers limit the greatest useful magnification to about 15. Types of magnifiers are shown in Fig. 115.

109. The Compound Microscope

The compound microscope consists essentially of two positive lenses, the first of which, the **objective** or **object glass,** forms a real and magnified image of the object, and the second, the **eyepiece,** is used as a magnifier to examine the aerial image formed by the objective. In practice both objective and eyepiece will be systems of lenses.

Fig. 116 shows the path of light from one end of a small object

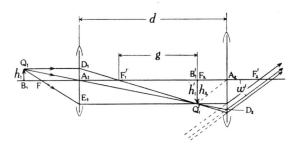

Fig. 116.–Compound microscope

B_1Q_1. A real inverted aerial image h'_1, having a lateral magnification m_1, is formed at B'_1 by the objective A_1. If the microscope is used with the unaccommodated emmetropic eye, B'_1 coincides with the first principal focus F_2 of the eyepiece A_2 which, as a magnifier, gives a magnifying power $m_2 = \dfrac{-q.}{f'_2}$ The total magnification M therefore equals $m_1 \times m_2$. From expression (6.10) $m_1 = \dfrac{-g}{f'_1}$ where g (distance from F'_1 to F_2) is called the **optical tube length.**

Therefore $$M = \frac{-g}{f'_1} \times \frac{-q}{f'_2} = \frac{-g}{f'_1} \times \frac{250}{f'_2 \text{(mm)}} \quad \dots \quad \dots \text{(10.03)}$$

In practice, microscope objectives are usually specified in terms of their focal lengths, either in inches or millimetres, and are generally designed for an optical tube length of 160 millimetres. Eyepieces are marked in terms of their magnifying power, $\dfrac{-q}{f'}$.

110. The Telescope. Principle

Telescopes are used for the following three purposes:

1. To produce an image of a usually distant object which will subtend a larger visual angle than that subtended by the object, that is, to give an apparent magnification or magnifying power.

2. As an instrument of large aperture, collecting sufficient light to enable faint distant point sources of light, such as stars, to be seen, and of high resolving power (see Chapter XV). This is the purpose of the large astronomical telescopes.

3. To determine accurately a definite line relative to a certain part of an instrument, that is, as a sighting device, as in surveying instruments, spectrometer, etc.

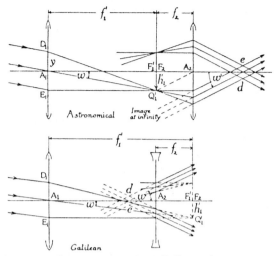

Fig. 117.—Astronomical and Galilean telescopes

The telescope consists, in principle, of two lenses, the first—the objective or object glass—being a positive lens which forms a real inverted image of the distant object. This image is observed through a second lens—the eyepiece—which acts as a magnifier. The eyepiece may be either positive (**astronomical telescope**), or negative (**Galilean telescope**). In practice, the objective will be an achromatic combination (see Chapter XIII) and the eyepiece of one of the forms described in § 115.

The path of the light from a point on a distant object through the two forms of telescope is shown in Fig. 117. An unaccommodated emmetropic eye sees an image at infinity subtending an angle w' instead of the angle w subtended by the object. As will be seen from the figure, this image will be inverted in the astronomical telescope and erect in the Galilean. When the telescope is in normal adjustment for an unaccommodated emmetropic eye and a distant object, both incident and emergent light are parallel, and the objective and eyepiece are separated by a distance equal to the sum of their second focal lengths. Such a system is known as an **afocal** or **telescopic system.**

111. Magnification or Magnifying Power

The magnification of the telescope is expressed as the ratio of the angle subtended by the image seen through the telescope to the angle subtended by the object at the eye, without the telescope. When the telescope is in "infinity adjustment", as in Fig. 117

$$\frac{\tan w'}{\tan w} = \frac{\dfrac{h'_1}{f'_2}}{-\dfrac{h'_1}{f'_1}} = -\frac{f'_1}{f'_2} = M \qquad \dots \qquad \dots \quad (10.04)$$

The magnification will be negative in the astronomical and positive in the Galilean telescope.

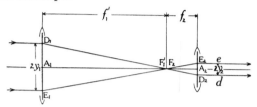

Fig. 118.—Astronomical telescope. Magnification as ratio of pupils

112. Entrance and Exit Pupils. Ramsden Circle

It will be seen from Fig. 117 that the light filling the effective aperture $D_1 E_1$ of the objective will, after passing through the instrument, pass through, or, in the case of the Galilean telescope, appear to come from a circular area $d\ e$ which is the image of the objective formed by the eyepiece. This is the **exit pupil, Ramsden circle** or **eye ring**. The objective is the **entrance pupil**.

The magnification may be expressed in terms of the diameters of the entrance and exit pupils. From Fig. 118

$$M = -\frac{f'_1}{f'_2} = -\frac{2y_1}{2y_2} \quad \dots \qquad \dots \qquad \dots \quad (10.05)$$

The magnification of a telescope is conveniently determined experimentally by a measurement of the diameters of objective and Ramsden circle.

The exit pupil has an important bearing on the field of view of the telescope as will be readily seen from Fig. 117; thus in the astronomical telescope, the exit pupil being real, it may be made to coincide in position with the pupil of the eye, and oblique pencils from the edges of a wide field can enter the eye. In the Galilean telescope, however, the exit pupil being virtual, the pencils from different points of the field diverge from the axis on leaving the eyepiece, and only those within a narrow angle can enter the eye pupil. The narrow angular field of view obtained is the chief drawback of this form of telescope.

Entrance and exit pupils and their significance will be more fully dealt with in Chapter XVII.

113. Graticules

If a scale engraved on a thin parallel glass plate is placed in the common focal plane of objective and eyepiece, it will be seen in focus at the same time as a distant object, and such an arrangement may be used to measure the angle subtended by an object. Such scales, which are frequently used in telescopes and binoculars, especially those used for military purposes, are known as **graticules.** The divisions of the scale are made of such a length that they subtend the required angle, say $\frac{1}{2}$ degree, at the nodal point of the objective and, since object and image subtend the same angle at the nodal points, the angle subtended by any object can be read off directly on the scale.

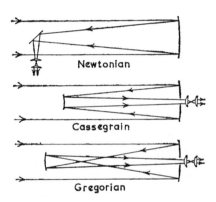

Fig. 119.—Reflecting telescopes

114. Reflecting Telescopes

Owing to the great difficulty and expense of producing objectives of very large aperture, the largest astronomical telescopes are of the reflecting form in which the objective is replaced by a surface silvered concave mirror. The mirror, which is "figured" to a parabaloidal form to correct spherical aberration, forms a real image. This is observed, usually after further reflection, by the eyepiece in the same way as in the refracting telescope. The image is, of course, free from chromatic aberration, and it was due to the apparent impossibility of producing an achromatic lens that Newton was led to construct the reflecting telescope. The largest telescope at the present time, the mirror of which has a diameter of 200 inches, is that at Mount Palomar in California. Various forms of the reflecting telescope are shown in Fig. 119.

115. Eyepieces

We have considered the eyepiece of the microscope and telescope as a single lens, but for several reasons it will generally be necessary to replace this single lens by a system of separated lenses. It will be seen from Fig. 117 that in order that light from a large field of view may pass through a single lens eyepiece this must be of large diameter. But the

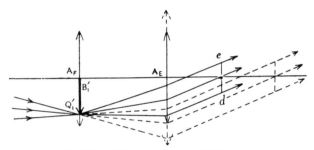

Fig. 120.—Effect of eyepiece field lens

lens cannot have a large diameter with the short focal length required for magnification. The difficulty may be overcome by using two separated positive lenses as the eyepiece. The use of two separated lenses will also permit of more control of the aberrations, particularly those affecting oblique pencils, and an example of this is discussed in § 162.

If a positive lens A_F (Fig. 120) is placed in the plane of the image $B'_1 Q'_1$ formed by the objective of the microscope or telescope, the image will be unchanged in position and size. The pencils of light from each point on the image will however be deviated towards the axis and will pass through a second lens A_E of smaller diameter than would otherwise be required. The paths of the rays without the lens A_F in place are shown by broken lines in the diagram. The first lens A_F is known as the **field lens** and the second A_E, nearer the eye, as the **eye lens.** The field lens is frequently used in optical systems where large fields are required. The introduction of the field lens will also bring the exit pupil closer to the eyepiece, and this may be an advantage, particularly with low power eyepieces, as the eye may be brought close to the eyepiece and extraneous light cut off by means of an eyecup. In the arrangement shown in Fig. 120 any imperfections of the field lens or dust on its surfaces would be seen magnified in the plane of the image. To avoid this, in most practical eyepieces, the field lens is placed a short distance either in front of or behind the image.

The two chief forms of eyepiece are the **Huygens** and the **Ramsden** (Fig. 121). The eyepiece most commonly used in the microscope and the astronomical telescope is the Huygens. This consists of two plano-convex lenses placed with their curved surfaces towards the incident light and separated by a distance equal to half the sum of their focal

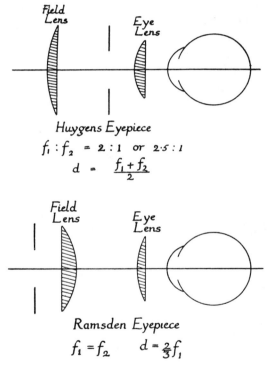

Huygens Eyepiece

$$f_1 : f_2 = 2 : 1 \text{ or } 2\cdot5 : 1$$

$$d = \frac{f_1 + f_2}{2}$$

Ramsden Eyepiece

$$f_1 = f_2 \qquad d = \tfrac{2}{3} f_1$$

Fig. 121.—Huygens and Ramsden eyepieces

lengths. The ratio of the focal lengths of field lens to eye lens varies somewhat, but is usually 2:1 in the microscope eyepiece and 2·5:1 in that of the telescope. When a microscope or telescope is to be used for making measurements by means of a scale or graticule placed in the first focal plane of the eyepiece, the Huygens eyepiece cannot be used.

It will be seen from Fig. 102 that the first principal focus of the eyepiece lies between the two lenses. Thus the image of a scale placed in this position is formed by the eye lens alone whereas the image to be measured is formed by the complete eyepiece, and therefore these two images are not equally affected by any distortion given by the lenses.

To overcome this difficulty the eyepiece must have its first focal plane in front of the field lens, so that the scale and the image formed by the objective are viewed through the complete eyepiece. The most usual eyepiece of this type is the Ramsden, which consists of two plano-convex lenses of equal focal length placed with their curved surfaces facing one another. To obtain the best colour correction the lenses, as in the Huygens eyepiece, should be separated by half the sum of their focal lengths, but as this brings the field lens into the focal plane, any dirt or imperfection on this lens will be sharply focused. The lenses are therefore separated by two-thirds the focal length of either.

116. Erecting Eyepiece. Terrestrial Telescope

A microscope or a telescope having either a Huygens or a Ramsden eyepiece gives an inverted image of an object. This is not of serious importance in the microscope or in the telescope used for astronomical or measurement purposes, but when the telescope is to be used for observing ordinary terrestrial objects, an erect image is essential. In order to take advantage of the large field of view given by the astronomical telescope some means must be incorporated in the telescope to erect the image. The method of erecting the image by means of reflection, as in the prism binocular, has been described in Chapter III. A further method is by the use of an **erecting eyepiece.**

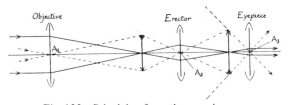

Fig. 122.—Principle of erecting eyepiece

If a short focus positive lens A_2 (Fig. 122) is placed in such a position that the image formed by the telescope objective is outside its first principal focus, a real inverted image is formed of the first image, that is, a real erect image of the original object. This image is observed with the ordinary eyepiece A_3. In practice, the erecting eyepiece usually consists of four plano-convex lenses, as in Fig. 123, the first two forming the

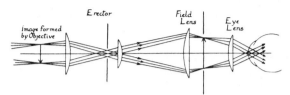

Fig. 123.—Four-lens erecting eyepiece

erecting system and the second two a modified Huygens eyepiece. The chief drawback to the terrestrial telescope with erecting eyepiece is its length, the length of the eyepiece alone being usually about ten times its focal length. By moving the erector along the axis, the size of the second image may be varied. This is the principle of the variable power telescope.

117. The Photographic Objective

This is essentially a positive lens forming a real image on a surface, sensitive to light, coated on either a thin glass plate or a sheet of celluloid. The requirements of such a lens are that it should form, on a plane surface, a well-defined image free from distortion and subtending a large angle, and that the aperture of the lens should be as large as possible. These difficult conditions can only be met by the use of carefully designed systems of a number of lenses. The modern photographic objective—the **anastigmat**—will give a good image subtending an angle of 50° or more with an aperture diameter which can be as large as one-half the focal length.

Photographic objectives are specified in terms of equivalent focal length, and a focal length about equal to the diagonal of the plate with which the lens is to be used will be suitable for general work.

When a large image of a distant object and therefore a long focal length is required, the long camera necessary with the ordinary type of objective can be avoided by the use of the **telephoto system.** This consists of a positive lens followed by a negative lens of shorter focal length, the lenses being separated by a sufficient distance to give a positive combination. Such a system will have its principal points outside the system in front of the positive lens, and its equivalent focal length is therefore considerably longer than its back vertex focal length. By suitably choosing the lenses and their separation an equivalent focal length several times the vertex distance can be obtained. Modern telephoto objectives usually have an equivalent focal length twice or three times the back vertex focusing distance.

For the purpose of controlling the illumination of the image and the

181

depth of focus (see Chapter XVII) the photographic objective is fitted with an adjustable aperture, usually placed between the lenses of the combination. The value of the aperture is expressed as the ratio of the focal length to the diameter of the effective aperture,* and is known as the **stop number** or **f number.** It is given as $f/8$, $f/16$, etc., where the effective aperture has a diameter $\frac{1}{8}$, $\frac{1}{16}$, etc., of the focal length.

It can be shown that the illumination of the image varies as the area of the aperture, that is, as the square of its diameter and inversely as the square of the focal length. The exposure required under any given conditions therefore varies directly as the square of the $f/$No. In order to simplify the calculation of exposures the marked apertures are arranged in a series in such a way that each one necessitates an exposure double that required with the preceding larger one, the series being as follows:—

	$f/4$	$f/5\cdot6$	$f/8$	$f/11\cdot3$	$f/16$	$f/22$	etc.
Exposure required	1	2	4	8	16	32	

118. The Projection Lantern

The projection lantern is also used to form a real image on a screen, but it differs from the camera in that the object is usually more or less transparent, the image being formed by the light from a source passing through the object; also the image is usually many times larger than the object. In order that light from each point of the object may pass through the objective or projection lens the source must subtend at least as large an angle at the objective as that subtended by the object. This would entail the use of a source of large area and therefore of low luminance, that is, the amount of light from each part of the source would be small.

Fig. 124.—Principle of projection lantern

* The effective aperture is the image of the actual aperture as seen from the front of the system, that is, the entrance pupil. (See Chapter XVII.)

By using a condenser, a large aperture short focus lens or system, between the source and object, an image of the source may be formed close to the objective and this will subtend a sufficiently large angle even when the source is small. The path of the light through the projection lantern is illustrated in Fig. 124. The source should be of high luminance, but small enough for the whole of the beam from the condenser to pass through an objective of reasonable aperture. The focal length of the objective is governed by the size of the image required, and the distance of the screen—the "throw," and the focal length of the objective governs that of the condenser, as can be seen from the figure. Further reference to the action of a condenser will be found in Chapter XVIII, § 228.

CHAPTER X   *EXERCISES*

NOTE—In the following exercises the lenses—objectives, eyepieces, etc.— are to be considered as thin.

1. Explain briefly the following terms, with diagrams
   Myopic eye.
   Hypermetropic eye.
   Crystalline lens.
   Far point.
Why is a negative lens used to correct a myopic eye and a positive lens a hypermetropic eye for distance? Give diagrams.

2. State what is meant by (a) the lateral magnification of an image, (b) the apparent magnification of an image as seen by an eye.

3. An object $\frac{1}{4}$ inch long is placed 6 inches from a positive lens of 8 inches focal length; what will be the visual angle subtended by the image as seen by an eye 4 inches from the lens?

4. What must be the position of an object in order that it may be seen distinctly through a + 10D lens placed 5 cm. in front of an eye, the eye being accommodated for a distance of 40 cm.? If the lens has a diameter of 1 cm., what length of object can be seen through the lens?

5. Show fully how a positive lens produces an apparent magnification of an object seen through it. What will be the magnification produced by a + 15D lens 5 cm. from an emmetropic eye when the eye is (a) unaccommodated, (b) accommodated 4D?

6. An object $\frac{1}{4}$ inch long is placed 3 inches from a positive lens of 4 inches focal length; what will be the apparent magnification of the image as seen by an eye 2 inches from the lens?

7. Show with a carefully drawn diagram how a magnified image is formed with a compound microscope. Explain exactly what is meant by the magnification of a microscope and derive an expression for calculating the magnification.

8. State exactly what is meant by the magnification of a microscope. What will be the magnification of a microscope having an objective of $\frac{2}{3}$ inch focal length, an eyepiece of $1\frac{1}{4}$ inch focal length and a tube length of 6 inches?

9. A microscope has an objective of $1\frac{1}{2}$ inches focal length and an eyepiece of 2 inches focal length, the distance between the lenses being $6\frac{1}{2}$ inches; what will be the magnification of the microscope?

A scale is placed in the first focal plane of the eyepiece and the length of the image of a certain object measured on this scale is found to be 0·16 inch, find the actual length of the object.

10. The objective and eyepiece of a compound microscope, each considered here as a thin lens, have focal lengths of 2 inches and 1 inch respectively and are separated by 9 inches. Calculate where the object must be to give distinct vision to an emmetrope, accommodated for a point 10 inches from the eyepiece. Draw a careful diagram, full size, showing the passage of a pencil of rays through the instrument from one point of the object.

11. A compound microscope has an objective of 1·5 cm. focal length and an eyepiece of 3 cm. focal length, the lenses being 18 cm. apart. What will be the position of the object and the magnification of the microscope when focused for an emmetropic person with accommodation relaxed?

Find the equivalent focal length of the microscope as a complete system and hence the magnification considering the system as a magnifier.

12. Explain why a compound microscope should consist of a short focus objective and a short focus eyepiece. (L.U.)

13. The microscope of question (11) is used to form a real image on a photographic plate 50 cm. from the eyepiece (photomicrography). How far and in which direction must the microscope be moved with respect to the object in question (11) and what will be the lateral magnification of the image?

14. Give a diagram showing the path of light through a compound microscope and deduce the formula giving the magnification of the instrument.

15. Two thin convex lenses when placed 10 inches apart form a compound microscope whose apparent magnification is 20. If the focal length of the lens representing the eyepiece be $1\frac{1}{2}$ inches, what is the focal length of the other?

16. Give a careful diagram showing the path of light through an astronomical telescope. A telescope 12 inches long is to have a magnification of eight times. Find the focal length of objective and eyepiece when the telescope is (a) astronomical, (b) Galilean.

17. Explain, with diagrams of the path of the light, two types of telescopes. State the advantages and disadvantages of each type.

18. Explain, with a diagram of the path of the light, the optical principle of the astronomical telescope. What is the disadvantage of this type of telescope? Describe a method by which this disadvantage may be overcome.

19. Show by means of a diagram how the magnification is produced in the case of a Galilean telescope. A Galilean field glass magnifying five times has an objective of 180 mm. focal length; find the focal length of the eyepiece and its distance from the objective when the field glass is focused for a myope of 4D.

20. A + 5D cyl. and a + 20D cyl. both with axis vertical, are separated by a distance of 25 cm. How will distant objects appear when viewed through the combination, the eye being close to the stronger lens?

21. An astronomical telescope is to be designed to give a magnification of × 8, a $\frac{3}{4}$ inch Ramsden eyepiece being used (separation = $\frac{2}{3}$ focal lengths); the aperture of the objective is to be one-sixth its focal length. Treating the objective and eyepiece

lenses as thin, show their positions on a scale diagram and trace accurately through the instrument a pencil from one extremity of a distant object which subtends an angle of 5°. Measure the magnification given by the diagram.

Discuss the diagram as drawn: diameter of tube, sizes of eye and field lenses, angular field, etc.

22. A horizontal telescope contains a pair of horizontal cross-wires one-tenth of an inch apart. The telescope is focused on a vertical staff 10 feet away from the object glass, which has a focal length of 10 inches. Find the length on the staff apparently intercepted between the wires.

23. Describe the construction and action of a simple telescope, and show how by adding (a) lenses, (b) prisms, an erect image may be obtained instead of an inverted one. (L.U.)

24. The objective of a reading telescope has a focal length of 10 inches and the eyepiece 2 inches. The telescope is focused on an object 40 inches away, and the image is formed at the distance of most distinct vision (12 inches). Draw a diagram showing the path of rays through the telescope. Determine the magnification and the length of the telescope. (L.U.)

25. A Galilean telescope has an objective of 120 mm. focal length and magnifies five times when used by an emmetrope to view a distant object. What adjustment must be made when the telescope is used by (a) a myope of 10D, (b) a hypermetrope of 5D? Find the magnification in each case.

26. A biconvex lens has surface powers of 5D and 20D respectively, is made of glass of refractive index 1·5 and the separation of the surfaces is 37·5 cm. Find the positions of the principal points of this lens.

What kind of system is this and what properties does it possess?

A mark is made near the axis on the 5D surface. Find the position and magnification of the image of this produced by the 20D surface.

27. A Huygens eyepiece consists of two lenses having focal lengths of $2\frac{1}{2}$ inches and 1 inch respectively, separated $1\frac{3}{4}$ inches; find the equivalent focal length and the positions of the principal foci. What will be the magnification of a telescope having an objective of 20 inches focal length with the above eyepiece?

28. Explain the function of the eyepiece in a telescope or microscope. Why is an eyepiece usually formed of two separated lenses? Describe the Huygens and Ramsden eyepieces giving, with reasons, examples of the use of each form.

29. Explain the principle of the telephoto lens. What is its particular advantage?

30. A telephoto lens is made up of two single thin lenses of focal lengths + 20 cm. and − 8 cm. respectively, separated by a distance of 15 cm. Find (a) by calculation, and (b) graphically, the focal length of the combination and the position of its principal points. What should be the camera extension and what will be the size of the image of a distant object that subtends an angle of 2°?

31. What is the significance of the f/No. of a photographic lens? Describe the system of numbering which is commonly adopted and explain the reasons for its choice.

32. A camera is fitted with a lens of 6 inches focal length. Three of the marked apertures have effective diameters of 1·09, 0·75 and 0·375 inches respectively. Express these as f/Nos. and find the exposures necessary with the first and last, the correct exposure with the 0·75 inch aperture being 2 seconds.

33. The finder on a camera consists of a 6D concave lens having a rectangular aperture of 25 × 20 mm. and a peep-hole 75 mm. behind it; what extent of object at 10 metres distance can be seen in this finder?

34. Explain the principle of the projection lantern. What are the conditions to be satisfied in obtaining the brightest possible image on the screen?

XI - Photometry

119. Introduction

In the study of geometrical optics which has been developed in the previous chapters we were very concerned with the *direction* of light rays. In the study of **Photometry** the *amount* of light is our first concern although obviously the direction is important when the amount in different places is being calculated. In § 1 the energy given out by the heated poker included heat as well as light. The heat energy does not stimulate the retina of the eye and is therefore not included in photometry which is solely concerned with the visual sensation produced by the radiated energy.

The visual sensation produced will vary greatly in two respects with the wave-length of the radiation. It has already been stated in § 7 that different wave-lengths give rise to the sensation of different colours, and this is considered more fully in Chapter XIII. Also the intensity of the sensation produced by equal amounts of emitted energy will vary throughout the visible spectrum. From the spectrum of white light (Plate 4A) it will be evident that the brightest portion is situated in the yellowish-green and that the brightness falls off rapidly towards both the red and violet ends. Assuming a source that is emitting equal energy at all wave-lengths and plotting the luminosity, that is the amount of visual sensation, against the wave-length, we obtain the relative luminosity or visibility curve shown in Fig. 125. The form of this curve varies somewhat with different normal observers and also with different values of luminance, particularly when the luminance is low. The curve shown, which is the result of a large number of observations, has been

187

adopted as the standard luminosity curve for ordinary values of luminance, that is over about one candela per square foot. It will be seen that the greatest visual effect occurs for a wave-length of 555 mμ (see § 139) in the yellowish-green portion of the spectrum.

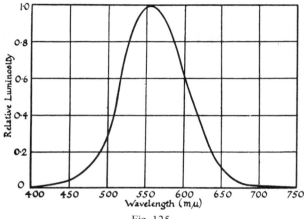

Fig. 125.

The eye is incapable of making an absolute measurement of the amount of light entering it; we can look at two sources and estimate that one appears "brighter" than the other if there is sufficient difference between them, but cannot form a reliable judgment as to by how much they differ. The eye can, however, decide with a fair degree of accuracy whether two adjacent surfaces appear *equally* bright; and this is the basis of all practical visual photometric measurements.

According to the law of Weber (1834) the smallest perceptible difference of apparent brightness or luminosity is a constant fraction of the luminosity. This fraction, known as *Fechner's fraction* (1858), is, over a large range of luminosities, about one per cent and the eye can therefore distinguish between two adjacent surfaces that differ in luminance by this amount. The eye is now replaced in many photometric observations by the photo-electric cell.

In Photometry we have to differentiate between the following quantities:

(1) The amount of light emitted by a point source, the *luminous flux* and *luminous intensity*.

(2) The amount of light received on a unit area of a surface in a given position; that is, the *illumination* or *illuminance* of the surface.

(3) The amount of light *emitted* or *re-emitted* per unit area of a surface; that is the *luminance* of the surface. This was formerly called the brightness of the surface.

In discussing these measurements it should be remembered that in practical photometry it is rarely necessary to strive for accuracies better than ± 1%, Fechner's Fraction. The exposure meters on cameras for instance may only be accurate to ± 10%. However, the range of the measurements can be over many orders of magnitude and it is often crucially important to eliminate *stray* light from the experimental area.

120. Luminous Flux and Luminous Intensity

There is a great difference in the amount of light emitted by different sources. Most practical sources are *incandescent* sources which emit energy over a wide range of wave-lengths, mainly as heat in the infra-red part of the spectrum (see Chap XIII). This is wasted energy as far as the emission of light is concerned and the *luminous efficiency* (see § 128) is very important to lamp manufacturers. In considering the amount of *light* given out by a source, the energy emitted must be evaluated or *weighted* according to its ability to stimulate visual sensation as given by the relative luminosity curve (Figure 125). The rate of flow of *light* from a source is the **luminous flux, F,** of the source, and the unit of this is the **lumen** (lm), which is defined later.

In no practical source is the distribution of flux uniform in all directions; hence it is necessary to define the amount of light radiated with respect to a given direction. In no practical source is the amount of luminous flux uniform over all parts of the source; hence it is necessary to define the amount of light radiated with respect to a point on the source or to assume that the whole source is contained at a single point. This last assumption works very well in the majority of cases (see § 121/122) and so the two requirements above can be met by considering the flux emitted by a point source into a cone of very minute solid angle constructed round a given direction. The concentration or density of this flux per unit solid angle is termed the **luminous intensity, I,** of the source in the given direction. The unit of luminous intensity is the **candela** (cd.) formerly the candle, and luminous intensity is often referred to as **candle-power.** If F is the flux in lumens emitted within a cone of solid angle w steradians, theoretically in this case infinitely small, then $\dfrac{F}{w} = I$ lumens per steradian = candelas in the given direction.

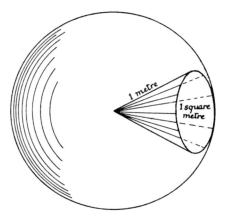

Fig. 126.—Unit solid angle. Steradian

NOTE. — The unit solid angle, or *steradian,* is the solid angle of a cone that, having its apex at the centre of a sphere, cuts off an area of the sphere's surface equal to the square of its radius. (Fig 126). The value of any solid angle in steradians is equal to the area of the sphere's surface included in the angle divided by the square of the radius of the sphere. The surface area of a sphere of radius r being $4\pi r^2$ the solid angle for a complete sphere is 4π steradians.

It can be shown that in a cone having a half plane angle θ the solid angle $w = 2\pi(1 - \cos\theta) = \pi \sin^2\theta$ steradians when θ is small.

While no source will be a point as required by the definition of luminous intensity, in many cases the size of the source is negligible at the distances at which it is used. (see § 121).

121. Illumination

A surface receiving light is said to be illuminated, the **illumination** E at any *point* of it being defined as the density of the luminous flux at that point, or the flux divided by the area of the surface, when the latter is uniformly illuminated. The metric unit of illumination is the **lux** (lx) or **lumen per square metre** (lm/m^2), the illumination of a surface, normal to the direction of the light, one metre from a source of one candle-power. The British unit is the **lumen per square foot** (lm/ft^2), the illumination of a surface normal to the direction of the light one foot from a source of one candle-power. (These units were formerly called the metre-candle and foot-candle respectively).

If F is the flux from a point source at the centre of a sphere of radius d, the illumination E on the surface of the sphere, from the definition, will be

$$E = \frac{F}{\text{area of sphere}} = \frac{F}{4\pi d^2}$$

$\dfrac{F}{4\pi}$ is the candle-power I, and therefore,

$$E = \frac{I}{d^2} \quad \ldots \quad \ldots \quad \ldots \quad (11.01)$$

This is the law of inverse squares (§ 3). It should be clearly understood that expression (11·01) applies only to the illumination received on a surface *normal* to the direction of the incident light from a *point* source, where I is the candle-power of the source in the particular direction.

It can be shown by applying the inverse square law to each part of a source larger than a point that the error introduced by assuming it is a point source will be less than 1% if the distance, d, is greater than 10 times the largest dimension of the source.

Fig. 127 represents the section of a pencil of light from a point source B of candle-power I falling on a surface LM perpendicular to the pencil. If F is the flux in the pencil of solid angle w, then the illumination of the surface at distance d from the source is given by

$$E = \frac{F}{\text{area of LM}} = \frac{F}{wd^2} = \frac{I}{d^2}$$

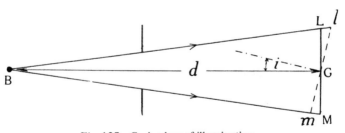

Fig. 127.—Cosine law of illumination

If the surface is rotated about G into a new position lm, the same flux is distributed over the larger area, and the illumination of the surface is now

$$E' = \frac{F}{\text{area of lm}}$$

As the illumination refers to any one point on a surface, we may consider the area lm as very small compared with its distance from the source, and then,

$$\text{area of lm} = \frac{\text{area of LM}}{\cos \text{LGl}} = \frac{\text{area of LM}}{\cos i}$$

where i is the angle of incidence.

Therefore

$$E' = \frac{F \cos i}{\text{area of LM}} = \frac{I}{d^2}.\cos i \qquad \dots \qquad \dots \quad (11.02)$$

Thus, *the illumination of a surface varies as the cosine of the angle of incidence;* this is known as the cosine law of illumination. The law of inverse squares and the cosine law are the two fundamental laws of photometry.

In practical applications of photometry we frequently require to find the illumination at different points on a plane surface, as for example,

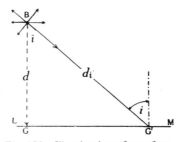

Fig. 128.—Illumination of a surface

the surface of a street, an illuminated test chart, etc. Fig. 128 represents a source B of a candle-power I above a plane surface LM; the illumination at any point G′ on the surface will be

$$E = \frac{I_i \cos i}{d^2_i}$$

where i is the angle of incidence and I_i the candle-power of the source in that direction. From which we have

$$E = \frac{I_i \cos^3 i}{d^2} \qquad (11.02A)$$

$$= \frac{I_i d}{d^3_i} \qquad \dots \qquad \dots \qquad \dots \qquad (11.02B)$$

where d is the perpendicular distance of the source from the surface.

Example:

A room 15 ft. by 12 ft. is illuminated by a 100 candle-power lamp suspended from the centre of the ceiling and 8 ft. above the floor. Find the illumination on the floor (*a*) directly under the lamp and (*b*) in a corner of the room, assuming the candle-power to be 100 in both these directions.

(*a*) $\qquad\qquad$ I = 100 c.p. $\qquad$ d = 8 ft.

$$E = \frac{I}{d^2} = \frac{100}{64} = 1{\cdot}56 \text{ lumens per sq. ft.}$$

(*b*) Distance GG′ (Fig. 128) from point on floor directly under lamp to corner of the room.

$$GG' = x = \sqrt{7{\cdot}5^2 + 6^2} = \sqrt{92{\cdot}25} \text{ ft.}$$

Distance d_i from lamp to corner

$$d_i = \sqrt{d^2 + x^2} = \sqrt{64 + 92{\cdot}25} = 12{\cdot}5 \text{ ft.}$$

Hence from (11.02B)

$$E = \frac{100 \times 8}{12{\cdot}5^3} = 0.41 \text{ lumens per sq. ft.}$$

Example:

Light from a point source of 50 candle-power, which radiates uniformly in all directions, falls on a + 1D spherical lens of 5 cm. aperture placed 50 cm. from the source. What fraction of the total flux from the source is incident on the lens? What will be the illumination in lumens per square metre on a screen held one metre beyond the lens, on the side opposite from the source, and perpendicular to the optical axis of the lens? (S.M.C.).

At 50 cm. from the source the total flux is spread over a sphere of 50 cm. radius, the area of which will be $4\pi \times 50^2$. The area of the lens aperture is $\pi \times 2{\cdot}5^2$. Therefore the fraction of the total flux received by the lens

$$= \frac{\pi \times 2{\cdot}5^2}{4\pi \times 50^2} = \frac{1}{1600}$$

As the source is 50 cm. from a + 1D lens a virtual image of the source is formed one metre in front of the lens. Therefore the diameter of the patch of light on the screen will be twice the diameter of the lens and its area four times the area of the lens.

Total flux emitted from the source $= 4 \pi I = 200\pi$ lumens. Flux received by the lens is $\dfrac{1}{1600}$ of total flux.

$$= \frac{\pi}{8} \text{ lumens.}$$

Area of patch of light on screen $= \pi \times 5^2$ sq. cm. Therefore, neglecting any loss of light at the lens,

$$\text{illumination of screen} = \frac{\pi}{8 \times \pi \times 5^2} = \frac{1}{200} \text{ lm/cm}^2.$$

$$= 50 \text{ lm/m}^2$$

Alternative solution:

$$\text{illumination on lens} = \frac{I}{d^2} = \frac{50}{50^2} \text{ lm/cm}^2$$

Area of illuminated patch is four times that of lens and therefore illumination is one-quarter; hence

$$\text{illumination of screen} = \frac{50}{50^2 \times 4} = \frac{1}{200} \text{ lm/cm}^2.$$

122. Luminance

Illumination refers only to the amount of light *received* by a surface regardless of the nature of that surface, but in many cases we are more concerned with the amount of light *emitted* in a given direction by a surface. The light emitted per unit area of a surface is the **luminance, L,** of the surface. It will depend, in the case of a non-self-luminous body, both upon the illumination and the fraction of the incident light diffusely reflected by the surface. This latter is known as the **reflection coefficient** or albedo of the surface. Luminance may also refer to a diffusely transmitting substance, such as the diffusing globe of a lamp, or to an actual source, such as a flame or an electric lamp.

In its perception of luminance the eye interprets the flux to be coming from an area projected at right angles to the direction of vision,* and the luminance of a surface is defined as follows: *The luminance in a given direction of a surface emitting light is the luminous intensity measured in that direction divided by the area of this surface projected on a perpendicular to the direction considered.* Luminance is expressed in candelas per unit area of surface, the metre units being the **nit** (nt.),

---

* The sun or a lamp surrounded by a spherical diffusing globe appears to the eye as a flat disc.

one candela per square metre and the **stilb** (sb.), one candela per square centimetre. Some typical approximate values of luminance are given below:

Clear blue sky	2·5	candelas per square inch
Candle	3	,, ,, ,, ,,
Incandescent gas mantle	20–25	,, ,, ,, ,,
Tungsten lamp filament (100 watt)	1,500	,, ,, ,, ,,
Pointolite lamp	12,000	,, ,, ,, ,,
Arc lamp crater	200,000	,, ,, ,, ,,

The reflection coefficients for a number of surfaces are given in Table 1 in the Appendix.

Another method of expressing luminance assumes that the surface is a perfect diffuser, i.e. one that appears equally bright from whatever direction it is viewed. This will mean that the luminous intensity per unit area of apparent or projected area is constant for all angles. Since for a plane surface the projected area is proportional to the cosine of the angle between the direction of viewing and the normal to the surface, the luminous intensity of the surface must vary in the same proportion. This is stated in the *cosine law of emission of* Lambert (1727–1777) as follows: *For a perfectly diffusing surface the candle-power per unit area of the surface in any direction varies as the cosine of the angle between that direction and the normal to the surface, so that the surface appears equally bright whatever be the direction from which it is viewed.* No surface completely satisfies this requirement, but a few, such as a coating of magnesium oxide, sand-blasted opal glass and scraped plaster of paris, have diffusion closely approaching the ideal. Many surfaces closely resemble the perfect diffuser when the angle of incidence is small and the direction of viewing is nearly normal. Such surfaces are said to be "matt."

Treating the surface as a perfect diffuser luminance is expressed in terms of the total luminous flux in lumens emitted by a unit area (actual not projected) of surface. A surface emitting or reflecting one lumen per square centimetre has a luminance of one **lambert,** but a luminance of one-thousandth of this value, the **millilambert,** is a more generally useful unit. These units have now largely been replaced by the **foot-lambert,** the luminance of a surface emitting a flux of one lumen per square foot. One foot-lambert equals 1·076 millilamberts. The corresponding metric unit is the **apostilb** (asb.), the square metre being the unit of area.

It can be shown that for a perfect diffusing surface the luminance in lumens per unit area is π times the brightness in candelas per unit area, i.e. a surface having a luminance of one candela per square foot has a luminance of π foot-lamberts.

As the illumination, E, of a surface is the flux in lumens received by a unit area of the surface and of this flux a fraction, r, the reflection factor, is reflected, the luminance of an illuminated perfectly diffusing surface will be given by rE lumens per unit area or $\dfrac{rE}{\pi}$ candelas per unit area.

A further measurement of luminance, recognising that photometry is based on visual parameters, makes allowance for the variations in diameter of the pupil of the eye. The concept of *retinal illuminance* from an extended source assumes that the retina is a smooth surface. The value is given by the equation:

$$E = \frac{Bt\,\alpha \cos\theta}{k}$$

where B = luminance in direction of viewing (nits),

t = transmittance of eye,

α = area of pupil (m^2)

θ = angle of incidence of principle ray through the eye

k = constant equal to the area of the retinal image (m^2) divided by the solid angle of the visual field (steradians).

As $\cos\theta$ is commonly unity and t and k are constants, the retinal illumination may be found by multiplying the number of nits by the area of the pupil in square millimetres. This gives a value in **Trolands**, a unit often used in visual experimentation.

123. Standard Sources of Light

Because the measurements of photometric quantities are largely comparisons between sources it is necessary to define a standard source in the same way that a metre or yard is a standard length. The original standards of luminous intensity was the **International Candle**. This, despite careful specification, was far from constant in its intensity and its colour differed considerably from that of modern sources. The candle as originally specified is now only of historical interest, although the term *candle-power* still survives. The candle was replaced by various flame lamps, the chief of which was the Vernon Harcourt lamp burning pentane vapour with a hollow cylindrical flame and having an intensity of 10 candle-power. Later the standard was maintained with a series of specially constructed electric carbon filament lamps kept at the standardising laboratories of the various countries.

In 1948 an entirely new primary standard was adopted. This consists of a very small cylinder containing molten platinum. The surface of the platinum at the temperature of the "freezing point" of platinum has a constant luminance. One sixtieth of the luminous intensity from one square centimetre of this surface is the **candela,** which now replaces the candle as the unit of luminous intensity.

As working standards, specially constructed electric filament lamps are now generally used. In these the filament is mounted in a single plane being perpendicular to the photometer bench. The glass bulb is considerably larger than would be used in the ordinary way in order to minimise the blackening caused by the deposition of metal particles on the glass, and the filament is "aged" by being run for at least 100 hours before being standardised. In use the lamp is run at a constant current and voltage slightly below that at which it was "aged".

124. Comparison Photometers

Photometry is concerned with the visual effects of emitted radiations so the primary instrument of measurement is the human eye. As this can only judge differences in luminous intensity qualitatively it must be assited if quantitative results are to be obtained. The method adopted consists in arranging that the two sources produce equal luminance on two similar adjacent surfaces, each surface being illuminated by one only of the sources. A simple arrangement is shown in Fig. 129.

Fig. 129.–Photometer principle

The diffusing surfaces at P lie at equal angles to the light paths but can both be seen simultaneously. If E_1 and E_2 are the illuminations on the two surfaces produced by sources B_1 and B_2, whose candle powers are I_1 and I_2 respectively, then

$$E_1 = \frac{I_1 \cos i_1}{d_1{}^2} \qquad \text{and} \qquad E_2 = \frac{I_2 \cos i_2}{d_2{}^2}$$

where i_1, i_2 are the angles of incidence of the light on the surfaces. If now d_1 or d_2 is changed the eye can observe the surfaces and judge when they are equally bright, that is, when no difference exists and the system is *balanced*. Under these circumstances

$$\frac{I_1}{d_1{}^2} = \frac{I_2}{d_2{}^2} \qquad \text{as} \qquad i_1 = i_2 \dots \qquad \dots \qquad \dots \quad (11.03)$$

Thus the relative candle-powers of the two sources are given by the ratio of $d_1{}^2$ to $d_2{}^2$.

In order that the judgement of equality shall be as accurate as possible the two surfaces must be seen simultaneously and have a sharp dividing edge. Various methods have been devised to obtain these conditions. An early arrangement merely consisted of a rod in front of a screen. When the two sources are properly located the two shadows on the screen are just touching and are equally dense. The ratio of intensities is then given by the ratio of the square of the source-to-screen distances as in the expression (11.03).

A more complex system replaces the V-screen of Fig. 129 with a screen of opaque white paper normal to the line between the lamps having at its centre a spot made translucent by treating the paper with oil or wax. Unequal illumination will show a dark central spot on a bright surround or vice versa. Although the precise derivation of the brightness is a little complex the balance position when the spot disappears still occurs when the illuminations are equal and the expression (11.03) applies.

A more complex and sensitive approach is that of the **Lummer-Brodhun Photometer** shown in Fig. 130. The light from each source passes through an aperture in the metal case and falls on either side of a white diffusing screen S; this is formed of plaster of Paris or of two sheets of ground white opal glass with a plate of metal between them. Two right-angled totally reflecting prisms P_1 and P_2 reflect the diffused light from the screen surfaces to the comparison prism P_3 and P_4. The prism P_3 has its hypotenuse face ground to a curve except for a small circular area in the centre, which is plane and polished, and this is pressed into optical contact with the plane hypotenuse of the prism P_4.

Thus light from the left hand side of the screen reaches the eye via the central part of P_3P_4, while the surrounding area receives light from the right hand side. The edge of the central area may be sharp but will disappear when a balance is obtained as before. A more complex working of the prism P_3 can produce a more complex pattern to be erased at balance. One such pattern is given in Fig. 130b. and systems incorporating this are known as **contrast photometers**.

These systems give adequate accuracies provided the two sources are of identical colour. When a colour difference is apparent an alternative method can be used in which the two illuminated surfaces are viewed

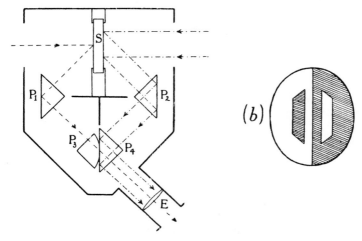

Fig. 130.–Lummer-Brodhum photometer. (b) contrast pattern

sequentially rather than side-by-side. Any difference in illumination now shows up as a flicker and the basis of this method is that flicker due to colour difference disappears at a lower speed of alternation than flicker due to luminosity difference so that the judgement of equal luminance is not affected by the presence of a colour difference.

Such a system due to **Guild** is shown in Fig. 131. D is a disc which can be rotated at various speeds. D, T and S are coated with diffusing material such as magnesium oxide. The eye is placed at E and sees the inside of the tube T uniformly illuminated to 2.5 ft-lamberts while the hole A subtends an angle of $2°$.

As the disc is rotated the eye sees the aperture filed alternately illuminated by the light from S and D and when the disc is rotating slowly the field will appear to flicker even when the luminance at S and D is equal. On increasing the speed of rotation the flicker becomes less pronounced and at a high enough speed will disappear. The test source B_2 and the photometer are fixed and measurements are made by moving the comparison lamp B_1 while reducing the speed of rotation of the disc until flicker almost disappears. The luminance of S and D is then equal and the relative intensities of B_1 and B_2 can be found as with other photometers.

To obviate any difference there might be between the reflection factors of the two surfaces of the photometer and any effects of stray light, a substitution method is generally employed. In this a third lamp, the comparison lamp, which need not be of known candle-power, is used with

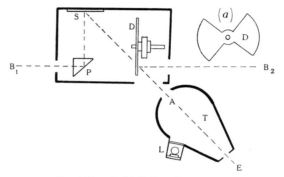

Fig. 131.—Guild flicker photometer

the standard and the test lamp, i.e. the lamp to be tested. Balance is first obtained between the comparison lamp and the test lamp. The distance between the comparison lamp and the photometer is then kept fixed and the standard substituted for the test lamp. By moving the standard lamp balance is again obtained. Thus the same side of the photometer is illuminated in turn by both standard and test lamps. Then if I_1 and I_2 are the candle-powers of test lamp, and standard respectively and d_1 and d_2 their respective distances from the photometer when balance is obtained,

125. Photoelectric Photometers

The limitations of comparison photometers are never more evident than when a simple portable instrument is required for measuring the luminous intensity of actual lamps in offices, on roadways etc. These are assisted by reflectors which may become dirty or misaligned. A photoelectric cell (or photo-cell) offers a solution provided:

(1) It responds to light in the same way as the eye.
(2) It has an output proportional to the luminous intensity.
(3) It does not change from day to day.
Very few cells satisfy the first requirement and the nearest is the **selenium barrier-layer photo-cell.** The response of this device to light of different wave-lengths is shown in Fig. 132. compared with the visual response. If a correction filter which absorbs ultra-violet light is placed in front of the cell the response can be made very close to that of the eye.

The cell produces a current through a resistance connected across it and

Fig. 132.—Spectral response of EEL cells (courtesy, Evans
Electroselelenium Ltd)

Fig. 133.—Typical output curves for 45 mm EEL
cell connected to various external resistances
(courtesy, Evans Electroselelenium Ltd)

the variation with light falling on it is shown in Fig. 133. It can be seen
that this is linear for low values of the resistance and a micro-ammeter in
the circuit could be calibrated directly in foot-candles or lux.

201

In manufacture pure molten selenium is poured on to a steel base plate. A very thin transparent metal layer is spluttered over the selenium to form a conducting surface. The light reaches the selenium through the metal layer and the current is generated between this layer and the base plate. Rather like standard sources of light, these cells are "aged" after manufacture by exposing them to a very strong light under short-circuit conditions. This helps to ensure constant values to the curves of Fig. 133 throughout their life.

Figs 132 and 133 refer to products of Evans Electroselenium Ltd and are reproduced with their permission.

126. Photometry of Screens

In the section on illumination the concept of the perfectly diffusing surface was discussed and defined as a surface which appears equally bright from all directions. If such a surface is used as a screen for the projection of a colour transparency or cine-film the projected image would appear

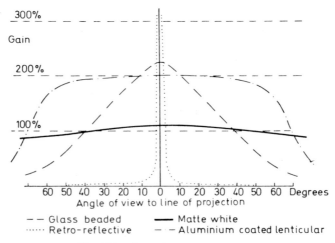

Fig. 134.—Gain of various screens

equally bright in all directions. Most surfaces are not perfect and return the light in a partially specular reflection. As the audience tends to sit between the screen and the projector this can give a brighter image. Provided no objectionable high spots occur this can be a good thing.

The ratio by which a given screen appears brighter than a perfectly diffusing surface with the same illumination is often referred to as the **gain** of the screen in a particular direction. It is important to realise that there is no gain in the sense of extra light being generated. The additional brightness in a given direction is obtained at the expense of less light in another direction. The reflectivity of the screen can affect the apparent gain particularly in the case of a dirty surface.

Curves of the gain of typical screens are given in Fig. 134. The curve for retroreflective screens can reach gain figures above 1000% and these have application in road signs and vehicle number plates where the driver sits within a few degrees of the projection axis of his headlamps.

127. The Photometry of Sources, Integrating Spheres

Luminous intensity or candle-power refers to the quantity of light emitted by the source in one given direction. The standard sources described have their intensities specified in a particular direction. In

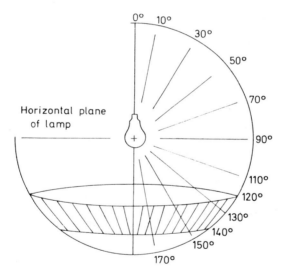

Fig. 135.—Measurement of mean candle-power

ordinary commercial work, however, the given candle-power is usually an average of the intensities in all directions when it is known as the **mean spherical candle-power.**

The determination of this value for a lamp or lighting system may be done by one of two methods. The intensity may vary in all directions although a proportion of systems may be symmetrical about a vertical

203

axis. In this case the total luminous flux may be measured by supporting the lamp in its operating position and measuring the intensity at different vertical angles as in Fig. 135.

If the luminous intensity is measured at equal angles as shown, the results cannot simply be averaged as the light emitted near the horizontal plane of the lamp is also being emitted over a considerable solid angle all round the fitment. Similarly the light measured directly above or below the lamp is being emitted nowhere else.

This geometrical feature leads to the adoption of zone factors which are proportional to the area swept out by the region around the angle chosen. These zone factors are given by the difference in the cosines of the *bounding* angles multiplied by 2π. For zones of $20°$ the factors are given in the table:

Angle of Measurement	Zone	Zone Factor
$10°$	$0°-20°$	0·378
$30°$	$20°-40°$	1·091
$50°$	$40°-60°$	1·671
$70°$	$60°-80°$	2·051
$90°$	$80°-100°$	2·182
$110°$	$100°-120°$	2·051
$130°$	$120°-140°$	1·671
$150°$	$140°-160°$	1·091
$170°$	$160°-180°$	0·378

The total of each luminous intensity reading multiplied by its zone factor gives the total flux, F, from the lamp. When this is divided by 4π the mean sphereical candle-power is obtained in lumens per steradian or candela. If the zone factors in the table are summed the result is 4π.

As a more approximate method, a series of angles may be taken which in themselves give zones of equal solid angle. These are known as **Russell Angles** and those giving six equal zones would be $-$ $33·6°$, $60°$, $80·4°$, $99·6°$, $120°$, and $140·4°$. The average of luminous intensity values obtained at these angles gives the mean spherical candle-power directly.

When the source has large variations in its intensity in different **horizontal** directions then the same approach may be used but the lamp rotated completely about its vertical axis for each reading so that an average reading may be obtained. This is sometimes referred to as the mean **horizontal** candle-power. The values obtained in this manner may be used in the same way as for the symmetrical source described above.

An alternative method for total flux measurement involves the use of an **integrating sphere**. This device is a large sphere with a diffusely reflecting inner surface. The theory of its use depends on the application of the cosine law of illumination to the geometry of a sphere from which it can be shown that the surface illumination on the sphere wall *due to reflected light* is proportional to the total flux of a source inside. No direct light from the source must reach the point at which the illumination is measured.

This means that baffles must be inserted in the sphere and a hole cut in the wall to allow the illumination to be measured. These detract from the accuracy of measurement as do the presence of fittings inside the sphere. If the test lamp and standard source can be introduced successively into the same place within the sphere for measurements, these inaccuracies can be minimised.

128. Incandescent Lamps

Tungsten-filament lamps are so called because the source of light is a fine filament of tungsten metal through which is passed an electric current. The resistive heating of the filament raises its temperature to a maximum of about $3500°K$. At this temperature the tungsten would rapidly oxidise. To avoid this the filament is enclosed in a glass envelope which is evacuated or filled with an inert gas. Even with this precaution the tungsten is slowly evaporated from the filament on to the walls of the envelope reducing its transmission and eventually ending the life of the lamp after a few hundred hours of use. Longer life can be obtained by reducing the power through the lamp (underrunning) but this reduces the colour temperature (see § 156) and also reduces the efficiency of the conversion of electrical power into light. At its design voltage a tungsten filament lamp will normally give between 15 to 25 lumens per watt. This is the total amount of light radiated by the lamp. The candle-power in specific directions depends very much on the shape and type of the filament. The conversion figures given above represent an efficiency of conversion of electrical energy into visible energy of between 8 and 13%.

These efficiency figures improve with higher filament temperatures but the worse evaporation of the filament material delayed the exploitation of this feature until the invention of the "quartz-iodine" lamp (more correctly termed the tungsten–halogen lamp). In this lamp the inert gas contains the halogen, iodine or bromine which has the property of redepositing evaporated tungsten on to the filament. This action requires a high bulb temperature and as glass proved unsuitable quartz is used in a much smaller shape so that it becomes hotter, but the quartz does not play an active role.

129. Luminescent Sources

Although a source of light gives out energy which it must obtain some-how, the method of heating used in incadescent sources is not the only way. Substances called **phosphors** have the property of absorbing energy from electrons, electric fields, chemical reactions and even other light, particularly ultra-violet. They then re-emit this energy in the form of light. Phosphors are frequently sulphides and oxides or silicates and phosphates of such metals as zinc, calcium, magnesium, cadmium, tungsten and zir-conium. They are used in such commonplace items as fluorescent lamps and television picture tubes and more complex devices such as X-ray machines and electron microscopes. They are all luminescent sources but as they are often of large dimensions and flat the term luminescent screen is generally used.

Phosphors are used in fluorescent lamps to obtain a much broader range of colours than can be obtained with incandescent sources. They are also more efficient, with luminous efficiency values towards 90 lm/watt and less heat radiation. The light is generated by a two-stage process as the electric current is discharged through a mercury vapour which produces green and ultra-violet light. This ultra-violet light is then absorbed by the phosphor coating on the inside of the glass tube containing the gas and re-emitted as white light.

In television picture tubes the phosphor is selected not only for its colour and efficiency but also for its short response time. The light emission is stimulated by high-energy electrons striking the phosphor on the inside of the screen. As the picture changes it is important that the phosphor can start and stop emitting light very quickly as the electrons are switched on and off.

130. The Action of Filters

Unless the light is travelling in a complete vacuum, there is a continuous reduction in its intensity due to the media through which it passes. This reduction may be due to absorption such as occurs with dark glasses, scattering such as occurs with ground glass or reflection such as occurs with glass surfaces. In principle there is no difference between reflection and scattering but all three processes reduce the amount of light trans-mitted from one place to another. Even the atmosphere has finite values of absorption and scattering. These vary considerably with meteorological conditions but the transmitted light is unlikely to be better than 95% over a kilometre.

For optical and ophthalmic glass the absorption value is considerably more but a further loss occurs at the surfaces of any sample due to reflection. Thus the absorption of most optical glasses is below 2% per centimetre but curves of the transmission of 1 cm samples often show values about 90% due to the 4% reflection loss at each surface.

Transmission values may vary considerably with wavelength (Chapter 13) but filters can be obtained with very little transmission change with wavelength. Such filters are neutral to the colour of the light passing through them and are known as **neutral density filters.** If two such filters have transmission values of 40% and 60% then the transmission of both used together is found by *multiplying* these values, ie. 24%. This fact has led to the adoption of the concept of **optical density** which is the $\log_{10}$ of the reciprocal of the transmitted fraction. Thus a transmission of 50% means that half the light is transmitted so the density is $\log_{10} 1/0 \cdot 5$ i.e. 0·3. This is very useful when the transmission values are very low as the density values increase as in the table:

Optical Density	0·1	0·25	0·4	0·5	0·75	1·0	2·0	3·0	4·0
Transmission %	80	56	40	32	18	10	1	0·1	0·01

The usual advantage of a logarithmic scale is found in that when two filters of density 0·5 and 0·25 are used together they give a density of 0·75 by simple addition. In transmission terms this is: $0·32 \times 0·56 = 0·18$ or 18%. This calculation is only correct when the spectral absorptions are identical for the two filters. Thus the densities of two differently coloured pieces of glass cannot be added in this way.

131. The Measurement of Illumination and Luminance

The photometers described in § 124 and § 125 are used to measure the candle-power of sources, although they do this by measuring the illumination due to a particular source at a given distance. If more than one source is present (other than the standard) then the illumination due to both will be measured. The comparison photometers are rather too cumbersome and the photoelectric system is commonly used for this purpose being placed on desk-tops and roadway surfaces to measure the illumination due to many sources. Indeed it is not capable of differentiating between the sources.

If the photoelectric cell is used as a camera exposure meter it measures the total light reaching the camera from the scene. No account can be taken of whether this contains a few bright sources or an even distribution and exposure errors can be caused by this. The cell essentially measures the illumination on the camera due to the scene and not the luminance of the surfaces comprising the scene.

An instrument which does the latter is the **S.E.I. Photometer** (Fig. 136). The surface to be measured is viewed through a low-power telescope, T. By means of a modified Lummer-Brodun prism in the telescope a comparison spot, C, is seen superimposed on the centre of the image field.

Fig. 136.—The S.E.I. photometer

This spot subtends an angle of ½° at the eye and is reflecting light from a diffusing screen, S, illuminated by a small electric lamp fed from a dry battery, B, through a rheostat, R. The lamp also illuminates a ring-shaped photoelectric cell, E, which is connected to a micro-ammeter, M. By adjusting the luminous output of the lamp by means of the rheostat the needle of the microammeter can be made to coincide with a standardising mark and in this way the luminance of the diffusing surface, S, can be kept at a constant value.

Between the diffusing surface and the prism are two opposed neutral wedges, N, which can be moved in opposition to one another by means of a rack and pinion mechanism operated by rotating the base of the instrument. The light reaching the comparison spot can in this way be varied over a range of 1 to 100. By interposing neutral filters in the path of the light from the surface under test the range of the instrument can be greatly increased.

The surface to be measured is viewed through the telescope with the comparison spot appearing superimposed upon it. The luminance of the spot is then adjusted to match that of the surface by rotating the base of the instrument and so moving the wedges. Colour filters below the comparison spot allow for matching surfaces illuminated by daylight or artificial light. In addition to the usual scales for obtaining photographic exposures the instrument is scaled for luminance in log. foot-lamberts.

To measure illumination a block of magnesium carbonate is placed in the required position and the luminance of the surface of the block is measured. As the surface is an almost perfect diffuser and has a reflectance factor of nearly 100%, the luminance in foot-lamberts corresponds very nearly to the illumination in lumens per square foot.

The S.E.I. photometer still depends on the eye as the final arbiter and variations between observers often occur. A more sophisticated instrument is the **Spectra-Pritchard Photometer** (Fig. 137). This uses the eye solely to define the surface to be measured. The eye views the scene via a mirror

Fig. 137.—Aperture mirror disc system of the Spectra Pritchard Photometer, Model 1980, manufactured by the Photo Research Division of the Kollmorgan Corporation

having a hole in it which therefore appears as a black dot to the observer. The light from the scene is imaged on to the mirror so that the black dot is sharp and the light passing through the hole is coming only from the part of the scene covered by the black dot. This transmitted light reaches a specially calibrated photodetector via a series of filters. The mirror has holes of different sizes which can be rotated into the observed scene while the filters have a considerable range of optical densities so that a very versatile broad-range instrument is made.

1. Define the illumination of a surface. Explain exactly what is meant by the statement that the illumination due to a source of light varies inversely as the square of the distance from the source.

If an electric lamp of 16 candle-power is set up at 40 inches from a screen, find the illumination on the screen (a) in lumens per sq. ft., (b) in lumens per sq. metre.

2. Define the terms candle-power, foot-candle and metre-candle, and find the relation between the foot-candle and the metre-candle. An illumination of 6 foot-candles is required on a study table. What candle-power lamp should be installed at a height of 4 feet above the table if 70 per cent. of the light is received indirectly? (L.U.)

3. What is meant by the candle-power of a source of light? Describe carefully an experiment to determine this value.

4. Explain what is meant by (a) the illumination and (b) the luminance of a surface. Describe how these qualities may be measured.

5. Given that the illumination of the earth in full moonlight is 0·02 lumens per sq. ft. and that the distance of the moon is 235,000 miles, find its candle-power.

6. Express lumens per sq. ft. in lumens per sq. cm. and lumens per sq. metre. A 50 C.P. lamp is 12 feet above a surface: find the illumination in lumens per sq. ft. and lumens per sq. m.

7. Define "Candle-power," "Mean spherical candle-power," "Foot-candle" and "Lumen." What is the minimum illumination of the roadway in a street where the lamps are 2,000 candle-power on standards 50 feet high and 60 yards apart? (L.U.)

8. Explain exactly what is meant by intensity of illumination and deduce the expression for the intensity on a surface placed obliquely to the light reaching it from a small source.

Describe briefly one form of standard source. (S.M.C.)

9. How does the intensity of illumination at a point on a screen vary (a) with the distance of the point from the source, (b) with the angle of incidence of the light?

A small source of 100 candle-power is suspended five feet above a horizontal table. Draw a curve showing how the intensity of illumination on a table produced by direct light varies along a straight line passing directly under the source. (L.U.)

10. Distinguish source intensity and intensity of illumination. How can the latter be measured and in what units is it expressed?

A surface receives light normally from a source at a distance of 2·82 metres. If the source is moved closer so that the distance is only 2 metres, through what angle must the surface be turned to reduce the illumination to its original value? (L.U.)

11. A point source of light of 20 candle-power is situated 50 cm. from a plane mirror and on a normal to its centre. If the mirror reflects 90 per cent. of the incident light, find the illumination on a screen 3 metres from and parallel to the mirror.

12. A 50 C.P. lamp is suspended above a table. It is found that by lowering the lamp 2 feet the illumination is increased three times. What was the original height of the lamp?

13. A small source of light of 100 C.P. is placed 2 feet from a circular plane mirror of 8 inches diameter; find the quantity of light in lumens falling on the mirror, and find also the illumination due to the direct and reflected light on a screen 6 feet from the mirror—the mirror reflecting 90 per cent. of the incident light.

14. A 100 C.P. lamp is 15 feet above the ground. Find the illumination on the ground (a) directly under the lamp, (b) 20 feet to one side.

15. A lens of 3 inches diameter is placed 10 inches from a small source of light of 20 C.P. Find the total quantity of light falling on the lens and also the illumination on a surface 4 inches from the lens if the light be brought to a focus 6 inches from the lens.

16. A small source of 30 C.P. is placed 10 cm. from a + 5D lens of 10 cm. diameter. Find the illumination on a screen 1 metre from the lens, neglecting losses by reflection or absorption at the lens. What change in the illumination will be produced by reducing the aperture of the lens to 5 cm. diameter?

17. The illumination given out by a lamp is not necessarily the same in all directions. The candle-power of an arc lamp was measured in various directions; it was 250 C.P. directly downwards; 1,070 C.P. at 30° to the vertical. If this lamp is 25 feet above the pavement, find the illumination directly under the lamp, and also on a screen at the place on the pavement corresponding to the 30° direction and placed perpendicular to the direction of the light.

18. A room 20 by 15 feet and 10 feet high is illuminated by a 50 C.P. lamp suspended from the centre of the ceiling and 8 feet above the floor. Find the illumination on the floor (a) directly under the lamp, (b) in the corner of the room.

19. A diagram is illuminated for photographic copying by two lamps of 50 C.P., 4 feet apart, and 3 feet from the plane of the diagram. If the diagram is placed centrally between the lamps, find the illumination at the centre of the diagram.

20. A sight-testing chart 22 inches by 11 inches is illuminated by a 50 C.P. lamp placed 24 inches in front of the chart and in line with the centre of the top edge; what will be the illumination of the chart at the centre and at the bottom corners?

21. Two lamps, each of 200 candle-power, are hung 5 feet apart and 6 feet from the floor above a table 2 feet 6 inches high. Find the illumination on the table directly under each lamp and at a point midway between them. (S.M.C.)

22. A 30 C.P. lamp is suspended over the middle point of a rectangular table of 5 feet; find the height of the lamp above the table in order that the illumination on the table directly under the lamp may be 1·5 lumens per sq. ft. What will be the illumination at the corners of the table?

23. A newspaper is held in a railway carriage in such a position that the plane of the paper and the centre of the paper are 3 feet and 7 feet respectively from the lamp. (a) If the candle-power of the lamps be 32, calculate the illumination on the centre of the paper. (b) What would the candle-power of the lamp have to be in order that the illumination on the paper might be 1·5 lumens per sq. ft.

24. A room 15 feet by 12 feet is illuminated by a 50 C.P. lamp in the centre of the room, and 12 feet above the floor. Find the illumination in one corner of the room and 4 feet above the floor.

25. What are meant by the illumination and luminance of a surface? A test chart 3 feet by 2 feet is illuminated by two 30 C.P. lamps one on each side of the chart, in line with the middle of its longer edges and 3 feet in front of it. Find the illumination at the centre of the chart.

26. (a) A table is situated directly under a 64 C.P. lamp which is 7 feet above the table. What will be the illumination on the centre of the table?

(b) If the table is moved 10 feet from its original position, calculate the candle-power of the lamp required to give the same illumination as before on the centre of the table in the new position.

27. Light from a lens 40 mm. diameter is converging to a point 75 cm. from the lens: find the diameter of the pencil at 15, 30 and 80 cm. from the lens. Compare the illumination on a screen in these three positions.

28. A rectangular table 8 feet by 5 feet is illuminated by two lamps each of 50 candle-power suspended 3 feet above the opposite corners. Find the illumination on the table at its middle point. (S.M.C.)

29. State Lambert's law on the variation of the emission of light with direction.

A perfectly diffusing incandescent surface A of area 1 sq. mm. and brightness 20 candles per sq. mm. is set up parallel to and 50 cm. from a screen, the line AB being normal to the screen.

Find: (a) the luminous intensity of the surface A in candles along a direction AC inclined 30° to the normal AB.

(b) the illumination in lumens per sq. metre on the screen.

(i) at B; (ii) at C. C is the point where AC cuts the screen. (S.M.C.)

30. Describe an accurate form of photometer and explain the principles on which its action depends. How would you use it to determine the percentage of light reflected from a surface? (L.U.)

31. Describe two methods for measuring the mean spherical candle-power of a lamp in its fitting.

32. Two lamps, one of 10 candle-power, are placed at opposite ends of a 2 metre bench and a screen placed between them is found to be equally illuminated on its two sides when 80 cm. from the 10 C.P. lamp. Find the power of the other lamp. What form should the screen take in order that the illumination of the two sides may be easily compared?

33. In comparing two sources of light by means of a photometer, it is necessary to arrange that:

(a) The photometer head is on the line joining sources (axis):

(b) All extraneous light is excluded;

(c) Both sides of screen are equally clean;

(d) Both sides of screen are equally inclined to axis of photometer.

Explain *clearly* the need for each of these precautions.

34. State and briefly discuss Lambert's law on the emission of light.

A small source of light of area 2 sq. mm. is set up 20 mm. from a + 60D condensing lens of diameter 40 mm. The luminous flux falling on the lens from the source is 40 lumens. Find the luminance (brightness) of the source in candelas per sq. mm. and, assuming no losses, the illumination of the image of the source in lumens per sq. metre. (S.M.C.)

35. Two lamps of 30 and 20 candle-power respectively are placed 2 metres apart. Find the positions on a line joining them where a screen would be equally illuminated by these two lamps. How would you determine these positions experimentally?

36. Explain the meaning of the term optical density as applied to an absorbing medium. Illustrate by finding the optical density of a filter which transmits 20% of the light flux incident upon it.

What will be the transparency (i.e. percentage of light transmitted) and optical density of a filter of the same material and twice the thickness?

37. Describe a method of measuring the absorption of a tinted glass. What are the difficulties that are met with and how may they be overcome?

38. A photometer bench 5 feet long has a lamp at each end and a balance is obtained on a grease spot disc at 3 feet 6 inches in front of the left-hand lamp. On interposing a sheet of dark glass between this lamp and the disc, balance is obtained at 2 feet from the same lamp. What percentage of light is transmitted by the glass?

39. Two lamps are mounted at opposite ends of a 2 metre bench and a photometer screen is found to balance when placed 140 cm. from the brighter lamp. On placing a tinted glass in front of the brighter lamp it is found that the screen is balanced when the distances from the two lamps are reversed. What is the percentage absorption of the glass?

40. A photometer was placed between two lamps 2 metres apart and was found to balance when 80 cm. from one of the lamps. Where must it be placed to again balance if a tinted glass absorbing 55 per cent. of the light be placed in front of the brighter lamp?

41. Two lamps whose candle-powers are in the ratio of 2 to 1 are placed at opposite ends of a 2 metre bench. In front of the more powerful lamp is placed a screen which absorbs 30 per cent. of the incident light. What will be the position of a photometer between the lamps when equally illuminated?

42. What is an illumination photometer? Describe a simple form of such an instrument and explain how you would use it to measure the variation of illumination at selected points in a room as twilight draws on.

A lamp of 125 C.P. is suspended at a height of 8 feet above a horizontal table. At what distance from the foot of the perpendicular let fall from the lamp on to the table will the intensity of illumination of a sheet of paper laid on the table be one foot-candle? (L.U.)

43. State Lambert's cosine law of emission.

Explain what is meant by a perfectly diffusing surface.

The luminance (brightness) of a flat perfectly diffusing surface of area 2 sq. mm. is 25 candela per sq. mm. What is its luminous intensity along (a) the normal direction; (b) a direction inclined at 60° to the normal?

If the light leaving the surface normally falls on a circular screen 1 cm. in diameter and 50 cm. from the surface, what amount of luminous flux, expressed in lumens, will fall on the screen? (S.M.C.)

44. A *uniform diffuser* is a surface having the same luminance (L) in all directions. Show that, for such a source of light, the luminous intensity (I) in any direction varies as the cosine of the angle between that direction and the normal to the surface. (NOTE.—"luminance" is the new term for "brightness").

Such a source, small in area, is suspended parallel to and 2 metres above a horizontal table. The illumination on the table at a point A vertically beneath the source is 5 lumens per square metre. Find the illumination on the table at a point B distant 1 metre from A. (S.M.C.)

45. (a) Light from a point source of 50 candela enters an eye of effective pupil

diameter 3 mm., the source being 50 cm. from the eye. What amount of luminous flux, in lumens, enters the eye?

(*b*) Calculate the illumination on the ground midway between two lamp-posts 100 yards apart and 16 feet high, each lamp being of 450 candle-power in the direction considered. Express your result in lumens per square foot. (S.M.C.)

46. Two small lamps A and B give equal illuminations on the two sides of a photometer head when their distances from it are in the ratio 2: 5. A sheet of glass is then placed in front of lamp B and it is found that equality of illumination is again obtained when the distances of A and B are in the ratio 6:5. Find the percentage of light transmitted by the glass.

From this figure calculate the optical density of the glass. (S.M.C.)

XII – The Nature of Light

132. Introduction

So far we have considered that light can be represented by straight lines drawn on paper. Indeed this is why the name **Geometrical Optics** is applied to that branch of the subject. This description is adequate where we are concerned with the way light reacts with prisms, lenses and mirrors, but is no longer sufficient when we wish to understand how light reacts with itself or when it meets optical devices with very fine structure.

In these circumstances a more complete description is required, which includes the wave motion described in Chapters I and II and incorporates the way in which this wave motion is produced and propagated. This branch of the subject is called **Physical Optics**.

133. The Dual Nature of Light

Our understanding of the propagation of light began some centuries earlier than our understanding of how it is produced, and the same order will be followed here. Very early experiments with burning-glasses showed light to be a form of energy. In the seventeenth century **Römer's** observations on the moons of Jupiter revealed that the times between their eclipses by the planet steadily increased and decreased as the earth orbited the sun. Calculations showed that these changes could be accounted for by the different distances the light had to travel if it had a velocity of about 190,000 miles per second.

This finite velocity means that the energy has to travel in some sort of self-contained packet across the intervening distance. The possible ways in which this might occur reduce to two. In the first case energy may be propagated as moving matter as in the case of energy received when a ship is hit by a projectile; or it may be propagated as a wave motion travelling through a continuous medium which does not move as a whole, as in the case of the movement of a ship caused by water arising at a distant disturbance.

Each of these methods has formed the basis of a theory of light and indeed the modern concept of radiant energy propagation is based on a duality which utilises both ideas. A full description of these concepts is beyond the scope of this book and would be more than is needed to understand a large part of physical optics. The essential idea that light is propagated in discrete packets of energy whose actions *can be described* by a wave motion will be developed as required.

134. Historical Development

Isaac Newton considered the evidence of rectlinear propagation and the laws of reflection and refraction and championed the **corpuscular theory,** which assumed that light consisted of exceedingly minute particles shot out by the source. Refraction was explained by assuming that the particles travelled faster in the denser medium. Although this sounds inherently unlikely, a greater problem was the partial reflection and partial refraction which was found to take place on glass-air surfaces and similar boundaries.

It would seem that a particle is either reflected or refracted, and if so, why should identical particles be treated differently? Newton was forced to assume that the boundaries between media were subject to "fits of easy reflection and easy refraction" which were in turn determined by vibrations set up by the particles.

In this last hypothesis, Newton came surprisingly close to the modern concepts of the dual nature of light. However, the assumptions regarding the speed of the particles in the denser medium were the first to be challenged. Measurements made by Foucault and Michelson in the nineteenth century gave contrary results and the **wave theory** became totally accepted until the present century.

The original wave theory was propounded somewhat vaguely by both Grimaldi and Hooke: but to **Huygens** in 1690 is accorded the honour of the first written explanation of the principle described in Chapters I and II. This was not altogether accepted and its opponents maintained that waves on water and sound waves required a medium to carry them whereas light could pass through a vacuum. Huygens assumed the existence of a medium known as the *ether,* filling all space. All experiments have failed to detect the nature or even the existance of such a medium and modern theory regards it as a mathematical concept rather than an actuality.

A more pertinent objection to the wave theory was that waves could bend round objects while light gives sharp shadows and appeared to travel in straight lines. It was left to the work of Young and Fresnel to show that light did bend round objects but that its vibrations were of such short wave-length that the effect could be seen only in special experiments. This will be discussed more fully later in this and in succeeding chapters. An added virtue of the wave theory was that if a transverse vibration was assumed instead of longitudinal as in sound, the phenomenon of polarisation (Chapter XVI) could be explained.

These successes ensured the acceptance of the wave theory of light throughout the whole of the last century and Maxwell was able to integrate these ideas with those of electricity and magnetism. The observed actions of reflection and refraction, interference and diffraction and polarisation could all be explained.

It was only after the discovery of the electron as the fundamental particle of electricity and the ability of light to eject these from metals (photo-electricity) that problems were seen to arise. It was found by experiment that although the numbers of electrons ejected was proportional to the intensity of the light, the energy of the individual electrons was only proportional to the frequency of vibration of the light. In other types of vibrating energy the energy flow is proportional to the amplitude and it was difficult to see why the energy of the ejected electrons did not vary with this. The theories of Planck regarding the emission and absorption of radiant energy not as a continuous action but in discrete packets or quanta and Einstein's concept of **photons** carrying the concentrated energy of the light surrounded by a wave motion of finite length were found to be viable.

Further deductions from Einstein's theories led to the idea of stimulated emission which culminated in the invention of the **laser** in 1960 as described in § 144. Thus the history of man's ideas on the nature of light is a story of change and modification as new discoveries were made. We have no reason to suppose that the present concepts represent an ultimate description.

135. Wave Motion

Because the original concept of a continuous wave motion goes so far in explaining the basic phenomena of interference and diffraction, it is necessary to study the fundamental characteristics of a simple wave motion. In any case such a study is a prerequisite to understanding the more complex theories to follow. If we observe the ripples on the surface of a pond, two things are at once evident; first, that any floating object executes an up-and-down motion without appreciably altering its general position, and, secondly, that crests and troughs, places of maximum displacement of the surface above and below its normal position, are repeated at regular intervals. It is obvious that there is no general forward movement of the water as the ripples or waves continue to travel out from a disturbance.

Let AB, Fig. 138, represent a stretched flexible cord, If a regular up-and-down motion is given to the end A each particle will communicate its movement to the next particle in turn, and a wave will travel along the cord. The form taken up by the cord for different positions of A is shown in Fig. 138. Two particles such as b and c, having the same displacement and direction of motion, are said to be in the same **phase**, and the distance between two successive particles in the same phase is a **wave-length**. If the end A moves in such a manner that it retraces its path in regular intervals of time, its motion is said to be periodic, and the time between successive passages in the same direction through any point is termed the **period** or

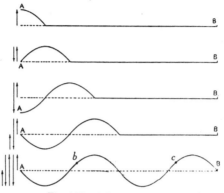

Fig. 138.—Vibrating cord

periodic time of the motion. From Fig. 138, it will be evident that during the time that A makes one complete up-and-down vibration, the disturbance has travelled a distance equal to one wave-length; during two complete vibrations it will have travelled a distance of two wave-lengths, and so on. Then if

λ = wave-length
V = velocity at which disturbance travels
T = period,

we have

$$\lambda = VT \qquad \dots \qquad \dots \qquad \dots \qquad \dots \qquad \dots \qquad \dots \quad (12.01)$$

which is the fundamental equation of wave motion.

The number of vibrations taking place per second is termed the **frequency,** N, of the vibration, and

$$\lambda = \frac{V}{N} \qquad \dots \qquad \dots \qquad \dots \qquad \dots \quad (12 \cdot 01A)$$

The maximum displacement from its normal position of a particle, i.e. when it is at the crest or trough of a wave, is termed the **amplitude,** *a*, of the wave.

The form of the wave will depend upon the nature of the vibration of the individual particles, and each particle will be vibrating in the same way according to the periodic motion imparted by the source, such as the end of the cord in Fig. 138. The motions of the individual particles in a wave motion may be very complex, but it can be shown that any complex vibration can be resolved into a number of simple to-and-fro motions of the type known as **simple harmonic motion** (S.H.M.).

136. Simple Harmonic Motion

If a point P (Fig. 139) moves around a circle with uniform velocity, the to-and-fro motion of the projection of P upon any diameter is termed a Simple Harmonic Motion; thus as P moves uniformly round the circle, the point R executes a S.H.M. along the line XOX' and the point Q, a S.H.M. along the line YOY'. The maximum displacement of Q and R is equal to the radius of the circular path of P; this is the **amplitude,** *a*, of the S.H.M. The **phase** of the S.H.M. at any instant is the angle turned through by the line OP measured from some fixed line, so that if O be the mean position of a particle moving with S.H.M. along the line YOY' its phase when at Q is given by the angle POX.

If *y* is the displacement of Q from O after an interval of time t then

$$y = a \sin POX$$

In the periodic time T the point P will have made one complete revolution and have turned through an angle 2π radians. Therefore in

Fig. 139.—Graphical representation of simple harmonic motion

any time t the point P will have turned through an angle $\dfrac{2\pi t}{T} = \stackrel{\frown}{POX}$, and we may express the displacement of the particle Q in terms of the periodic time as

$$y = a \sin \frac{2\pi t}{T} \qquad \dots \qquad \dots \qquad \dots \ (12.02)$$

where Q is in its mean position when t = o.

By considering Fig. 139 it can be seen that a particle executing a S.H.M. has, on passing through its mean position, its maximum velocity $\frac{2\pi a}{T}$, which is the velocity of the point P. Its velocity then diminishes as it recedes from its mean position, becoming zero at the extremity of its displacement. On returning, the velocity increases, again having a value $\frac{2\pi a}{T}$ on passing through the mean position and a zero value at the other extremity of the displacement of the particle.

A S.H.M. is characteristic of the vibration in a straight line of a particle, acted on by a restoring force, proportional to the displacement of the particle from a centre of attraction. An example is seen in the motion of a weight suspended from the end of a spring.

In the equation (12.02) we have assumed that the vibrating particle is in its mean position when t = 0. If, however, the particle is at say Q (Fig. 139) when $t = \theta$ and at S after t seconds, where P and P_1 are the corresponding points on the circle, then

$$y = OS$$
$$= a \sin SPO = a \sin (POP + POX)$$

$$= a \sin \left(\frac{2\pi t}{T} + \theta \right) \dots \qquad \dots \qquad \dots \ (12.03)$$
Where $\theta = \overset{\frown}{POX}$

Equation (12.03) is the general equation of an S.H.M. The angle $\left(\frac{2\pi t}{T} + \theta \right)$ is termed the **phase angle.**

Many of the phenomena of light, especially those of interference, diffraction and polarisation, may be explained by considering the resultant motion of a particle acted on simultaneously by two or more S.H.M. The displacements due to each S.H.M. at successive values of t are calculated, and the resultant displacement at each value of t is the algebraical sum of the separate displacements at the corresponding time. This is known as the **principle of superposition** and has been shown to be true when the displacements are small. The theory of interference is based on this principle.

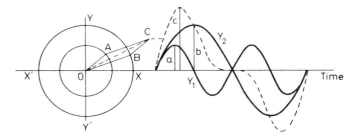

Fig. 140.–Composition of two simple harmonic motions

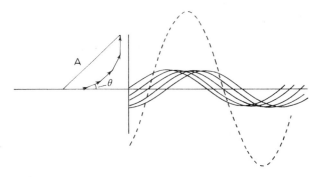

Fig. 141.–Composition of five SHMs of equal frequency and
amplitude but progressively out of phase by θ ($\pi/8$)

The composition of a number of S.H.M. acting in the same straight line
may be solved graphically by plotting, in the way shown in Fig. 139, each
S.H.M. to the same scale and with the same co-ordinates. The curve of the
resultant displacement is then the sum of the separate displacement curves.
An example is shown in Fig. 140 where the broken line is the curve of the
resultant displacement due to two S.H.Ms,

$$y_1 = a \sin \frac{2\pi t}{4} \text{ and } y_2 = b \sin \frac{2\pi t}{8} \text{ executed in the same vertical line.}$$

At any moment in time the displacement of the resultant can be found
by drawing the lines OA and OB at the correct phase angles and constructing, by the method of the parallelogram of forces, the vector addition to
obtain the resultant OC. As the vectors OA and OB rotate at different
frequencies in this example, OC has a varying length which when projected
on to YY′ produces the broken line wave motion.

221

If OA and OB are of the same frequency then OC is of constant length depending on AOB, the **phase difference** between OA and OB. With more than two S.H.Ms contributing to the resultant a polygon of forces can be built up to give the resultant. The resultant is of maximum length when the polygon reduces to a straight line, that is the contributing S.H.Ms are all *in phase,* or out of phase by multiples of 360°. Often the contributing S.H.Ms have equal amplitudes but are progressively out of phase by some angle θ. The vector addition in this case is shown in Fig. 141.

The resultant waveform of Fig. 140(a) occurs when OA and OB produce the S.H.M. by projection on to the same vertical line YY'. If the projections on to XX' were used a similar waveform would result. If, however, we combine OA projected on to YY' and OB projected on to XX' a more complex result occurs. If the analysis is restricted to cases where the frequency of the two contributing motions is the same a result is obtained of great importance in the study of polarised light.

Suppose the two contributing wave forms are represented by

$$x = 11 \sin \frac{2\pi t}{T} + \frac{\pi}{4} \text{ and } y = 7 \sin \frac{2\pi t}{T}$$

which are executed along XOX' and YOY' respectively of Fig. 142. If two circles are drawn with radii proportional to 11 and 7 they can be divided

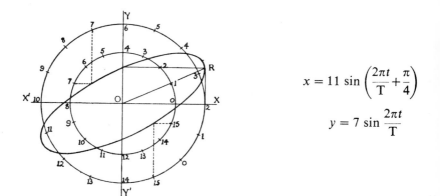

$$x = 11 \sin \left(\frac{2\pi t}{T} + \frac{\pi}{4} \right)$$

$$y = 7 \sin \frac{2\pi t}{T}$$

Fig. 142.—Two S.H.Ms executed in directions at right angles

into equal parts representing equal time intervals. As the two motions are of the same frequency the number of divisions is the same but the difference in phase is represented by the different starting points. Remembering that the resultant is derived from the components of the rotating vectors on *different* axes these must be taken *before* the parallelogram of forces is drawn. The result of these at any instant is found to lie on the ellipse shown. In fact the result of any two orthogonal S.H.Ms of equal frequencies is an ellipse except for the special case where the phase difference is zero or π when the result is a straight line. A further special case occurs when the phase difference is an odd multiple of $\dfrac{\pi}{2}$ and the amplitudes are also equal. These conditions result in a circular motion.

137. Wave Trains and their Superposition

It was seen in § 135 that the form of a wave will depend upon the motion imparted to the particles by the source. If this motion is a S.H.M. the simultaneous displacements of all the particles may be found from the expression (12.02) above.

Let us suppose that a source A, Fig. 143, is executing a S.H.M. given by the expression $y = a \sin \dfrac{2\pi t}{T}$, and that the vibration is transmitted from one particle to the next with a velocity V. Then any particle b at a

Fig. 143.—The form of a wave motion. – A wave train

distance x from A commences to vibrate $\dfrac{x}{V}$ seconds after the particle at A and its motion is given by

$$y = a \sin \frac{2\pi}{T}\left(t - \frac{x}{V}\right) \quad \dots \quad \dots \quad \dots \quad (12.04)$$

As $$\lambda = VT$$

then $$y = a \sin 2\pi\left(\frac{t}{T} - \frac{x}{\lambda}\right) \quad \dots \quad \dots \quad \dots \quad (12.05)$$

Equations (12.04) and (12.05) give the complete form of any simple wave motion, for by giving any constant value to x we have the equation for the motion of a particle at this point, and by giving any constant value to t the curve obtained gives the form of the train of waves at that instant. The curve so obtained should not be confused with that in Fig. 139, which shows the displacement of a *single* particle for various values of t.

When two or more waves are superposed, the resultant displacement at any point may be found from the separate displacements by the methods of § 136. As the resultant displacements will still be periodic, a new wave is formed. If the waves are travelling with equal velocities in the same direction and the vibrations are taking place in the same plane, the resultant wave form can be found graphically. Fig. 144 shows the resultant wave, as a full line, when two waves, shown as broken lines, are superposed. The two waves with the same amplitude are travelling in the same direction with equal velocities and one has twice the wave-length of the other. Any point on the resultant wave has a displacement OR equal to the algebraic sum of the separate displacements OP and OQ, where distances above the x axis are taken as positive and below as negative.

This process may be repeated for any number of simple waves, and the most complex wave forms may be built up of a number of simple waves of suitably chosen wave-length and amplitude. Fourier (1768–1830) has shown mathematically that any form of wave can be expressed as the sum of a series of simple waves.

The difference in phase at any instant of two particles in a wave train may be found from equation (12.05). Thus, if x_1 and x_2 be the distances of the two particles from the source, the displacement of the particle at x_1 is

$$y_1 = a \sin 2\pi \left(\frac{t}{T} - \frac{x_1}{\lambda} \right)$$

and of the particle at x_2

$$y_2 = a \sin 2\pi \left(\frac{t}{T} - \frac{x_2}{\lambda} \right)$$

Therefore the difference in phase at any instant

$$\alpha = 2\pi \left(\frac{t}{T} - \frac{x_1}{\lambda} \right) - 2\pi \left(\frac{t}{T} - \frac{x_2}{\lambda} \right)$$

$$= \frac{2\pi}{\lambda} (x_2 - x_1)$$

$$= \frac{2\pi}{\lambda} \times \text{path difference} \quad \dots \quad \dots \quad \dots \quad (12.06)$$

This will also express the difference in phase between two disturbances arriving at a single particle from two sources at distances x_1 and x_2 providing the two sources are producing exactly similar vibrations. The expression (12·06) is of great importance in problems on interference.

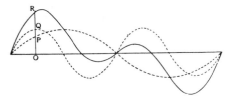

Fig. 144.—Superposition of waves

Referring to § 136 we saw that the maximum velocity of the particle executing S.H.M. was $\frac{2\pi a}{T}$ where a is the amplitude of the motion and T the period. The energy of the particle, being all kinetic at the moment of maximum velocity, is given by $\frac{1}{2}mv^2$ where m is the mass and v the velocity. Thus energy is proportional to $\frac{2\pi a}{T}^2$ or, more importantly, to the amplitude squared. This result is also important in interference and diffraction.

138. Wave Fronts

The two previous sections have analysed the wave motion as a one-dimensional vibration. In Chapter I we loosely considered a wave front moving out from a source and saw its propagation in terms of wavelets generated by each point on the wave. The wave itself had travelled from the source in a given time and this time is the same for all parts of the wave front. If we isolate different points on the wave front they will all be in the same phase and the wavelets generated will be in phase. Thus if exactly similar vibrations be travelling out uniformly in all directions in a homogeneous medium from the point source B (Fig. 1) any sphere DAE having its centre at B, will be a surface of equal phase generating wavelets of equal phase. The only way to obtain wavelets of different phase is to take them from points at different distances from B.

139. The Electro-Magnetic Theory

For many years after the wave theory of light had become generally accepted it was supposed that the vibrating atoms of a luminous body set up waves in an elastic-solid medium which filled all space, and most of the more important phenomena of light have been satisfactorily explained on this theory.

In 1873, however, Clerk Maxwell (1831–1879) put forward a new form of the wave theory in which light is considered to consist of alternating electro-magnectic waves. This **electro-magnetic theory,** which conceives an alternating condition of the medium in place of an oscillation of position of its particles as in the elastic-solid theory, satisfactorily explains most of the known facts about light and overcomes the difficulties met with in the older theory. Maxwell showed mathematically that the electro-magnetic vibration would be transverse to its direction of travel and would travel in free space with a velocity of 3×10^{10} cm. per sec., which, as we have seen, agrees with the measured velocity of light.

About 1886 Hertz (1857–1894) succeeded in producing electro-magnetic waves by purely electrical means, and this was a most important discovery in confirmation of Maxwell's theory. These waves, which are those now used for radio, have many of the properties of light waves, but their wave-lengths extend to many thousand times those of light. The discovery in 1895 of the Röntgen or X-rays by Röntgen (1845–1923) and later of the gamma rays emitted by radio-active-substances, gave us another series of electro-magnetic waves with the same characteristics as light waves, but of very much shorter wave-length.

Later researches have succeeded in filling up the gaps that existed between the various groups of radiation and we now have a complete range of radiations, differing only in frequency and therefore in wave-length, extending from the longest electrical waves of some thousands of metres wave-length, as used in wireless, through the infra-red or heat radiations to the extremely small portion of the complete range, the visible spectrum, which we know as light, where the wave-lengths vary from 7×10^{-4} mm. in the red to 4×10^{-4} mm. in the violet. Beyond the violet we have the still shorter wave-lengths, the ultra-violet radiations, with wave-lengths extending from that of the extreme violet down to about 14×10^{-6} mm., and beyond these again the X-rays, the wave-lengths of which extend down to 6×10^{-9} mm. In the gamma rays the shortest wave-length found is about 6×10^{-10} mm. More recent work has shown the existence of radiations of very great penetrating power and having wave-lengths as short as 8×10^{-12} mm. To these radiations the name cosmic rays has been given.

The various regions of this enormous range of radiations have been shown to possess many properties in common; they all travel at the same speed, and all give rise to certain phenomena, such as interference and diffraction. The different regions are, of course, produced by different sources and detected in different ways. The known range of electro-magnetic waves is shown diagrammatically in Fig. 145.

The wave-lengths of light and other short wave-length radiations are usually expressed in terms of one of the following units:

(1) The **angstrom** (Å). This is equal to 10^{-10} metre or a ten-millionth part of a millimetre.

(2) The **millimicron** (mμ) or **nanometre** (nm), which is equal to 10Å, 10^{-9} metre or one-millionth part of a millimetre.

Fig. 145.—The electromagnetic spectrum

(3) The **micron** (μ), or micrometre (μm) which is equal to one-thousandth part of a millimetre, 10^{-6} metre.

Thus the wave-length of the yellow light of the sodium flame is 5,890 Å = 589 nm = 0.589 μm = 0.000589 mm. We shall adopt the nanometre in referring to wave-lengths. Although the nanometre is the internationally recommended term, the millimicron is still commonly used and some exercises will use the old unit.

140. Incandescence and Characteristic Colours

When the heated poker was discussed in Chapter I, the change in colour with temperature was noticed. As the poker becomes hotter it gives out more light and appears brighter. At the same time the colour of the light emitted changes from red only towards blue. In other words the poker finds it more easy to emit blue light the hotter it gets. This is the first indication that different colours are associated with different amounts of energy.

If instead of a solid poker a pure element such as sodium is heated in a flame, the light it gives out is an intense orange-yellow. The same colour appears when an electric discharge is passed through a sodium vapour as in a common type of street light. If the sodium vapour is replaced by mercury vapour blue and green colours are generated to give a bluish-white appearance. Thus it is seen that particular colours are associated with particular elements.

141. The Quantum Theory and Atomic Structure

The Huygens conception of light travelling out from a luminous point as a *continuous* wave surface is equally applicable to both the elastic-solid and the electro-magnetic forms of the wave-theory, and this conception can satisfactorily account for the observed facts of reflection, refraction, interference, diffraction and polarisation. There are, however, certain phenomena, notably those concerned with an interchange of energy between radiation and matter, absorption, and the photo-electric effect, which cannot be explained in terms of the radiation being transmitted as a continuous wave surface.

At the beginning of the present century Planck successfully explained certain facts regarding the emission and absorption of energy on the hypothesis that energy is not emitted or absorbed continuously, but in definite minute units or **quanta.** Later Einstein extended the quantum theory to light, and considered that light travelled, not as a wave surface over which the energy was equally distributed, but in minute bundles of energy—**light quanta** or **photons***—the energy of which remains concentrated as they travel through space.

Fig. 146.—Emission spectrum of the hydrogen atom

When the light emitted by energised gases is examined by the spectroscopic methods described in Chapter XIII a bewildering array of spectral colours is obtained. The simplest arrangement is found when the gas is hydrogen, Fig. 146. This shows a series of lines which become closer together with decreasing wave-length in a very regular manner. It was these lines which led to the establishment of the whole subject of Atomic Physics when Niels Bohr postulated a structure of matter in which light-

---

* The term photon is also used for a unit of retinal illumination; this is defined as the illumination of the retina produced when the brightness of the observed object is one candle per square metre and the aperture of the eye pupil one square millimetre.

weight electrons orbit a heavy nucleus. The electron orbits are restricted in number and each orbit requires a specific amount (or level) of energy in the electron which occupies it. Bohr postulated that light is emitted when an electron moves from a high energy level orbit to one having a lower energy level. The difference in energy is carried away by the emission of a single photon of light, the wave-length of which is defined by the equation

$$E = h\nu$$

where E is the energy difference

h is Planck's constant

ν is the frequency of the light.

This equation is of fundamental importance in physics and the constant h (6.63×10^{-13} joule. second) is a fundamental constant of nature. Thus the energy radiated by the heated poker increases with increasing temperature as more atoms are emitting energy more frequently but the energy of each photon of light emitted depends solely on the change in energy of the electron actually emitting the light.

The particular photon and the electron energy levels are very closely related. This is evidenced by the fact that if the photon passes another electron in another atom in the lower of the energy levels there is a high probability that the photon will be absorbed by the electron which jumps to the higher energy level. This is further discussed in § 142 and § 157 (Chapter XIII).

142. Stimulated Emission and Optical Pumping

When an incandescent filament emits light we can imagine many atoms having their electrons raised to high energy levels by absorbing energy from the free electrons constituting the electric current through the filament. These electrons subsequently fall to lower levels and emit light. No specific colour is generated as many different electrons and energy levels are involved. The actual moment when a particular electron emits light is not predictable as it does so on a *spontaneous* basis although the time spent in the higher (or excited) level may be very short. However, for transparent sources such as gas discharge lamps a second process exists.

It is now known that if a photon of the same wave-length as that about to be emitted encounters an electron in the excited state a high probability of *stimulated emission* occurs. In this case the emitted photon is naturally of the same wave-length as the stimulating photon but more importantly it is emitted in the same direction and in phase. However, this is not very easy as there are many electrons in the lower state ready to absorb the stimulating photon and the excited electron may also emit spontaneously

229

before the stimulating photon arrives. Thus the processes of stimulated emission, spontaneous emission and absorption can be seen as competing actions within the luminous volume. Very careful arrangements are needed if the stimulated emissions are to predominate over the normal luminescence of the material.

One approach is to ensure an over-abundance of electrons in the excited state. This can be achieved without upsetting the stimulation process by indirect *optical pumping*. With this method intense light is focused into the material so that its particular wave-length is absorbed by the electrons which enter a higher energy level than that required. A proportion of these will then emit light spontaneously and descend to the required high energy level. From here they may spontaneously or by stimulation descend to the lower state from which they can be again raised by the incoming light beam. If this light is sufficiently intense it is possible to maintain more electrons in the high level than in the lower. This *inverted population* of electronic energy levels is essential in the design of **lasers**.

143. Coherence and Resonant Cavities

It is seen from the preceeding sections that light is normally emitted when an electron in a higher energy level *spontaneously* jumps to a lower level and emits a photon of light. This hardly relates to the analogy of ripples on the surface of water used in Chapter I.

If, however, the ripples are caused by a single stone dropped into the water then they are seen to have a definite duration. A circle of a finite number of ripples moves out from the point of impact. When these have passed no other ripples occur until another stone has been thrown in. If the second stone entered the water before the first set of ripples had died away it would be possible, by carefully judging the correct moment, to make the second set of ripples in phase with the first. Indeed a whole series of stones could be dropped at specific times to generate a continuous ripple system moving out from the point of impact. If the stones were dropped on a random basis over a small area then the ripple system would be a complex superposition of several ripple trains even though they would have the same wave-length and speed of propagation.

The difference in these two cases is similar to that between **coherent** and **incoherent light**. The latter case is typified by the incandescent filament where the atoms are spontaneously emitting at random times. In this example many different frequencies are also involved. The former case requires stimulated emission by a controlling wave. This wave can be generated only in a **resonant cavity**. Resonance occurs in sound when a travelling wave is reflected at one end of an enclosed space and returns to the other end for a second reflection. If it now finds itself in phase with the remainder of the wave not yet reflected the two superpose to give a greater amplitude. If they are out of phase zero amplitude will result.

It is easy to show that the length of the cavity is the important dimension. For a fixed length 1, any wave for which $n\lambda = 2l$ will resonate, n being an integer. If the material with an inverted population (§ 142) is contained in an optical resonant cavity a controlling light wave reflected at each end will stimulate emission from the material to build up its amplitude as it travels backwards and forwards.

144. Gas Lasers

The first gas laser was built in 1961 by Javan, Bennett and Herriott at Bell Telephone Laboratories. It consisted of a tube, 100 cm. long, filled with a mixture of helium and neon. At both ends of the tube reflectors were fitted to make the resonant cavity. These were flat mirrors having a flexible coupling to the tube so that they could be adjusted parallel to each other to within a few seconds of arc. In the original model the discharge in the gas was generated by radio-frequency radiation from electrodes wrapped round the tube. Electrodes can also be used inside the tube between which a current can be passed through the gas. Fig. 147 shows the general arrangement.

The stimulated emission occurs in the neon but the helium helps to absorb the electrical energy and generate the inverted population in the neon. When spontaneous emission occurs one of these photons is reflected normally by an end mirror and proceeds down the tube stimulating other excited electrons to emit. On reflection the same action occurs on the return journey and again after a further reflection.

When the stimulating wave passes over an excited electron the stimulated emission may not start until most of the wave has passed and need not occur at all. This means that after two reflections of the stimulating wave the stimulated action may still be in process and so the need to remain in phase is paramount. Energy is continually lost in the absorption of the stimulating wave by electrons in the lower energy levels and in the less than perfect reflection at the mirrors. A lasing action will only occur and be maintained if these losses are smaller than the gains due to stimulated emission. Once this has been achieved one of the mirrors may be made partially transparent so that the beam of coherent light emerges from the cavity. This also represents loss to the system.

In general terms the efficiency of the system is of the order of 0·1% and the energy radiated a few milliwatts — far less than a simple flash lamp. On the other hand the beam is extremely well collimated so that its luminous intensity is very high, forming a safety hazard. (see § 147).

Since the first helium-neon laser many other gases have been found to lase although this original combination has proved to be the most tractable. Low powered types can now be purchased for a few tens of pounds while specially stable types are used in measuring systems (§ 188). The light emitted is red, 632·8 mμ but other electronic levels will also lase giving better efficiencies although the light emitted is infra-red. For visual work a green colour is preferred and after some delay the argon ion laser was invented. Now a large number of gases are known to lase giving a large choice of wave-lengths.

W – Windows   B – Bellows
R – Reflectors   E – Electrodes

Fig. 147.–Layout of the first gas laser

In all cases the light emitted has an extremely narrow spread of wave-lengths. The need to remain in phase after a reflection at each of the end mirrors means that the cavity length must contain an integral number of wave-lengths. Whereas photons spontaneously emitted may have a small range of wave-lengths as the various atoms are influenced by collisions with other atoms which give rise to slight differences in the energy levels; photons in the light beam from a laser being related to the resonant cavity length are the most monochromatic known to man.

145. Pulsed Lasers

Although the He-Ne laser described in the previous section was the world's first *continuous* source of stimulated radiation it was preceded by some months by the pulsed ruby laser of **Maiman** at Hughes Research Laboratories. The major problem to be overcome was the too great a loss of energy by spontaneous emission, non-optical losses and absorption in the end mirrors. Previous exploratory work by **Townes, Basov and Prokhorov** for which they received the Nobel Prize had shown the extent of these other means of energy dissipation compared with the required method.

Maiman's solution (Fig. 148) was to wrap a xenon flash tube around the working element which was a cylinder of pink ruby containing 0·05% chromium. The end faces of the cylinder were polished flat and parallel and coated to provide the resonant cavity, although one face is only partially reflecting to allow the emergence of the laser beam. The flash lamp surrounding the ruby is operated by discharging a bank of capacitors across it. A large proportion of this energy is dissipated as heat but a fraction is emitted as blue-green light which is absorbed by the ruby. For a short time the ruby has an inverted population in its chromium atoms. With high-speed detectors it is possible to observe that spontaneous incoherent emission from the ruby starts almost immediately radiating in all directions but after about 0·5 milliseconds the coherent radiation emerges from the partially reflecting face of the ruby rod. This laser radiation may last for only a few milliseconds but peak powers in the order of tens of kilowatts have been recorded.

Once again the precise wave-length emitted depends on the length of the rod, the resonant cavity. During the discharge this heats up and expands and it is possible to observe the change in wave-length due to this.

Fig. 148.—Layout of the first ruby laser

The wave-length is about 694·3mμ, which is governed by the energy levels in the atoms of chromium known as the *active* element or *working* element. Again other wave-lengths can be found although the red line first discovered is the most efficient.

An alternative lasing material is neodymium which can be contained in a glass rod or a crystal. This gives out light at 1·09 μm in the infra-red part of the spectrum. Solid lasing materials such as these can now be made to lase continuously as gas lasers but any defects in the material absorb energy and can heat up and fracture. Later research turned to liquid lasers where any local hot-spot can be healed by the movement of the liquid.

146. Semi-Conductor Lasers

The population inversion so necessary to coherent light generation can be obtained in the junctions of semi-conductor devices. Instead of inverting the population by putting energy into the working volume from other lamps, the direct passage of a current across the junction produces this effect. The excited electrons may emit light spontaneously and such devices are often called **light emitting diodes** or **LEDs.** If the ends of the junction region are plane and well polished a resonant cavity is formed as the refractive index of the material is so high that substantial reflectivities occur without any reflective coatings.

The most common material is gallium arsenide and this emits in the infra-red part of the spectrum. LEDs are available which emit red and green light. The efficiency is about 50% but large powers are difficult to obtain. For the infra-red emitters a number of suitable detectors are available and such devices have been used on eye-movement experiments where the infra-red beam is reflected off the sclera but not from the cornea. There is very little heat generated and at the low power levels which suffice, emitters and detectors can be mounted close to the eye on a spectacle frame without danger and without the subject being able to see the light as it is infra-red.

147. Photometry of Lasers — Vision Hazards

If plane mirrors define the resonant cavity of a laser then the emitted laser beam will be plane as far as diffraction effects allow (see Chapter XV). Plane mirrors can be difficult to adjust and often concave mirrors are used but with a further curved surface to give a parallel output beam having a divergence of about $^1/_3$ minute of arc. This means that the total solid angle containing the energy is about 10^{-8} steradians while more common sources of light radiate into 4π steradians.

Because of the large loss of energy by other means which occurs in laser systems the efficiency is usually about 0.1%. Thus a watt of energy must be supplied to a He-Ne laser to obtain a milliwatt of light. Most small gas lasers produce between one and 30 milliwatts of red light. In photometric terms this is equivalent to little more than 5 lumens. However, candle-power is measured in lumens per steradian and as these few lumens are concentrated into such a narrow beam the intensity of the source seen looking along the beam is 500 million candela! Because this is in a parallel beam the eye focuses this source on to the retina. In all but the very weakest laser such an intensity will cause burn damage on the retina before the muscles of the eyelid can close to protect the eye. It is therefore very important that operating lasers are not directed towards observers or on to mirrors which could reflect the beam into unprotected eyes.

With pulsed lasers the hazard is even greater as the chance reflection of the pulse from a metal surface such as an optical bench or equipment case could direct sufficient energy into an eye to cause blindness. Workers in laser laboratories are normally supplied with protective goggles containing glass which absorbs those wave-lengths at which the laser is operating.

Other parts of the spectrum are transmitted so that the experimentor can see what he is doing.

The use of lasers for interference and holography (Chapters XIV and XV) normally requires the laser beam to be divergent and a mirror or lens is used for this purpose. This gives the beam a solid angle of about one-tenth a steradian and the apparent intensity of the source is now only 500 candela — similar to tungsten filament lamps. Nevertheless all lasers should be treated with the utmost respect and any examination of the beam is best done by allowing it to fall on a piece of white card or paper even if the result looks quite dim.

148. Application of Lasers

The potential of lasers was recognised from their beginning. Many uses were proposed but as these did not come to fruition within a few years some cynicism has been generated culminating in the tag that lasers were "a solution looking for a problem". This is far from the truth. Two major uses of lasers occur in Optical Metrology and in Holography which are discussed in Chapters XIV and XV respectively. Other uses involve one or more of the three most obvious attributes of this unique source:

(1) Narrow wave-length range
(2) High collimation
(3) Large radiant energy density (high intensity)

Referring to the last attribute first, the use of pulsed lasers for photo-coagulation in cases of retinal detachment is a case of a hazard being put to controlled use. Other energy based applications include the cutting of metals, glass, cloth and even the erasing of typewritten letters by the vaporisation of material. The absorption of the laser energy can be so immediate that no burning is caused although in the case of metal cutting some systems incorporate a jet of oxygen to cause combustion.

The high degree of collimation available particularly from gas lasers allows aerial straight-edges to be created. A laser beam shone into a tunnelling or road grading machine can be used to ensure a straight line of travel. In setting up optical bench experiments a laser can be very useful but stray reflections must be carefully avoided.

The narrow wave-length range permits lasers to be used as sources in various types of spectrophotometer (Chapter XIII); of great value in the chemical and allied industries. Various laser systems have been proposed for monitoring pollution.

The major attribute of lasers is, of course, their coherence as described in § 143. The applications which use this are Optical Metrology and Holography, which, as mentioned above, are dealt with in subsequent chapters; Optical Communications and Optical Processing are largely beyond the scope of this book.

149. The Measurement of the Velocity of Light – Historical Methods

The deduction by Romer (§ 133) in 1675 that light travelled at a finite velocity was based on his observations of Jupiter's moons when the earth was at opposite locations in its orbit so that the light was travelling over different distances. A second astronomical method was used by Bradley (1728) on stars located perpendicular to the earth's orbit. When these were viewed at six-monthly intervals the orbital velocity of the earth is in opposite directions so that the vector addition of the earth's velocity and the light velocity results in a slight angular displacement of a telescope aligned with the star. This was called the **aberration of light** although the word is now applied primarily to image defects.

Galileo had already tried a century earlier to measure the velocity by terrestrial observations. Two observers A and B, with lanterns, were stationed on hills some miles apart. A uncovered his lantern and B uncovered his as soon as he saw the light from A. A noted the time which elapsed between showing his light and seeing that of B, as this would be the time taken for the light to travel from A to B and back to A. Obviously such a crude method would not give a sensible result when the transit velocity is as fast as that of light. The same general principle, however, was used by Fizeau in 1849 in the arrangement shown in Fig. 149.

Light from the source S passes through a lens L_1, is reflected at a plane glass plate P and brought to a focus at F. It is then made parallel by the lens L_2 and travels a great distance to the reflector L_3M. This reflector consists of a concave mirror placed at the principle focus of the lens L_3 and having its centre of curvature at the optical centre of the lens. This arrangement ensures that the principle ray of any pencil reaching the lens is reflected back along its original path*. The reflected light again passes through L_2 and some of it passes through the plate P and is received by an eyepiece E and the eye. In the plane of the focus F is placed a toothed wheel W which can be rotated at a high speed, and is in such a position that light from S alternately passes through the space between two teeth or is cut off by a tooth as the wheel rotates.

Light is sent out when the space is coincident with the image F. Depending on the speed of the wheel the time taken for the light to go out to the reflector and back will be such that a tooth or a space is in the

*A similar arrangement is used in road signs and markings to reflect the light from vehicle lamps back towards the driver.

position F at the moment of its return. In Fizeau's original experiment the distance MF was 5·36 miles (8·63 kilometres). By steadily increasing the speed of the wheel which had 720 teeth the image seen in the eyepiece

Fig. 149.–Measurement of the velocity of light. Fizeau's method

grew alternately brighter and darker and by measuring the number of revolutions of the wheel at the brightest and darkest conditions the velocity can be calculated.

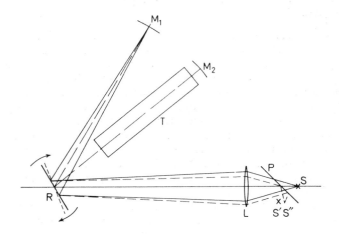

Fig. 150.–Measurement of the velocity of light: Foucault

Foucault, one time collaborator of Fizeau, published a different method in 1850 using a rotating mirror. The principle of this is shown in Fig. 150.

In this method the light is sent out via a rotating mirror R. During the transit time of the light between RM, and back the mirror rotates so that the image of S is no longer at S′ but shifted to S″. In Foucault's experiment RM was 20 metres and the displacement S′S″ was 0·7mm; the velocity was found to be 298,000,000 metres per second. This apparatus was also used by Foucault to make the crucial determination as to whether light travelled faster or slower in a denser medium. For this the second mirror M2 (equidistant from R) and the tube T were added and two displaced images could be seen when the tube was filled with water. The image from the water beam was displaced more than the other showing that the rotating mirrors had had time to rotate further. Thus the velocity of light in water is less than that in air.

A more accurate experiment was done by Michelson in 1885 using white light when he found that the ratio of the velocity in air to that in water was 1·330. A further experiment using carbon disulphide gave a short spectrum at S″ showing that the light had different velocities for different wave-lengths. The ratio of velocities for yellow light was 1·75 while the refractive index of carbon disulphide is only 1·64 only discrepancy is dealt with in § 150.

150. The Measurement of the Velocity of Light — Modern Methods

The measurements made with liquids give a refractive index different from that calculated by refraction measurements on prisms, etc. This is because in any medium except a vacuum light has two velocities. Returning to § 141 it was shown that most light is emitted in rather short trains of waves when an electron changes its energy. Light emitted in this way, although monochromatic, is far from the very narrow range of frequencies found when a laser is used to generate a continuous beam from a resonant cavity using simulated emission. It was shown by Fourier in the early part of the last century that the length of the wave train and the range of frequencies within it are mathematically linked together. A short wave train is formed by using a wide range of frequencies as discussed in Chapter XIII.

If the different frequencies have different velocities it can be shown that the velocity of the wave train or **group** of waves is considerably slower than the wave or **phase** velocity of the component wave-lengths. Only in a vacuum does the group velocity equal the phase velocity. In modern determinations of the velocity of light it is important to distinguish which velocity is being measured, although most measure the group velocity. When this is measured in air a correction must be made for the refractive index of the medium. Michelson devised a method using a more sophisticated rotating mirror system for which the light path was in an evacuated pipe one mile long. By successive reflections at mirrors the total distance traversed amounted to 10 miles. The pipe could be evacuated to a pressure of 0·5mm. of mercury and values at various pressures were obtained. The rotating mirror had 32 sides and rotated at such a speed that the subsequent face of the mirror replaced the first face during the transit time of the light. The mean of almost three thousand measurements was found to be 299,774 km/sec ± 11 km/sec.

Other methods involving Kerr-cells and ultrasonic waves have been used as well as radio waves. These latter have been used in radar systems and also in cavity resonators where the frequency as well as the wave-length can be measured.

CHAPTER XII   EXERCISES

1. What is meant by the wave-length of light, and how is it related to its frequency of vibration and velocity of propagation? (S.M.C.)

2. State the fundamental equation of wave-motion. Given the velocity of light in air as 3×10^8 metres per second, determine the wave-lengths for red, yellow and blue light, the frequencies being

Red	395 billions per second.	
Yellow	509 ,, ,, ,,	
Blue	617 ,, ,, ,,	

Give the wave-lengths in Angstrom Units, nanometres, microns and inches.

3. A series of waves of wave-length 100 cm. is travelling across a pond on which three corks A, B and C are floating at a distance of 1·5, 2·25 and 3·8 metres respectively from a fixed post. In which direction will each of the corks be moving when a trough of a wave is passing the fixed post? State also whether each cork is above or below its normal position.

4. What is meant by a simple harmonic motion? Define the terms, period, amplitude and phase.

5. Plot the graph of the S.H.M. represented by

$$y = 3 \sin \frac{2\pi t}{8}$$

6. A particle is acted on simultaneously in the same straight line by two S.H.M. represented by

$$y = 3 \sin \frac{2\pi t}{6} \text{ and } y = 5 \sin \left(\frac{2\pi t}{8} - \frac{\pi}{2} \right)$$

Plot a graph showing the resultant motion.

7. Find graphically the resultant motion of a particle acted upon by two perpendicular S.H.M. of equal period and amplitude, and differing in phase by (a) 0, (b) $\frac{\pi}{4}$, (c) $\frac{\pi}{2}$, (d) $\frac{3\pi}{4}$, (e) π, (f) $\frac{3\pi}{2}$.

8. Repeat the graphical constructions of question 7 for the case when one S.H.M. has twice the amplitude of the other.

9. A particle **B** executing a simple harmonic motion given by the expression $y = 8 \sin 6 \pi t$ is sending out waves in a continuous medium travelling at 200 cm. per second. Find the resultant displacement of a particle 150 cm. from B, one second after the commencement of the vibration of **B**.

10. Explain the significance of the work of Römer, Foucault, Maxwell and Planck in the establishment of the theory of light.

11. List and describe the prime requirements for a system to emit coherent light.

12. Describe a laser and one application of it.

13. In using Fizeau's method of determining the velocity of light the disc has 200 teeth and revolves 2,800 turns per minute to cut off the light when the mirror is 5 miles distant; find the velocity of light.

14. In measuring the velocity of light by Fizeau's methods the distance from the toothed wheel to the distant mirror is 6 kilometres. If the wheel has 750 teeth, what must be the least number of revolutions per second in order that the light returning from the mirror shall be cut off?

15. The maximum displacement of a very distant star from its mean position is found to be 20·44 seconds. If the orbital velocity of the earth be 30,557 metres per second, what is the velocity of light?

16. In astronomy stellar distances are measured in "light years," the distance which light travels in a year. The star Sirius is 8·55 light years away, what is its distance in miles.

17. Write a short account of the complete radiation spectrum extending from the long electrical waves at one end to the short gamma and cosmic rays at the other end. Indicate by name the various regions of radiation in their order on a diagram, inserting figures giving their approximate wave-lengths.

Why is the spectrum sometimes called the electro-magnetic spectrum?

Make a brief mention in your essay of the work of Hertz and Röntgen in developing our knowledge of radiation. (S.M.C.)

XIII – Dispersion and Colour

151. Dispersion

From the observations described in the last chapter it is known that light of all colours travels with the same velocity *in vacuo*. As was stated in Chapter II, this is no longer the case when light enters a denser medium. The light of each different colour or frequency then has a different velocity and in most cases the red light travels the fastest and the violet the slowest. This effect is always evident when the light passes obliquely through a refracting surface; the different colours are refracted different amounts according to their velocities, and are therefore separated or dispersed.

152. The Production of a Pure Spectrum

Newton, in his original experiments on the spectrum, allowed light from the sun to pass through a small circular aperture into a darkened room, and received the small circular patch of light on a white screen. On interposing a prism this circular patch was broadened out into a band of colour, each of the colours, which together constitute white light, being deviated a different amount by the prism. Such a spectrum is, of course, very far from pure, since, as can be seen from Fig. 151, the different coloured patches overlap considerably. If the screen is removed and the eye placed in the light leaving the prism, the aperture is seen as a spectrum in the position V′G′R′. This spectrum, while being reasonably pure if the aperture is quite small, will have very little width. Substituting a narrow slit parallel to the refracting edge of the prism for the circular aperture, a wide fairly pure spectrum will be seen on looking through the prism, the spectrum consisting of virtual images of the slit in each of the constituent colours.

Usually we shall require a spectrum that can be received on a screen or examined with an eyepiece, and in order that the spectrum shall be as pure as possible, that is, that there shall be little overlapping of the

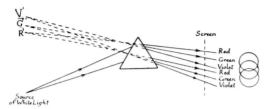

Fig. 151.—Dispersion by a prism

various colours, certain conditions are necessary. These are as follows:

(1) The light must pass through a narrow slit with its length parallel to the refracting edge of the prism.

(2) A well-defined real image of this slit must be formed by means of a positive lens.

(3) Each pencil of light must be parallel on passing through the prism, otherwise various parts of a pencil will be deviated by different amounts and a sharp image cannot be formed.

The arrangement shown in Fig. 152 satisfies these conditions. A narrow slit S, usually adjustable in width, is illuminated by light from

Fig. 152.—Optical arrangement of spectroscope

the source of which the spectrum is required. The slit is situated at the first principal focus of a positive lens A_1 and slit and lens then form a collimator from which emerge parallel pencils from each point on the slit. The light passes through the prism P where it is dispersed, each different colour being refracted in a different direction, and is then brought to a focus by a second achromatic lens A_2. A well defined spectrum is therefore formed in the second focal plane of the lens A_2, each different colour being a fine line, an image of the slit S. When the source is an incandescent solid, the spectrum will be a continuous band of colour consisting of an infinite number of these slit images. The image in the focal plane of A_2 can best be examined by means of an eyepiece E, and it will be seen that the lens A_2 and the eyepiece form an astronomical telescope. The complete instrument constitutes a **spectroscope.**

153. The Spectrometer

We frequently require to measure the angle through which light of various colours is deviated by a prism, as, for example, in the measurement of refractive index (Chapter IV), and the instrument used for this purpose is the **spectrometer** (Fig. 153). The arrangement of the instrument is shown in plan in Fig. 154. Its essential parts are a collimator, an astronomical telescope, a table for carrying a prism and a divided circle. The collimator, with an adjustable vertical slit, is mounted with

Fig. 153.–Modern spectrometer (courtesy P.T.I. Co. Ltd.)

its axis directed towards the axis of the divided circle and is usually fixed. The telescope, which has a Ramsden eyepiece with cross-lines at its focus, rotates in a horizontal plane about the axis of the circle and its position is read on the circle by means of a vernier. The prism table, which is provided with levelling screws, also rotates about the axis of the circle, and in some instruments its position can likewise be read on the circle by means of a vernier. In adjusting the spectrometer the telescope eyepiece is first focused on the cross-lines and the telescope then focused on a distant object. The slit of the collimator is illuminated and the telescope having been brought into line with the collimator the latter is adjusted by moving the slit in or out until a sharp image of the slit is formed in the plane of the cross-lines of the eyepiece.

Fig. 154.—Plan of spectrometer

154. Measurement of the Refractive Index with the Spectrometer

Measurement of the Angle of a Prism

The angle of a prism may be measured on the spectrometer by either of the following methods:

Fig. 155a.—Measurement of angle of prism

(1) The prism is placed on the table of the spectrometer with its refracting faces vertical and the angle to be measured facing the collimator (Fig. 155a). A certain amount of the light is reflected from the two faces of the prism, and the angle between the two reflected beams is measured by reading the position of the telescope when an image of the slit is received on the cross-lines, first by reflection at the face AC and

secondly at the face AE. It can easily be proved that the angle between these two reflected beams is equal to twice the angle of the prism.

Fig. 155b.—Measurement of angle of prism

(2) The telescope is fixed at an angle of roughly 90° with the collimator (Fig. 155b). The prism is rotated until an image of the slit formed by reflection at the face AE is received on the cross-lines, and the position of the prism table is read on the circle. The prism table is then rotated until the image of the slit reflected from the face AC falls on the cross-lines, and the reading of the table again taken. The angle through which the table has turned will then be equal to $180° - a$, where a is the angle of the prism.

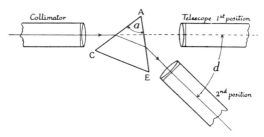

Fig. 156.—Measurement of minimum deviation

Measurement of Minimum Deviation

To determine the refractive index of the prism it will be necessary to measure the angle of minimum deviation; this may be done as follows: The collimator slit is illuminated by a source giving the light for which the refractive index is wanted—the sodium flame will be used when mean refractive index is required—and the telescope is brought into line with the collimator, so that an image of the slit falls on the cross-lines. The reading of the telescope in this position is taken. The prism, which will usually have a refracting angle of about 60°, is then placed on the table with one refracting face making an angle of about 45° with the beam from the collimator, Fig. 156. The telescope is turned until an image of the slit is seen. To find the position of minimum deviation the prism is slowly rotated in the direction that causes the image to move towards the first position of the telescope, i.e. towards the undeviated direction. Following the image with the telescope a position is found where the image commences to move in the opposite direction and this is therefore the direction of minimum deviation. The image of the slit in this position is brought on to the cross-lines, and the reading of the telescope again taken. The difference in the readings gives the angle of minimum deviation d, and the refractive index is found from the expression (4.06A)

$$n = \frac{\sin\left(\dfrac{a+d}{2}\right)}{\sin\left(\dfrac{a}{2}\right)}$$

155. Refractometry

Although the refractive index of a material may be found by grinding and polishing it into a prism shape and using it on a spectrometer to give an angle of minimum deviation the method is time consuming and expensive as a high degree of flatness is required on the incident and emergent

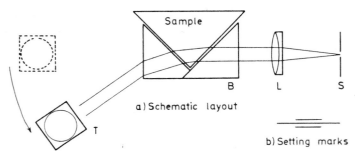

a) Schematic layout

b) Setting marks

Fig. 157.

faces for good accuracy. The methods described in Chap. IV § 42 are insufficiently accurate when the precise value of index is required at different wave-lengths so that achromatic lenses (§ 161) may be designed and manufactured. For this purpose the Hilger-Chance V-block Refractometer was developed and a side view of this instrument is shown in Fig. 157.

The V-shaped block B is made as indicated of two prisms of medium-index glass. The angle between the faces of the V is exactly 90° and the sample of material is ground and polished to about 90°. The quality of the polish and the accuracy of the angle is not of great importance as a film of liquid is placed between the sample and the block to ensure optical continuity. Monochromatic light from the source S is collimated by the lens L and passes through the loaded V-block which forms an Amici prism system (§ 164). The deviation of the emergent beam is measured by the horizontal telescope T which can be rotated as shown. This deviation depends on the refractive index of the sample and may be in either direction. In use the instrument is usually calibrated with samples of known materials.

The setting system received special attention during the design and utilises two short parallel lines in the eyepiece of the telescope. These can be set equidistant about the image of the slit seen in the telescope as in Fig. 157b. The eye is very sensitive to any non-symmetry and maximum accuracy can be achieved. This instrument is commonly used for quality checking by optical and ophthalmic glass manufacturers.

156. Wave-length, Colour and Temperature

By methods described in Chapters XIV and XV, it is possible to measure experimentally the wave-length and hence, as the velocity is known, the frequency of any light. It is found that when light is dispersed to form a spectrum, each portion of the spectrum consists of light of a particular wave-length and therefore frequency. Thus in the case of a continuous spectrum of white light, there is a continuous range of wave-lengths extending, in the portion visible to the eye, from about 390 nm at the violet end to 760 nm at the red end. We may therefore specify any *spectrum colour* or any position in a spectrum in terms of the wave-length or frequency of the light dispersed to that position.

Each colour temperature can be stated in terms of its 'mixed' value (micro-reciprocal degree) which is given by 10^6 divided by the colour temperature in degrees Kelvin. The reciprocal nature of this unit allows the filters described above to be assigned a mired shift value, which may be positive or negative. Lamps of different colour temperatures will be 'shifted' by the same number of mireds by a given filter. This analysis is used in photography but care must be exercised as it assumes that the lamps are nearly black-body radiators. Some fluorescent lamps depart considerably and photographic results show these differences.

The continuous spectrum of *white* light is usually obtained from an incandescent solid or liquid. The word incandescent means glowing with heat and the hotter the source the more it glows. As well as becoming more intense the source also changes colour from red through white to blue-white. When the intensity for each wave-length is plotted for a perfect (or black-body) source the curve varies with temperature as in Fig. 158.

It is therefore possible to describe the red-white or blue-white colour

Energy per unit of wavelength

Fig. 158.

seen in terms of the temperature and this is known as the colour-temperature usually given in degrees Kelvin. This need not be the same as the actual temperature for if a blue filter is placed in front of a tungsten lamp the red parts of its spectrum will be absorbed more than the blue and through the filter the colour-temperature will appear to be higher even though the lamp itself remains the same. Due to atmospheric absorption the colour-temperature of the sun is about $6000°K$ even though its actual temperature is higher than this. The colour temperature of most lamps ranges from $2000°K$ to $4000°K$. (Plate 4C).

157. Spectra and Spectral Analysis

There are two classes of spectra; *emission spectra* and *absorption spectra*. Within these classes there are three types; *continuous spectra, line spectra* and *band spectra*. Emission spectra are obtained when light coming from a source is examined by a dispersive instrument such as a spectro-scope. Absorption spectra occur when light from a source having a con-tinuous spectrum is passed through an absorbing material before entering the spectroscope. The electrons in the absorbing material absorb specific wave-lengths and then re-emit these by spontaneous emission. This occurs in all directions and so the beam passing through the material is depleted of these wave-lengths and dark areas or lines appear on the spectrum seen. Absorption spectra carry as much information about the structure of materials as emission spectra and as the sample does not have to be heated it is usually preferable to use this class of spectra in absorption spectro-photometers (§ 158).

Continuous Spectra

Solids and liquids tend to have continuous spectra in which every possible wave-length is emitted within a broad range without any sharp edges or details. Absorption spectra of liquids especially solutions are also broad but often show some structure particularly in the infra-red part of the spectrum. The absorption spectrum of an alcoholic solution of chloro-phyll, the green colouring matter of plants, for example, shows a fairly narrow absorption band in the orange-red and another broad band extend-ing from the blue to the violet end of the spectrum.

Line Spectra

When a gas is made incandescent, the spectrum consists of bright lines on a dark background. These lines are, of course, images of the slit. Any gas, therefore, when made incandescent, emits radiations of certain defi-nite wave-lengths, and each gas gives its own characteristic spectrum, although this may differ somewhat according to the means used to render the gas incandescent. Thus the spectrum of incandescent sodium vapour, which may be obtained by placing a small quantity of common salt in the flame of a bunsen burner, will be found to contain two bright yellow lines. These two lines are very close together, that is, they differ little in wave-length, and with a small spectroscope will appear as one.

The usual method of rendering a gas luminous, in order to examine its spectrum, is by means of a discharge tube. A glass or quartz tube is filled with the gas or vapour at low pressure. Two metal electrodes pass through the walls of the tube so that an electric current may be passed through the gas. This usually requires a high voltage to start with until the gas ionises. If the gas is hydrogen the tube will glow with a purplish coloured light and on passing this through a spectroscope the visible spectrum will be found to consist of four bright lines, the first red, the second greenish-blue, the third deep blue, and the fourth violet. These are shown in Fig. 146, Chap XII. The other lines are outside the visible region. The spectra of other gases and some of the more volatile metals, such as mercury and cadmium, can be obtained in the same way. Sometimes the electrodes used to pass the current through the gas can give rise to impurities and extra spectral lines. It is possible to excite the atoms of the gas by an electromagnetic field at radio frequencies. The gas is then sealed in a complete glass envelope. These are known as electrodeless lamps and are being increasingly used in spectroscopy.

It should be understood that with a given spectroscope the line corresponding to any wave-length will always occupy the same position; thus the yellow line of the sodium spectrum will fall in the same position as the corresponding yellow in the continuous spectrum. The same thing applies if the lower temperature discharge tube is inserted into a beam of white light falling on to a spectroscope slit when absorption lines are found in exactly the same position as the emission lines due to the absorption and re-emission of particular wave-lengths. A special case of absorption lines is considered in § 159.

Band or Fluted Spectra

In certain cases a spectrum will consist of broad luminous bands, which are sharply defined at one edge, and gradually fade away at the other giving a fluted appearance. When examined with a spectroscope giving great resolving power, it is found that these bands are composed of a great number of lines closely packed together towards the bright edge, and becoming more and more openly spaced towards the other. The band spectrum is characteristic of molecules and hence is produced by chemical compounds in the gaseous state. Line spectra are associated with atoms.

The line spectrum of hydrogen (Fig. 146, Chap XII) can be seen to be a band spectrum except that it is spread out over a very wide range of wave-lengths. The progression shown is due to the very regular construction of the hydrogen atom.

158. Spectrophotometers

Because of the very great importance of spectral analysis to modern science, the design and manufacture of instruments to give accurate and repeatable results has also become important. These measure the *intensity* of each spectral line or band as well as its wave-length so the term **spectrophotometer** is used. Although many of these use a prism to disperse the light a larger number use a diffraction grating (Chap XV) which has a more regular dispersing action. Interest in spectral lines extends well outside the visible spectrum and it becomes impossible to find materials for prisms and lenses. It is then necessary to construct a system using a reflecting diffraction grating and concave mirrors to collimate the light. In the extreme UV it is necessary to evacuate the instrument so that the atmosphere does not absorb the radiation.

Careful mechanical design is needed to rotate the prism or grating so that different wave-lengths are imaged on to an exit slit just in front of a photo-detector. It is commonly arranged that the same mechanical system drives a pen over a chart so that the x-axis of the pen position is related to wave-length.

The output of the photo-detector is used to drive the pen in the y-axis so that a graphical plot of the emission or absorption spectrum is recorded. Photo-detectors vary considerably in their response to different wave-lengths. This response may have to be carefully calibrated so that the pen records the true values. Alternatively, absorption instruments may be designed with two alternate beams, one of which passes through the sample while the other is direct. If the detector is used to measure intensities of these two beams alternately any differences due to detector response can be cancelled out. It is often necessary to switch in different detectors for different regions of the spectrum as some are satisfactory only in the UV to visible while others respond in the IR.

159. Fraunhofer Lines

The sun and most of the stars provide important examples of absorption spectra. The spectrum of sunlight is seen to be crossed by a great number of fine dark lines distributed throughout the length of the spectrum. These lines were first observed by Wollaston (1766—1828) in 1802 and were more fully studied by Fraunhofer (1787—1826); they are known as the **Fraunhofer lines.** They occupy definite constant positions in the spectrum and indicate that certain wave-lengths are absent in sunlight when it reaches the earth. Fraunhofer observed about 600 of these lines in the solar spectrum, and denoted the most prominent ones by letters from A in the extreme red to H in the violet; the wave-lengths corresponding to them are given in Table II in the Appendix. Many more Fraunhofer lines have since been discovered not only in the visible, but also in the ultra-violet and infra-red portions of the solar spectrum.

The explanation of these lines was given by Kirchoff who assumed the sun to consist of an incandescent solid or liquid core giving a continuous spectrum, surrounded by an envelope, the *reversal layer,* at a lower temperature, in which the various elements are in the form of gases. Most of the lines in the solar spectrum have been identified as due to elements also found on earth, while a few of the lines are due to absorption in the earth's atmosphere. The existence of the element helium was discovered in this way in the sun, before it was found on the earth.

The Fraunhofer lines are extremely useful in enabling us to specify particular colours or regions of the spectrum, for, as has been seen, each one occupies a perfectly definite and constant position corresponding to a particular colour or wave-length.

160. Dispersive Power

It has been shown in Chapter II that, for practically all transparent substances, the dispersion is such that the refractive index is greater the smaller the wave-length. The change in refractive index with change in wave-length will, however, differ with different media, and it is necessary that the dispersion of the various glasses and other media, of which lenses and prisms are to be made, shall be accurately known. The refractive indices of a glass are usually determined for the following colours:

The yellow line given by the sodium lamp, corresponding to the Fraunhofer line D.

The red and green-blue lines given by hydrogen, coinciding with the Fraunhofer lines C and F respectively.

In addition refractive indices are given by the glass manufacturer for the violet, G', line of the hydrogen spectrum and a number of other lines in the spectra of helium and mercury. These are the red, b, and the yellow, d, lines of the helium spectrum and the green, e, and the violet g and h lines of the mercury spectrum. The various indices will be denoted by n_D, n_F, n_C, n_d, etc.

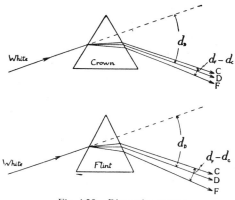

Fig. 159.—Dispersive power

The refractive index of a medium for light corresponding to the D line is known as the **mean refractive index** and is the value understood by the term unless otherwise stated. The value $n_F - n_C$ is termed the **mean dispersion** and the **partial dispersions** are given by n_C-n_b, n_d-n_C, n_e-n_d, etc.

The sodium D line has a broad structure making its use in measurements difficult. Optical glass companies adopted the helium d line but lately values have also been quoted for the mercury e line (546 nm) as being nearer the peak of the visual luminosity curve.

The mean dispersion of a glass is usually greater for glasses of high mean refractive index, but the increase in mean dispersion is not proportional to the increase in mean refractive index. This is illustrated by the following figures for the deviation of the C, D and F lines produced by two prisms, one of crown glass and the other of flint glass, the refracting angles being such that the minimum deviation for the D line is equal in each case.

	n_C	n_D	n_F
Hard Crown Glass	1·5150	1·5175	1·5235
Dense Flint Glass	1·6176	1·6225	1·6349

	d_C	d_D	d_F	$d_F - d_C$
Crown Prism: $a = 60°$	38° 31′	38° 42′	39° 14′	0° 43′
Flint Prism: $a = 52° 6′$	38° 27′	38° 42′	39° 42′	1° 15′

It will be seen that while the mean deviation, d_D, produced by the two prisms is equal, the dispersion between the F and C lines, $d_F - d_C$, for the flint prism is almost double that for the crown. The effect is shown diagrammatically in Fig. 159, the dispersion being exaggerated.

From expression (4.08) the deviation of a prism of small refracting angle for the D line is given by

$$d_D = (n_D - 1)\, a.$$

Similarly
$$d_F = (n_F - 1)\, a.$$
$$d_C = (n_C - 1)\, a.$$

The dispersion between the F and C lines is therefore

$$d_F - d_C = (n_F - n_C)\, a = \frac{n_F - n_C}{n_D - 1} \cdot d_D.$$

This ratio of the dispersion of the C and F lines to the mean deviation

$$\frac{d_F - d_C}{d_D} = \omega = \frac{n_F - n_C}{n_D - 1} \qquad \ldots \qquad \ldots \quad (13.01)$$

is called the **dispersive power** or **relative dispersion** of the medium.

In the calculation of achromatic lenses and prisms the reciprocal of dispersive power is frequently useful and is generally denoted by the symbol v and called the **v-value** or **constringence**.

$$v = \frac{n_D - 1}{n_F - n_C} \qquad \ldots \qquad \ldots \qquad \ldots \quad (13.02)$$

The **relative partial dispersions** give the ratios of the several partial dispersions to the mean dispersion, e.g. $\frac{n_C - n_b}{n_F - n_C}$, etc.

Lists are published by glass manufacturers giving these values for each type of glass manufactured, and as each melting will vary slightly from the type, exact figures for the actual melting are usually also supplied.

The two common types of optical glass are the crown glasses which will have values for n_D of a little over 1·5 and for v of about 60, and the flint glasses having n_D values of about 1·6 and v values of about 36. In addition to these many other types of glass are now made having special properties with respect to their dispersive powers and relative partial dispersions. A selection of varieties of optical glass is given in Table IV in the Appendix.

161. The Achromatic Lens

We have seen in Chapter VI that any thin lens will have chromatic aberration—light of different colours focusing in different places—and such a lens will not, therefore, produce a sharp image when light of different wave-lengths is coming from the object. It will be shown below that by combining two lenses made of different glasses it is possible to produce a combination that will focus light of two different colours at the same place; such a combination is said to be **achromatised** for these two colours. When the lens is to be used in conjunction with the eye, as for example the objective of a telescope or microscope, it will be achromatised for light corresponding to the C and F lines, but for photographic purposes it will usually be achromatised for light corresponding to the D and G′ lines, since the ordinary photographic plate is chiefly sensitive to the blue and violet end of the spectrum.

For a thin lens we have

$$F_D = (n_D - 1)(R_1 - R_2)$$
$$F_C = (n_C - 1)(R_1 - R_2)$$
$$F_F = (n_F - 1)(R_1 - R_2)$$

Therefore

$$F_F - F_C = (n_F - n_C)(R_1 - R_2) = \frac{n_F - n_C}{n_D - 1} \cdot F_D$$

or representing the mean power F_D by F,

$$F_F - F_C = \omega F = \frac{F}{V} \quad \ldots \qquad \ldots \qquad \ldots \quad (13.03)$$

In an achromatic combination it will be necessary that the difference in power for the two colours in the one lens shall be neutralised by that in the other and as the power of two thin lenses *in contact* is equal to the sum of the separate powers, the condition to be satisfied is that

$$\frac{F_1}{V_1} = -\frac{F_2}{V_2} \qquad \ldots \qquad \ldots \qquad \ldots \quad (13.04)$$

From (13.04) it is obvious that the two lenses must be made from glasses having different v values as otherwise F_1 will equal $-F_2$ and the combination will have no power. Also it will be seen that for a positive combination the glass from which the positive lens is made

Fig. 160.—Forms of achromatic lenses

must have the higher v value, and the positive lens will usually be of crown glass and the negative lens of flint.

Combining the expressions

$$F_1 + F_2 = F$$

and

$$\frac{F_1}{V_1} = -\frac{F_2}{V_2}$$

we have

$$F_1 = \frac{V_1}{V_1 - V_2} \cdot F \quad \ldots \qquad \ldots \quad (13.05\text{A})$$

and

$$F_2 = -\frac{V_2}{V_1 - V_2} \cdot F \quad \ldots \qquad \ldots \quad (13.05\text{B})$$

Having found the powers of the two lenses that when combined will give an achromatic combination, the curves can be chosen to correct or reduce other aberrations, especially the spherical aberration. It is usual, where possible, with lenses up to two or three inches diameter to arrange that the adjacent surfaces of the two lenses shall be of equal curvature, and the lenses are cemented together with Canada Balsam. Figs. 160a and 160b show two usual forms of small telescope objectives; b is frequently used as the objective of the prism binocular, and when made of suitably chosen glasses, will give correction for oblique aberration over a small angle.

Example:

An achromatic lens of + 3D power is to be made from the glasses of which the constants are:

						n_D	$n_D - n_C$	v
Crown	...	...	...	...		1·5175	0·0085	60·9
Flint	...	...	...	...		1·6225	0·0173	36·0

From (13.03)

for crown lens $\qquad F_F - F_C = \dfrac{F_1}{V_1} = \dfrac{F_1}{60\cdot9}$

for flint lens $\qquad F_F - F_C = \dfrac{F_2}{V_2} = \dfrac{F_2}{36}$

In order that the two lenses shall achromatise when in contact

$$\frac{F_1}{60\cdot9} = -\frac{F_2}{36}$$

or $\qquad\qquad F_2 = -\dfrac{36}{60\cdot9}\,F_1$

Substituting in $\qquad F_1 + F_2 = F$

we have $\qquad\qquad F_1 - \dfrac{36}{60\cdot9}\,F_1 = 3$

therefore $\qquad\qquad F_1 = +7\cdot34\ \text{D}$

$$F_2 = -4\cdot34\ \text{D}$$

The crown lens is to be equi-convex and cemented to the flint lens, then $R_1 = -R_2 = -R_3$

$$R_1 - R_2 = \frac{F_1}{n_{D1} - 1} = \frac{7\cdot34}{0\cdot5175} = +14\cdot18\ \text{D}$$

$$R_3 - R_4 = \frac{F_2}{n_{D2} - 1} = \frac{-4\cdot34}{0\cdot6225} = -6\cdot97\ \text{D}$$

Thence we have

$$R_1 = + 7{\cdot}09D \qquad r_1 = + \;\; 14{\cdot}1 \text{ cm.}$$
$$R_2 = - 7{\cdot}09D \qquad r_2 = - \;\; 14{\cdot}1 \text{ cm.}$$
$$R_2 = - 7{\cdot}09D \qquad r_2 = - \;\; 14{\cdot}1 \text{ cm.}$$
$$R_2 = - 0{\cdot}12D \qquad r_2 = - \;\; 833{\cdot}3 \text{ cm.}$$

Newton maintained that the dispersion of any medium was proportional to its refractivity, that is, that the dispersive powers of all media were equal and therefore the construction of an achromatic lens was impossible. For this reason he and his followers turned their attention to perfecting the reflecting telescope. Later it was argued, notably by Euler (1707–1783) in 1747, that the achromatic lens was a possibility since the eye was achromatic. That this is not the case can be very easily proved by looking at a bright object whilst half covering the pupil of the eye with a card, and it is now known that the eye has considerable chromatic aberration. In 1733 Chester Moor-Hall (1703–1771) had succeeded in producing two or three telescope objectives which gave images fairly free from colour defects, but it was not until 1757 that, through the work of John Dollond (1706–1761), the regular construction of these achromatic lenses was commenced.

162. Chromatic Difference of Magnification

Unless the achromatic combination is quite thin, as was assumed above, it is not possible to get both the principal foci and the principal points respectively coinciding for the different colours. In the case of the telescope objective it is necessary that the foci for different colours shall coincide, that is, the back focusing distances shall be equal. In eyepieces, however, it is necessary that the images in the different colours shall be the same size, and hence the equivalent focal lengths must be equal. If this be not so, images seen through the eyepiece will be fringed with colour, the defect increasing towards the edge of the field. When this second condition is fulfilled the system is said to be corrected for **chromatic difference of magnification** or **lateral chromatic aberration.**

The Huygens eyepiece is an example of a system where this condition is satisfied by the use of two lenses of the same glass, the blue and red rays from any point on the image emerging parallel, and the images in these two colours therefore subtending the same angle.

For two thin separated lenses in air we have

(1) $$F = F_1 + F - d\,F_1\,F_2$$

putting $$F_F - F_C = \delta F$$

(2) $$F + \delta F = (F_1 + \delta F_1) + (F_2 + \delta F_2) - d(F_1 + \delta F_1)(F_2 + \delta F_2)$$

257

Subtracting equation (1) from equation (2) and neglecting the term involving the product of the small quantities δF_1 and δF_2 we have

$$\delta F = \delta F_1 + \delta F_2 - (F_2\, \delta F_1 + F_1\, \delta F_2)d$$

In order that
$$\delta F = 0$$

$$d = \frac{\delta F_1 + \delta F_2}{F_2\, \delta F_1 + F_1\, \delta F_2}$$

$$\delta F_1 = \frac{F_1}{V_1}, \quad \delta F_2 = \frac{F_2}{V_2}$$

$$d = \frac{V_2 F_1 + V_1 F_2}{(V_1 + V_2)F_1 F_2} = \frac{V_1\,'_1 + V_2 f'_2}{V_1 + V_2}$$

If the two lenses are made of the same glass, $v_1 = v_2$ in order that there shall be no difference between F_C and F_F we have

$$d = \frac{f'_1 + f'_2}{2}$$

as the condition for achromatism of magnification.

163. The Achromatic Prism

For certain purposes it is sometimes required to use a combination of prisms to produce a deviation of the light without, as far as possible, any dispersion. Such an **achromatic prism** may be formed by combining

Fig. 161.—Achromatic prism

two prisms giving equal dispersions with the base of one against the apex of the other. Rays of light of the two colours, for which the prism is achromatised, will then emerge in parallel directions as shown in Fig. 161.

If we consider prisms of small angle for which the approximate expression (4.08) may be used, we have

$$d_D = (n_D - 1)a$$

and the dispersion
$$d_F - d_C = (n_F - n_C)a$$

In order that two prisms when combined shall achromatise, $d_F - d_C$ for the two prisms must be equal.

Therefore

$$(n_{F1} - n_{C1})a_1 = (n_{F2} - n_{C2})a_2 \quad \ldots \quad \ldots \quad (13.06)$$

If the two prisms be made of different glasses, the mean deviations d_D will be unequal when expression (13.06) is satisfied and their difference is the resultant mean deviation.

Example:

A prism of 15° refracting angle is made of the crown glass, particulars of which are given in § 160. What prism of the flint glass will be required to give achromatism for the C and F lines, and what will be the resulting mean deviation?

For the crown prism the dispersion

$$d_{F1} - d_{C1} = (n_{F1} - n_{C1})a_1 = 0 \cdot 0085 \times 15°$$

For the flint prism the dispersion

$$d_{F2} - d_{C2} = (n_{F2} - n_{C2})a_2 = 0 \cdot 0173\, a_2$$

and this must be made equal to $d_{F1} - d_{C1}$

Therefore $\qquad 0 \cdot 0173a_2 = 0 \cdot 0085 \times 15$
$$a_2 = 7 \cdot 37°$$
Then $\qquad d_{D1} = 0 \cdot 5175 \times 15 = 7 \cdot 77°$

and $\qquad d_{D2} = 0 \cdot 6225 \times 7 \cdot 37 = 4 \cdot 59°$

Therefore, resultant mean deviation, $d_D = 3 \cdot 18°$.

164. The Direct Vision Prism — Amici Prism

If the refracting angles of two prisms made of different glasses be such that the mean deviations they produce are equal, the two prisms when placed base to apex will produce no deviation on light corresponding to the D line, but the other colours will be dispersed. Thus, with the two prisms for which the deviations are given in § 160 combined in this way, the D line will be undeviated, while the C line is deviated 4′ to one side, and the F line 28′ to the other side of the D line. Such a combination forms a **direct vision prism,** and is frequently used in the spectroscope, particularly in the small pocket instrument.

To obtain sufficient dispersion the direct vision prism is usually composed of three or five prisms, the usual form being that shown in Fig. 162. This was first made by Amici (1786–1863) in 1860, and consists of a prism of very dense flint glass cemented between two prisms of crown glass. The glasses should be chosen to have the greatest possible difference in their v values. In the construction of some D.V. prisms liquids of high dispersive power, such as carbon disulphide and cinnamic ether, are used.

Fig. 162.—Amici direct vision prism

165. The Constant Deviation Prism

A very convenient form of spectroscope can be constructed by making use of a **constant deviation prism.** This may take the form shown in Fig. 163 and may be considered as being composed of two 30° prisms

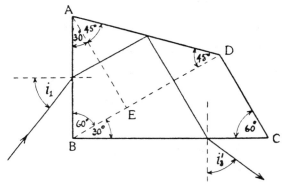

Fig. 163.—Constant deviation prism

ABE and BCD and a right-angled reflecting prism AED, although in practice it will be made in one piece. For a certain angle of incidence i_1 the ray refracted at AB is normal to AE and is totally reflected at the surface AD undergoing a deviation of 90°. As the angle ABC = 90° the angle of incidence at BC is equal to the angle of refraction at AB. Therefore $i'_3 = i_1$ and the emergent ray is at right angles to the incident ray. The prisms ABE and BCD may be considered as two halves of a 60° prism and when the angle of incidence i_1 at the surface AB is equal to the angle of refraction i'_3 at the surface BC the deviation is a minimum. Thus any ray that is deviated through 90° by the constant deviation prism is passing through at minimum deviation. The collimator and telescope of the spectroscope are permanently fixed at right angles to one another and the prism is rotated about a vertical axis. As the angle of incidence of the light from the collimator changes the spectrum moves across the field of the telescope and light of any wave-length is at minimum deviation when it falls in the centre of the telescope field. The prism can be rotated by means of a screw attached to which is a divided drum, which gives the wave-length for the particular portion of the spectrum falling on the cross-lines.

166. Irrationality of Dispersion. Secondary Spectrum

An achromatic lens composed of two different glasses can be designed to focus any *two* colours in the same place, but all other colours will focus in slightly different positions. The reason for this small residual colour defect, which is known as the **secondary spectrum,** will be seen if we examine the dispersions of various media between different parts of the spectrum. Thus, while the mean dispersion is higher in flint glass than in crown, the dispersion towards the red end of the spectrum is relatively higher in crown glass than in flint and relatively lower at the blue end as will be seen from the figures for the relative partial dispersions, Table III in the Appendix. These relative differences in the dispersions in different parts of the spectrum are still more marked if the dispersion of water is compared with that of flint glass.

Fig. 164a shows the relative lengths of the spectra formed at the same distance from prisms of equal refracting angle made of water, crown glass and flint glass. In Fig. 164b the spectra of the crown glass and water prisms have been made the same length as that of the flint glass prism between the B and H lines; this could be done either by suitably choosing the refracting angles, or by receiving the spectra at different distances. It will be seen that the other lines do not coincide. This effect is known as **irrationality of dispersion.**

The amount by which the power or focal length for any colour differs from the combined power or focal length for the C and F lines for which the lens has been achromatised may be found as follows:

For the D and F lines we have

$$F_D = (n_{D1} - 1)(R_1 - R_2) + (n_{D2} - 1)(R_3 - R_4)$$
$$F_F = (n_{F1} - 1)(R_1 - R_2) + (n_{F2} - 1)(R_3 - R_4)$$
$$F_F - F_D = (n_{F1} - n_{D1})(R_1 - R_2) + (n_{F2} - n_{D2})(R_3 - R_4)$$
$$= \frac{(n_{F1} - n_{D1})F_{D1}}{n_{D1} - 1} + \frac{(n_{F2} - n_{D2})F_{D2}}{n_{D2} - 1}$$

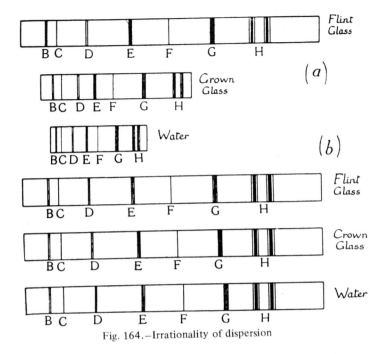

Fig. 164.—Irrationality of dispersion

From (13.05A and B)

$$F_F - F_D = \frac{n_{F1} - n_{D1}}{n_{D1} - 1} \cdot \frac{V_1}{V_1 - V_2} \cdot F_D - \frac{n_{F2} - n_{D2}}{n_{D2} - 1} \cdot \frac{V_2}{V_1 - V_2} \cdot F_D$$

$$\frac{F_F - F_D}{F_D} = \frac{n_{F1} - n_{D1}}{n_{D1} - 1} \cdot \frac{V_1}{V_1 - V_2} - \frac{n_{F2} - n_{D2}}{n_{D2} - 1} \cdot \frac{V_2}{V_1 - V_2}$$

$$= \frac{n_{F1} - n_{D1}}{n_{F1} - n_{C1}} \cdot \frac{1}{V_1 - V_2} - \frac{n_{F2} - n_{D2}}{n_{F2} - n_{C2}} \cdot \frac{1}{V_1 - V_2}$$

Calling the relative partial dispersions $\dfrac{n_D - n_C}{n_F - n_C}$, $\dfrac{n_F - n_D}{n_F - n_C}$

and $\dfrac{n_G' - n_F}{n_F - n_C}$ a, b and c respectively, we have

$$\frac{F_F - F_D}{F_D} = \frac{f'_D - f'_F}{f'_F} = \frac{b_1 - b_2}{V_1 - V_2}$$

Similar expressions may be obtained for the other spectrum lines. Thus it is seen to reduce the secondary spectrum glasses should be chosen having their respective relative partial dispersions as nearly equal as possible or having a large difference in their v values.

Lenses in which the secondary spectrum has been practically eliminated are called **apochromatic** and form the highest class of microscope and telescope objectives.

167. Types of Optical Glass

In designing a lens system to correct as far as possible the various aberrations considered in Chapter XVIII, advantage has to be taken of all the possible variables. These are the curvatures of the surfaces, the thicknesses and separations of the component lenses and the refractive indices and dispersive values of the various media. It is therefore obviously an advantage to have available a large variety of glasses with different characteristics. The elimination of the secondary spectrum is an example of the need for glasses with characteristics differing from those of ordinary crown and flint.

At the time of the introduction of the achromatic lens, the only optical glasses available were the ordinary crown and flint types. These are composed of silicates combined with soda and potash or both, and oxide of calcium in the case of crown glass and oxide of lead in the case of flint glass. During the first half of the nineteenth century systematic efforts to extend the range of available types of glass were made by Fraunhofer and Guinand in Germany and by Faraday, Harcourt and Stokes in England, but the great advance in this direction was made by Schott and Abbe, who in 1885 issued the first catalogue of the Jena glasses. This contained particulars of a number of glasses, the dispersive powers and partial dispersions of which differed considerably from those of earlier types of crown and flint. Other types of glass have since been produced and all these glasses and many others are now made by any manufacturer of optical glass.

The effect of the lead in flint glass is to increase the mean refractive index, the relative dispersion and the relative partial dispersion at the blue end. The substitution of barium for lead gives a much lower dispersion; thus we have in a dense barium crown a glass with a mean refractive index as high as 1·61 with a v value the same as that of hard crown. The introduction of boron into a glass lengthens the red end of the spectrum as compared with the blue, as will be seen from the figures in Table III for the boro-silicate crown. Other additions, such as phosphates and fluorine also produce important modifications in the optical characteristics of the glass, but a number of these glasses have been found to be unstable, and their manufacture has therefore been discontinued.

Fig. 165.–Glass types (Courtesy, Chance-Pilkington Ltd.)

CROWNS: BSC, Boro Silicate Crown; DBC, Dense Barium Crown; HC, Hard Crown; MBC, Medium Barium Crown; SBC, Special Barium Crown; ZC, Zinc Crown.

FLINTS: BF, Barium Flint; BOR.F, Borate Flint; EDF, Extra Dense Flint; DEDF, Double Extra Dense Flint; DF, Dense

Fig. 165 shows a diagram of various glass types plotted by refractive index and reciprocal dispersion (v-value). The double line somewhat arbitrarily divides the types into crowns and flints. Each area is commonly referred to by the descriptions indicated but for particular types six-figure reference numbers are also used from the first three significant figures of the term $n_d - 1$ and the first three significant figures of the v-value. Thus glass type BSC 510 644 has a refractive index of 1·50970 and a v-value of 64·44.

168. Other Optical Materials

As an alternative to glass, organic polymers (plastics) have advantages and disadvantages. Although they can be formed into lenses by moulding, which is cheaper than grinding and polishing, the accuracy of curve is not as good. Although having about half the weight of the equivalent glass lens, their resistance to scratching is usually lower. The homogeneity attainable with plastics is considerably poorer than that with optical glass and the choice of index and v-value is restricted to two locations for readily obtainable plastics:

Polycarbonate	1·58	29·5
Polystyrene	1·59	30·5
Polymethyl methacrylate	1·49	57·5 (Perspex)

Apart from spectacle lenses, optical plastics are used in cheap cameras and binoculars and also in some display systems where weight is important. The limitation to two main types is a very serious restriction on lens design and the extra elements required to give equal performance often offset the weight and cost advantage.

Most glasses have very high transmission between 450 nm and 1000 nm although some of the special types tend to absorb in the blue. For work outside these wave-lengths, in the ultra-violet and infra-red, other materials have been used — crystals of calcium fluoride, lithium fluoride and quartz are commonly used to enhance transmission in the ultra-violet. While for the infra-red oxide glasses such as calcium aluminate are used for wave-lengths up to 5 μm. In the so-called thermal region infra-red (8–13 μm) germanium and silicon are used as well as the polycrystalline materials in the Kodak 'Irtran' range.

169. Anomalous Dispersion

For most substances the refractive index increases as the wave-length decreases, the increase becoming more rapid with the shorter wave-lengths. In some substances, such as cyanine and fuchsine which are strongly coloured, the refractive index for some of the longer wave-lengths is greater than that for some of the shorter. If a prism is made by pressing some fused cyanine between two glasses inclined at a very small angle, the spectrum of white light seen through it will be found to have the colours in the order green, blue, red and orange, the green being the least deviated. The yellow is completely absorbed. This effect is known as **anomalous dispersion.** As cyanine absorbs a very great deal of the light, only a small amount of light will be transmitted, even at the edge of a prism of very small angle.

Substances giving this so-called anomalous dispersion have a strong absorption band in the visible spectrum. On the blue side of this absorption band there is an abnormal drop in the refractive index which gradually rises again towards the blue end of the spectrum. On the red side of the absorption band there is an abnormal increase of refractive index on approaching the band. Thus with cyanine, which gives an absorption band in the yellow, the refractive indices will have increasing values in the following orders of colours, green, blue, red and orange, as is seen in the spectrum with the cyanine prism. As with ordinary transparent substances such as glass there is a well marked absorption band in the ultra-violet and another in the infra-red, it would appear that the dispersion of such substances is of the same general form as that of substances having an absorption band in the visible spectrum. That is, the refractive index rapidly decreases for wave-lengths slightly less than those for the absorption band in the infra-red and rapidly increases for wave-lengths slightly greater than those for the absorption band in the ultra-violet. It is therefore supposed that there is no real anomaly in the so-called anomalous dispersion, and the general nature of dispersion has been explained by considering the interaction of the particles of the medium and the light vibrations passing through it.

170. The Ultra-Violet

Vision is dependent only on radiations having wave-lengths from about 760 nm to about 390 nm. Most sources of light, however, emit a much larger range of wave-lengths than this, comprising in addition to the visible spectrum, the shorter wave-length ultra-violet radiations at one end and the longer wave-length infra-red radiations at the other end of the visible spectrum.

Ultra-violet light may be detected by fluorescent materials and by photography. The latter was used by Becquerel (1842) to show the existence of Fraunhofer lines in the ultra-violet region of the solar spectrum. Fluorescence is still used today. Phosphors such as sodium salicylate re-radiate in the visible spectrum when excited by ultra-violet light and in some cases by x-rays.

Nowadays a large range of photo-electric detectors are available in the form of photo-multipliers or image tubes. As the energy of ultra-violet photons is higher than visible photons it is generally easier to eject electrons (photo-emission) from materials than in the visible region. Photo-detectors exist for wave-lengths down to 50 nm (500 Å). The shortest wave-length reaching the earth from the sun through a clear atmosphere is about 300 nm as wave-lengths shorter than this are absorbed by the air. For experiments at wave-lengths shorter than about 200 nm the apparatus must be evacuated leading to the term, **vacuum ultra-violet** which distinguishes these shorter wave-lengths from the **near ultra-violet**.

Ordinary glass is opaque to wave-lengths shorter than about 330 nm and materials such as quartz, sapphire and, for very short wave-lengths, lithium fluoride must be used.

Various sources of ultra-violet light are available although ordinary electric lamps emit ultra-violet radiations to some extent. For more intensity, particularly at the shorter wave-lengths, special sources are required. These radiations are chiefly emitted by sources in which the temperature is extremely high or in which the atoms are subject to intense electrical excitation. Xenon and mercury arcs are commonly used.

171. Biological Effects and Applications of Ultra-Violet

For a material to be affected in any way by radiations falling on it, these radiations must be absorbed by the material; radiations reflected or transmitted by the material will produce no effect.

Ultra-violet radiations, particularly those of shorter wave-length, are completely absorbed by living tissues, even in extremely thin layers, and photo-chemical or *abiotic* changes are produced. The extent of these changes increases as the wave-length decreases and is also dependent on the intensity of the radiation and the time of exposure.

The abiotic effects caused by the absorption of ultra-violet may be sufficient to destroy minute organisms, such as bacteria; and water, milk, etc., may be sterilised by exposing them to these radiations. In the case of large animals the effects will be confined to the surface owing to the rapid absorption of the ultra-violet by the tissues, and these effects may be beneficial or harmful according to the extent of change produced.

The health-giving properties of sunlight are well known, and the outward signs of sunburn and bronzing of the skin have been found to be caused almost entirely under natural conditions by the longer wave ultra-violet between 320 nm and 290 nm, the limit of the solar radiation reaching the earth. Extensive use is being made of the shorter wave the cadmium arcs, in the treatment of disease conditions, both general and local. In the latter the action is such as to cause slight damage to the tissues sufficient to set into being or to stimulate the processes of repair.

Much has been written regarding the action of ultra-violet on the eye. It can generally be accepted, that no deleterious effects are caused on the eye by exposure to the radiations present in sunlight to which the eye is adapted. In special circumstances, e.g., on snowfields particularly at high altitudes, radiations of shorter wave-length and of greater intensity than those to which the eye is normally accustomed may be present, and the wearing of protective glasses which absorb these radiations will be necessary. The shorter wave radiations produced by artificial sources, such as the arcs previously mentioned, are, however, readily absorbed by the eye tissues, and damage to these tissues, usually of a temporary nature, is produced.

Various chemical effects, for example the bleaching of dyes and the hardening of paints and varnishes, which are dependent on light action, have been found to be chiefly affected by the ultra-violet portion of the solar spectrum, and the commercial testing of these articles is considerably hastened by exposing them to a source rich in ultra-violet radiations. Another commercial application that has become of considerable importance during recent years is the exposure of certain foodstuffs to ultra-violet sources with the object of increasing their vitamin content.

Ultra-violet has been successfully applied to the microscope, the short wave-length in this case leading to considerable increase in resolving power. (See § 206).

172. The Infra-Red

The infra-red radiations are characterised by their heating effect, and their presence may therefore be detected by any sensitive heat measuring device. It is obvious that heat is radiated from any source of light, but the discovery that these heat radiations form a continuation of the visible spectrum was first made by Sir William Herschel (1738–1822) in 1800. Using a delicate thermometer with blackened bulb in order to find the heating effect of different colours of the spectrum, he found that while the temperature rose towards the red end of the spectrum, the maximum was reached in the region beyond the visible red end. In 1840 Sir John Herschel (1792–1871) succeeded in showing the existence of Fraunhofer lines in the infra-red region of the spectrum.

The detection method used was the drying of lampblack wet with alcohol. This measures the heating effect of the radiation and the **bolometer** uses the same effect to alter the resistance of platinum wire. If two wires of different materials are joined together a voltage difference can be generated by heating the junction and this forms the basis of the **thermopile.** When a gas is heated its volume or pressure must change and this is the principle of the **Golay cell**, which uses optical interference methods to

PLATE 1

FIGURE 1.—BI-PRISM FRINGES

FIGURE 2.—SHADOW OF NARROW WIRE

PLATE 2

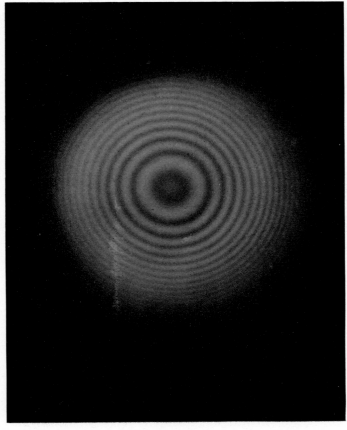

NEWTON'S RINGS
(The light centre to the ring system is due to the two glasses not being
quite in contact)

PLATE 3

SHADOW OF STRAIGHT EDGE

B

NEWTON'S RINGS

(§ 184)

Dark disc in centre indicates optical contact
Glass surfaces by Mr. D. Jones, Pilkington P-E
North Wales

A

NORMAL VISUAL SPECTRUM
Figures indicate nm wavelength
(§§ 119, 140, 156, 208)

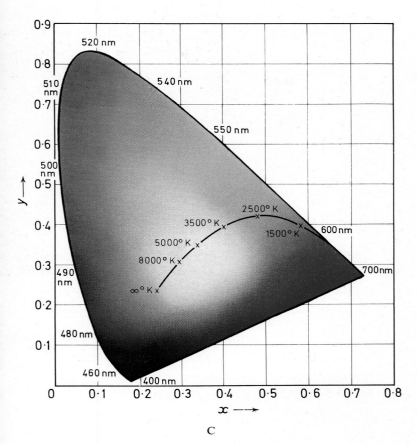

C

THE CIE CHROMATICITY DIAGRAM
(§§ 156, 175)

Based on the Thorn colour master published in
Lamps and Lighting (Edward Arnold) and
reproduced by courtesy of Thorn Lighting
Ltd., Research and Engineering Laboratories

In these pictures it is not possible with printing inks to reproduce exactly the colours of the optical
effects described. The particular illuminant used when viewing will also affect their appearance. The
intention is to indicate the major effects of the spectral range of light

PLATE 5

FIGURE 1.—LIGHT AT CENTRE OF CIRCULAR SHADOW

FIGURE 2.—AIRY DISC

PLATE 6

A

B

C

A. ONE SLIT
B. TWO SLITS
C. FOUR SLITS

FRAUNHOFER DIFFRACTION WITH ONE, TWO AND FOUR SLITS

PLATE 7

MULTIPLE-BEAM INTERFERENCE FRINGES
(By courtesy of Pilkington P–E Ltd)

achieve high sensitivity. All these detectors are known as **Thermal Detectors** as they depend on the heating effect of the absorbed radiation. They have a generally slow response of the order of milliseconds and a wide spectral range.

The transmission of the infra-red part of the electro-magnetic spectrum by the atmosphere shows considerable variation with wave-length. The region from visible red to 5 μm is generally transparent but a region three micrometres wide is heavily absorbed before reasonable transmission is again obtained between 8 μm and 14 μm. These transparent regions are known as windows and are commonly referred to as the near infra-red, and the far infra-red although the latter can be used for wave-lengths up to one millimetre.

Because the energy of individual photons decreases with increasing wave-length the ability to eject electrons (photo-emission) reduces and photo-emissive materials are difficult to obtain for wave-lengths below about 1·2 μm. However, semi-conductor materials do have the ability to absorb photons with a resultant change in conductivity (photo-conductivity) when their wave-length is as low as 20—30 μm. These **photo** — or **quantum detectors** have a much faster response (typically microseconds) but they also generate a lot of electrical noise due to random changes in the electron energies. In practice they are often cooled by liquid nitrogen or helium to reduce this noise.

Ordinary incandescent lamps radiate far more in the infra-red than in the visible regions. At normal temperatures every object has appreciable radiation in the thermal infra-red region. The human body, car exhausts and tyres can be easily detected as can any object at a different temperature (or contrast) with its surroundings.

173. Biological Effects and Applications of Infra-Red

Absorption of infra-red radiation produces a rise in temperature of the absorbing material; the radiant heat or infra-red from the sun and other heat sources is utilised in this way.

Excessive exposure of living tissue to infra-red radiations will cause damage to the tissue in the form of burns or *thermal lesions,* and this damage will often be more deep-seated than the abiotic changes produced by ultra-violet, because of the greater penetrability of the longer wave-length. Radiant heat is used in the treatment of disease conditions existing in positions too deeply seated to be reached with ultra-violet.

It has been found that serious and permanent damage may be caused to the eye by excessive exposure to the short-wave infra-red, such as may occur in certain industrial processes (glass and furnace work, welding, etc.)

and thy eyes of workers in these processes must be adequately protected from these radiations. Glass containing iron in the ferrous state very largely absorbs the short-wave infra-red.

On the other hand, body tumours have a slightly higher temperature than normal cells and so **Thermography** can be used, under controlled conditions, to detect their existence.

174. Colour

It is convenient to refer to light of any particular frequency or wave-length in terms of the colour sensation to which it gives rise when received by the normal eye, but it should be understood that colour is a purely visual sensation. Light, which we call white, consists of a continuous band of frequencies, as is seen from its spectrum, and this complex vibration gives rise to the sensation of white when received by the eye. Portions of this band of frequencies or a predominance of any portion of it give rise to the sensation of colour as distinct from white. Thus light of wave-lengths from about 760 nm to about 620 nm gives the sensation of red, from about 580 nm to about 510 nm the sensation of green, and so on.

That the type of sensation is not specifically determined by the wave-length of the radiation can be shown in a number of ways. Thus the sensation of yellow as produced by a radiation of wave-length 589 nm may also be evoked by presenting to the eye a suitable mixture of pure red and pure green, that is, by light containing no true yellow radiation. Also the sensation of white will result from the mixture in suitable proportions of certain pairs of pure colours known as complementaries.

Any colour can be specified in terms of three variables, its **hue**, its **saturation** or **purity** and its **luminosity**. The hue is the property which depends on the frequency of the light, the saturation depends on the amount of white present in addition to the light giving the hue, the less white light the more saturated the colour is said to be. Thus for the same hue we have the intense saturated yellow of the spectrum, the paler yellows when this hue is mixed with white and the various browns as the luminosity is reduced.

In the spectrum there are five very distinct hues, red, yellow, green, blue and violet. There is also purple, not found in the spectrum, but formed by a mixture of red and blue. The following divisions of the spectrum with the corresponding wave-lengths were given by Abney (1844—1920):

Red	...	...	...	End to 620 nm
Orange	...	...	...	620 nm ,, 592 nm
Yellow	...	...	...	592 nm ,, 578 nm
Green	...	...	...	578 nm ,, 513 nm

Blue-green	...	...	513 nm ,, 500 nm
Blue ...	...	...	500 nm ,, 464 nm
Ultramarine	...	...	464 nm ,, 446 nm
Violet ...	...	...	446 nm ,, End

175. Colour Mixing and Measurement*

When a mixture of light of two or more colours reaches the eye, new colour sensations are produced, and the eye is incapable of recognising the simple colours contained in the mixture. Light of different colours

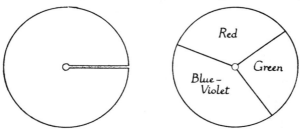

Fig. 166.—Maxwell's discs

may be mixed in a number of simple ways. Two or more projection lanterns with coloured filters may be arranged to project overlapping patches of coloured light on a white screen. A disc may be coloured with sectors of the colours to be mixed, and on quickly rotating the disc the colours blend and produce the sensation due to the mixture. An improvement on this method is the Maxwell's colour disc. Three discs coloured say red, green and blue-violet, and having a radial slot cut in each, as in Fig. 166, may be fitted together to form a single disc, with three sectors, the angles of which can be varied by sliding one disc over the others.

The results of adding colours may be stated in a general way as follows:

Red + Green + Blue-violet = White
Red + Green = Yellow
Green + Blue-violet = Green-blue (Peacock Blue)
Red + Blue-violet = Purple
Therefore
Yellow + Blue-violet = White
Green + Purple = White.

* For a full account of colour and colour measurement see *Visual Optics* Vol II by H. H. Emsley, Hatton Press, 1953, and *The Measurement of Colour* by W. D. Wright Hilger and Watts Ltd. 1958.

By altering the proportions of the colours one to another, all the hues are produced; thus starting with red and adding increasing quantities of green, the hue changes from red to orange-red, orange and yellow and then through the greenish-yellows to green. Adding increasing amounts of blue-violet to green the hue changes from green through the various hues of green-blue and blue to blue-violet. In the same way by adding the blue-violet to the red we get magenta, the various purples and violet.

It is generally more useful, however, to study the effects of mixing purer colours than can be used in the above methods. Three suitably chosen spectrum colours, usually red, green and blue, when mixed in certain proportions produce the *sensation* of white and any colour sensation can be matched in *hue* by varying proportions of the three primaries or **matching stimuli.** Although any colour can be matched in hue a complete match is not always possible because the colour produced by the mixture may be less saturated than the colour to be matched, the *test colour*. To obtain a complete match the test colour will have to be desaturated by the required amount by adding to it either a known amount of white or, as can be shown to produce the same effect, a certain amount of one of the matching stimuli.

Any colour can therefore be expressed in terms of the three matching stimuli, R, G, and B and this is the basis of **colorimetry,** a subject of considerable importance in industry. A number of instruments, known as **colorimeters,** have been devised for colour measurement and these differ mainly in the methods by which the matching stimuli are produced, mixed and varied in intensity. Coloured light, desaturated if necessary, from the specimen under test fills one-half of the observation field of the instrument and the mixture of the matching stimuli the other half.

In some instruments, such as the Wright Colorimeter, three narrow selected bands of the actual spectrum are mixed together in varying proportions, but in others the matching stimuli are obtained by the use of three colour filters giving highly saturated colours. This enables a simpler instrument, more adapted for industrial use to be produced. An example of an instrument using colour filters is the Donaldson Colorimeter (Fig. 167). The light source L is a 250 watt projection lamp; this is run at 90% its rated voltage, which gives a longer life and more uniform light output. The light is received on the plate A (Fig. 167a) in which are three rectangular apertures R, G and B, the dimensions of which can be independently varied by shutters sliding in grooves. Each shutter is operated by a separate rack and pinion and the extent of the opening is read on a scale on the shutter. Behind the apertures are mounted the three colour filters.

An image of the lamp filament in each of the three colours is formed by means of the lens C on the aperture D in the integrating sphere S. This sphere has a diameter of 6 inches and its interior surface is silver

Fig. 167.—Donaldson colorimeter

plated and coated with magnesium oxide. By repeated reflections in the sphere the light from the three filters is completely mixed and this mixed light from the sphere wall at I passes through a second aperture F in the sphere. The light from F after reflection by the right angle prism H is reflected by the photometer prism P to the eye at E. The eye sees one-half of the field evenly illuminated by light which is a mixture of the light passing through the three filters and the proportion of the three component colours is adjusted by means of the shutters in A. Light from the test surface T fills the other half of the field.

When it is necessary to desaturate the test colour light from the source is deflected by the prism K through a small aperture M fitted with a diffusing screen. A disc O carries three filters similar to the three main filters which can be brought in turn in line with the aperture and N is a circular neutral wedge which can be rotated over the aperture to regulate the amount of desaturating stimulus. The desaturating light is then reflected into the field of the test colour by the plane parallel glass plate Q.

When a given colour has been matched by the mixture of the three matching stimuli its colour can be expressed in the form of an equation, as was first suggested by Clerk-Maxwell in 1856.

Thus $c = r(R) + g(G) + b(B)$, where c is the colour and the numerical coefficients r, g and b represent the amounts of the red, green and blue stimuli required for the match. The units in which the matching stimuli are measured are arbitrarily chosen so that an equal number of each stimulus will match a standard white when added together. These units are called **Trichromatic Units** and are units of *colour* as distinct from units of *light*. In the **Unit Trichromatic Equation** the **Trichromatic Coefficients.** r, g and b will always add up to unity, thus for standard white we have

$$1W \equiv \tfrac{1}{3}(R) + \tfrac{1}{3}(G) + \tfrac{1}{3}(B).$$

For some colour, say a yellowish-green, the equation might be

$$C^* \equiv 0{\cdot}3(R) + 0{\cdot}6(G) + 0{\cdot}1(B)$$

or in the case of a colour, such as a pure spectrum green which must be desaturated before a match can be obtained, the equation might read

$$C \equiv 0{\cdot}480(R) + 0{\cdot}543(G) - 0{\cdot}023(B)$$
or $$C + 0{\cdot}023(B) \equiv 0{\cdot}480(R) + 0{\cdot}543(G)$$

that is when the given amount of blue is added to the colour a match can be obtained by the mixture of appropriate amounts of red and green.

The fact that most colour sensations could be produced by the mixture of three suitably chosen primaries led to the formulation of the **trichromatic theory** of colour vision. This theory, first put forward by Young in 1802 and later elaborated by Maxwell and Helmholtz, is usually known as the Young-Helmholtz theory, and supposes the existence in the retina of three types of receptors which when stimulated give rise to the sensations of red, green and blue respectively.

Plate 4C shows a colour reproduction of the colour triangle. The edge colours of this shape represent the spectral hues while the central 0·33, 0·33, 0·33 point is pure white. A straight line from this point to the edge represents the locus of saturations from 0 to 100% at that hue. It is seen that the values for which r = 1; or g = 1; or b = 1 are three different colours with saturations above 100%. They are therefore not real colours and cannot be duplicated by any dyes. Essentially they are theoretical points which allow the whole of the observable colour diagram to lie within the positive axes of the graph.

---

* The coefficient 1·0 for the test colour in unit trichromatic equations is usually omitted.

A plot of colour temperatures may also be done on this diagram and this is shown on Plate 4C.

Any two colours which when added together give the *sensation* of white are known as **complementary colours;** thus, as was seen above, yellow and blue-violet are complementaries, as are also green and purple. In many cases pairs of pure spectral hues, that is monochromatic light, having appropriate relative luminosities may be complementaries. Some of these complementary pairs of wave-lengths with their required relative luminosities as determined by Sinden (1923) are shown in the following table.

Complementaries (nm.)

λ_1	650	609	586	578·5	574	573	570·5
λ_2	496	493·5	487·5	480·5	472	466·5	443

Relative Luminosities

L_1	42·1	52·4	73·8	85·5	92·2	94·2	97·5
L_2	57·9	47·6	26·2	14·5	8·0	5·8	2·5

176. Coloured Objects

The colour of most objects is due to selective absorption. When white light falls on a coloured opaque substance, some of it passes into the substance, and certain frequencies are diffusely reflected close to the surface. A red object, for example, illuminated with white light, is reflecting some of the white light from the surface, and for a short distance inside the surface is diffusely reflecting the red and possibly also the orange and yellow, while the green, blue and violet are absorbed. The same object illuminated with green or blue light will appear almost black, as except for a small amount from the surface it is reflecting no light of this colour. Coloured transparent substances, coloured glasses, dyes, etc., owe their colour to a similar cause, a red glass transmitting only red light and absorbing light of other colours.

The colour of an object will obviously depend on the colour of the illuminating light, and in speaking of the colour of any object we are referring to its colour when illuminated with white light.

The colours obtained by reflection and transmission are usually far from pure, and examined with a spectroscope will be seen to contain a band of frequencies on either side of the predominant hue. There is also usually a considerable amount of white light reflected from the surface, and if this is diffusely reflected and mixes with the coloured light from below the surface, the colour will be less saturated than when the surface is polished, and the white light is reflected in a definite direction. This effect is seen in the difference in colour in polished and unpolished wood and marble.

The effects of mixing coloured substances, such as pigments, or of combining colour filters—dyes, coloured glasses, etc.—are quite different from those obtained in adding coloured lights. It is well known, for example, that the mixture of blue and yellow water-colours gives green, whereas we have seen that the addition of blue and yellow light gives the sensation of white. The explanation is simple: in the case of the pigments, the blue particles, when illuminated with white light, are absorbing the longer wave-lengths and reflecting the violet, blue and green; the particles of yellow pigment are absorbing the violet and blue and reflecting the green, yellow and red. A mixture of the two pigments will therefore absorb or subtract everything from the white light but the green which is reflected. If we place a red filter in one lantern and a green filter in another and project two overlapping patches on a white screen, the additive mixture produces the sensation of yellow, but if the two filters are placed together in one lantern, no light passes through the combination, as the light transmitted by one filter is absorbed by the other.

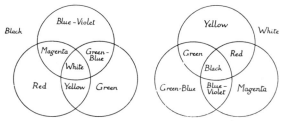

Fig. 168.—Additive and subtractive primaries

A great range of colours can be obtained by the mixture of pigments or dyes of three carefully chosen colours; these are known as the **subtractive primaries,** as distinct from the **additive primaries** dealt with earlier. The subtractive primaries are magenta, yellow and green-blue. These colours are in themselves mixtures, being composed of broad bands of the spectrum. They are such that, of white light, the magenta reflects all but green, the yellow all but blue and violet, and the green-blue all but red and yellow. The general results of additive and subtractive mixtures are shown diagrammatically in Fig. 168.

The subtractive primaries are the colours used in three-colour printing. Three photographs are taken, one through a red, one through a green and one through a blue filter. Prints are made from the three negatives in colours complementary to those of the filters, and the three prints are superposed. As the red filter transmits only red light the dark portions of a print from the negative taken through the red filter represent absence of red, and a print from this negative is therefore made in white minus red, i.e., green-blue or cyan. Similarly the print from the negative taken through the green filter must be made in magenta, and the print from the negative taken through the blue filter must be made in yellow. If the colours of the filters and inks are carefully adjusted the superposed prints can give excellent reproduction of the colours of the original object. Both the additive and subtractive systems of colour mixing have been used in colour photography but all the present day processes are subtractive.

177. Colour Due to the Scattering of Light

When light passes through a space filled with minute particles, some of it is scattered, chiefly due to diffraction. The percentage of light so scattered is found to be inversely proportional to the fourth power of its wave-length, so that the scattered light will consist mainly of the shorter wave lengths, the blue and the violet. This is known as the **Tyndall effect** after Tyndall (1820–1893) who investigated the effect experimentally.

The blue of the sky is due to this selective scattering. A certain amount of the shorter wave-lengths of the sun's light is scattered by the particles of dust and water and the molecules of gas in the earth's atmosphere, while the longer wave-lengths are transmitted. The increasing redness of the sun, as it gets nearer the horizon, is due to the same cause. As the light is then travelling through a great thickness of atmosphere more of the shorter wave-lengths are scattered, and the remaining transmitted light is therefore redder.

CHAPTER XIII   *EXERCISES*

1. Describe fully a method of obtaining a pure spectrum. What difference will be seen between the spectra of:

    (*a*) the sun, (*b*) the electric arc, (*c*) a tube containing incandescent hydrogen.

2. Explain clearly how you would use a spectrometer to determine the refractive index of the glass of a prism. Give the necessary theory and describe the experimental details.

3. What is meant by the dispersive power of glass? Describe carefully a practical method of measuring this value.

4. In measuring a prism with refracting angle of 60° 4′ on the spectrometer the following values were obtained for the minimum deviation for different colours:

Red	(C Line)	38° 23′
Yellow	(D Line)	38° 37′
Blue	(F Line)	39° 8′

Find the dispersive power of the glass. What kind of glass is this?

5. Manufacturers' lists of optical glasses give particulars of refractive index. dispersive power and partial dispersions. Briefly explain these quantities.

6. Explain clearly the terms:

Dispersion
Mean dispersion from C to F
Partial dispersions
Chromatic aberration.

7. Explain carefully how to determine the mean refractive index n_D and the dispersive power of a sample of glass in the form of a 60° prism. Describe the adjustments of the instrument used and show the method of calculation.

8. A 60° prism is made of flint glass having the following refractive indices:

n_C	n_D	n_F
1·615	1·620	1·632

Find the angle between the emergent light corresponding to the C and F lines when a beam of white light is incident on the prism in the direction of minimum deviation for the D line.

9. Explain the construction of an achromatic lens, showing exactly why it is necessary to use two different glasses. Why does a lens of this type not focus all colours in the same position?

10. Given that for a certain flint glass $\dfrac{n_D - 1}{n_F - n_C} = 36$ and $n_D = 1·6$, calculate the total curvature of a concave lens made from the glass in order when placed in contact with a convex lens having a difference of power for the F and C lines of 0·2167 D, the combination may be achromatic.

11. A positive lens is made of crown glass the refractive indices of which are as follows: red 1·5150, yellow 1·5175, blue 1·5235, the mean power of the lens being 5D; what will be the difference in power for the red and blue? What must be the mean power of a negative lens of flint glass, the refractive indices of which are red 1·6175, yellow 1·6225, blue 1·6348, in order that when combined with the positive lens the red and blue focus at the same point? What is such a combination called?

12. Given the following glasses, find the curves of an achromatic lens of 25 cm. focal length, the crown lens to be equi-convex and to be in exact contact with the flint.

Crown	$n_D = 1·517$	v = 60·5
Flint	$n_D = 1·612$	v = 37·0

13. An achromatic telescope objective is to be constructed of the following glasses:

	n_D	$n_F - n_C$	v
Crown	1·5188	0·0086	60·3
Flint	1·6214	0·0172	36·1

The radii of the crown lens are to one another as 2 to -3, while the flint curve next to the crown is at the same time represented by -2.815. Determine the radii of all four surfaces on the assumption that the lens is thin and is to have a focal length of $+25$ cm.

14. Explain why two different glasses must be used in the construction of an achromatic lens. A positive achromatic lens of 30 cm. focal length is to be constructed from the following glasses:

			n_D	$n_F - n_C$
Crown	...	...	1·519	0·0086
Flint	...	...	1·614	0·0166

The flint lens is to be cemented to the crown and its second surface is to be plane; find the radii of curvature of the surfaces.

15. Assuming that the optical system of the eye is roughly equivalent in its action to a spherical surface of water of radius 5·1 mm. and given the following refractive indices of water: for red light $= 1.332$, for blue-violet $= 1.340$, calculate the chromatic aberration as expressed by the interval between the corresponding foci. Assuming the eye-pupil to be 4 mm. in diameter, calculate the diameter of the least patch of confusion between these foci (where the retina may be assumed to be positioned) and comment on the result, remembering that the diameter of a foveal cone is about 0·0025 mm. (S.M.C. Hons.)

16. A thick lens is made of crown glass which has the following constants:
$$n_D = 1.52300, \quad n_F = 1.52924, \quad n_C = 1.52036$$
The lens is of meniscus form, its radii of curvature being $+55.0$ mm. and $+87.16$ mm., thickness 20 mm. For each of the three colours find:
    (a) Power and focal length of the lens.
    (b) Position of principal points.
    (c) Size of image of distant object subtending $10°$.

17. Show by diagrams how two prisms of crown and flint glass respectively may be combined to obtain:
    (a) dispersion without deviation, and
    (b) deviation without dispersion. (B.O.A.)

18. Explain the principles underlying the construction of (a) an achromatic prism, (b) a direct vision prism.

19. A thin prism with a refracting angle of $8°$ is made of crown glass having the following refractive indices: for red 1·527, for yellow 1·530 and for blue 1·536; a thin prism of flint glass is to be combined with the crown prism in order to neutralise the dispersion, the refractive indices of the flint glass being for red 1·630, for yellow 1·635 and for blue 1·648. What must be the angle of the flint prism and what will be the total mean deviation produced by the combination?

20. A direct vision prism is to be made of crown and flint glass having the following constants:

	n_C	n_D	n_F
Crown	1·527	1·530	1·536
Flint	1·630	1·635	1·648

The D ray is to be undeviated; what will be the resultant angle of dispersion between the C and F rays if the angle of the flint prism is 5°?

21. It is required to produce an achromatic prism combination to give a deviation of the mean ray of 1 in 70. The two component prisms are to be made of crown and flint respectively, which have the following constants:

$$\text{Crown } n_D = 1 \cdot 5173, \qquad v = 60 \cdot 2$$
$$\text{Flint } \quad n_D = 1 \cdot 6190, \qquad v = 36 \cdot 2$$

Find the angles of the component prisms.

22. Two thin prisms are to be combined to form a thin achromatic prism combination which is required to produce a deviation of the mean ray of 1 in 100. What are the angles of the component prisms? The first component prism is made of a crown and the second of a flint, the constants for these glasses being:

$$\text{Crown } n_F - n_C = 0 \cdot 0086, \qquad v = 60 \cdot 2$$
$$\text{Flint } \quad n_F - n_C = 0 \cdot 0171, \qquad v = 36 \cdot 2$$

23. An achromatic prism is to be constructed of glasses having the following values:

			n_C	n_D	n_F
Crown	...	...	1·527	1·530	1·536
Flint	...	...	1·630	1·635	1·648

The crown prism is to have an angle of 10°, find the angle of the flint prism and the deviation produced.

24. What is meant by the term **achromatism**?
Prove that a system of two lenses of the same material will be approximately achromatic if the lenses are separated by a distance equal to one-half of the sum of their focal lengths. State whether the system will be converging or diverging in character. (B.O.A.)

25. Two pieces of glass, one blue and the other yellow, are superposed. What colour is seen? If the two pieces are placed side by side in a lantern so as to illuminate a screen, will the effect be the same? Give reasons in each case. (S.M.C.)

26. Discuss the occurrence of the colour in the case of:
1. The blue sky.
2. A piece of blue glass.
3. A piece of blue paper. (B.O.A.)

27. How would you specify a given colour in order that it may be reproduced from a knowledge of the specification? (B.O.A.)

28. A constant deviation prism of the form shown in Fig. 163 is made of dense flint glass having the following refractive indices: red 1·630, yellow 1·635, blue 1·648. A narrow beam of white light is incident on the face AB in such a direction that the angle of refraction for the yellow at this surface is 30°. Find the angles between the red, yellow and blue light emerging from the surface BC.

29. Explain why it is possible with two pure spectrum colours suitably chosen (say a red and a green) to match daylight containing the whole range of spectrum colours. What does *match* mean here? Could such a light be used with which to paint a picture successfully? (B.O.A.)

30. The letters on a poster printed in red ink disappear altogether when looked

at through a red glass. So does the lettering on a poster printed in yellow ink when seen by gas light. Explain these results. (B.O.A.)

31. A better photographic rendering of objects, particularly of blue sky and clouds, is obtained when a yellow filter is used on the photographic objective; why is this?

32. Write a short essay on infra-red and ultra-violet radiation.

33.

	C	D	F	G
Crown	1·506	1·509	1·514	1·523
Flint	1·749	1·757	1·776	1·812

The table gives the refractive indices of two glasses for the stated Fraunhofer lines. Explain the meaning of the term irrational dispersion, using the quoted figures to illustrate your answer. Find the dispersive powers and the relative partial dispersions of the two glasses. (S.M.C.)

34. Explain each of the following:
    (a) Yellow and blue paints mixed together produce green paint, but yellow and blue lights projected on to a screen produce white light.
    (b) Smoke from the end of a burning cigarette appears blue whereas smoke from the mouth of the smoker appears grey.
    (c) Materials with bright saturated colours can be obtained in the red-orange-yellow range, but in the green-blue range the colours may be saturated but not bright. (S.M.C.)

35. Using appropriate diagrams, explain the following terms: (a) Dispersive power; (b) Secondary spectrum; (c) Anomalous dispersion. (S.M.C.)

36. The face AC of a prism ABC is silvered. A ray from an object is incident on the face AB and after two reflections, first at the silvered face AC and then at the face AB, it emerges from the base BC. Determine what relation must exist between the prism angles at A and B in order that there shall be no chromatism. (S.M.C.)

37. Describe the advantages and disadvantages of using plastics instead of glass in an optical system.

XIV - Interference and Optical Films

175. Introduction

The phenomenon of interference arises directly out of the wave theory of light. According to the principle of superposition of simple harmonic motions developed in Chapter XII interference should occur whenever two or more waves meet. That this is not so in practice is due mainly to the departure of ordinary light from a *continuous* wave-motion. To get steady interference effects with ordinary light sources we have to satisfy rather stringent conditions. On the other hand the light from a laser source approximates much more closely to a continuous wave motion and workers with these light sources will testify that interference phenomena occur so easily that they can become a nuisance.

If, however, the correct effects can be isolated they form a very precise method of measurement being able to assess accurately changes in dimension which are less than one thousandth of the wave-length of the light used. For green light this is 5 Angstrom Units or 2×10^{-8} inches. This branch of optics is now known as **Optical Metrology** and an outline of this technology will be given in this chapter.

On the other hand if optical components can be made with thicknesses less than the wave-length of light, the interference phenomena obtained can be used to modify reflection coefficients and obtain colour filters, very high transmission elements and polarising effects. This branch of optics is known as **Optical Thin Films** or **Optical Coatings** and this chapter will conclude with a review of the theory and applications of this subject.

176. Conditions for the Interference of Light

Two sets of waves of equal amplitude and frequency spreading out from point sources will neutralise the effects of one another in certain places and will give increased disturbance in others. This can be shown visually by generating two sets of ripples on the surface of mercury. If two small points are attached to a tuning fork and arranged so that they just dip into the mercury, two axactly similar sets of ripples will be created when the tuning fork is set into vibration. Each ripple system will consist of a series of concentric circles. The appearance will be as in Fig. 169

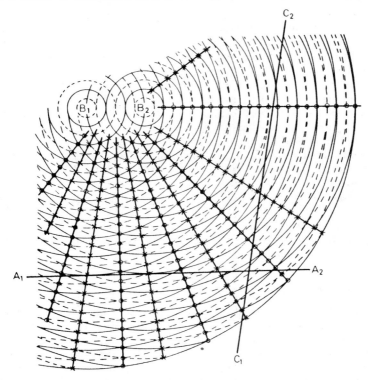

Fig. 169.—Interference between two sets of waves

where B_1 and B_2 are the sources and the crests of the waves are represented by solid lines and the troughs by broken lines.

Along the lines marked with crosses the crest of one wave coincides with the trough of another and since the amplitudes are equal the displacements will be neutralised in accordance with the principle of super-position. Along the lines marked with circles the crests coincide and the troughs coincide so that the resultant disturbance has twice the amplitude. As the energy of the wave-motion is proportional to the square of the amplitude there will be four times the energy of a single source along these lines and thus there is no loss of total energy from the two sources due to interference, but only a redistribution.

If a barrier were erected at A_1-A_2, it would receive no waves where the crossed lines meet it and double amplitude waves at the places where the circled lines meet it. In moving along the barrier from A_1 to A_2 one receives doubled and zero effects in a cyclic fashion. Similarly a barrier along C_1-C_2 would show the same effect but with the different regions of energy more spread out. Their location is crucially dependent on the distance between B_1 and B_2 *measured in terms of the wave-length of the ripples.*

As has been indicated earlier, two ordinary sources of light placed in positions such as B_1 and B_2 do not produce visible interference effects. If they did we should expect to find on a screen at A_1-A_2 bright patches where the crests coincided and darkness, because of zero amplitude, in the regions where a crest always coincides with a trough. These patches would stay much the same on the screen above and below the plane of the page and would be seen as alternately bright and dark **fringes**. If either of the two sources varies in phase with respect to the other sources these lines would move. With any two ordinary sources of light, the relative phase between them is constantly changing in a very fast random manner. The fringes formed are therefore blurred out to a uniform illumination. When the source has a finite size there is an equally random relationship in the phase between one part and another.

177. Coherence, Incoherence and Mutual Coherence

In § 144 and following the concept of coherent sources was developed and the various forms of laser described. However, interference phenomena do not require such complete coherence and indeed fringe patterns due to interference were being studied long before the laser was invented. This was achieved with largely incoherent sources because the presence of stationary fringes on screen A_1-A_2 of Fig. 169 depends only on obtaining a constant phase difference between sources. In other words it does not matter if the phase of B_1 changes as long as that of B_2 changes in an identical manner. In terms of superposition of waves at A_1 the wave-motion from B_1 and B_2 can vary from moment to moment provided that a crest from B_1 always coincides with a trough from B_2 and vice versa. Such a relationship between B_1 and B_2 is termed **mutual coherence**.

In all aspects of coherence it should be remembered that a continuous single-frequency simple harmonic motion cannot be perfectly achieved, nor can complete identity between two sources. It is proper therefore to speak of **degrees of coherence** which are less than 100%. In general terms the visibility of the fringe patterns described will vary with the degree of coherence obtained. However, some fading of the pattern can usually be tolerated.

In mathematical terms it may be stated that for two coherent sources the intensity on the screen is found by adding the respective amplitudes and squaring the result; whereas for incoherent sources the final intensity is found by adding the respective intensities. The latter method can never give darkness as the intensities are always positive being the square of the amplitudes shown in § 137.

178. Interference with Incoherent Sources

The methods of interference with incoherent sources are aimed at obtaining a sufficient degree of mutual coherence between two sources. This is done by making the two sources both dependent on the same source. For instance the reflection of light in a mirror provides an additional source apparently behind the mirror surface. It is possible to obtain interference between an incoherent source and its reflection provided there is a region where the two sets of waves overlap and provided the distance between the source and its reflection is not greater than a few wave-lengths. These conditions are easier to achieve if two reflected sources are used and Newton's rings (§ 184) is a particular case which constitutes the earliest study of interference phenomena.

The next example was devised by Young in the early nineteenth century. He illuminated a small aperture so that it became effectively a single small source. A short distance from this he placed a screen containing two small apertures which were illuminated only by the first aperture. These two apertures now constitute mutually coherent sources and provided that the illuminateing light is reasonably monochromatic, interference fringes may be observed on a screen in front of the two apertures as $A_1 - A_2$ in Fig. 169. The fringes are very faint as nearly all the original light has been obscured. Better results are obtained if narrow parallel slits are used in place of the three apertures.

Much better use of the available light was made by **Fresnel** in his **double mirrors, and bi-prism** methods. These are most easily explained in Fig. 170. The single source B is in each case converted to two sources B_1 and B_2, by reflection or refraction at AC and CD. So that the apparent sources are close together the angle ACD of the bi-prism and between the mirrors must be in the region of 179°. The interference takes place in the regions S where the beams overlap.

Considerably brighter finges may be obtained if B is an illuminated narrow slit parallel to the join of the mirrors or refracting edge of the prism. This gives clear parallel fringes in the region S with a regular spacing which may be measured experimentally using a travelling microscope. Fringes may be found anywhere in the shaded region S and are said to be non-localised.

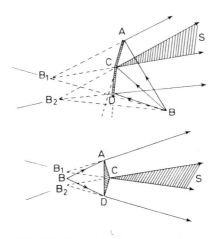

Fig. 170.—Fresnel's mirrors and bi-prism

Lloyd's Single Mirror method is different from the above as it uses only one reflected source. In Fig. 171 the illuminated slit B is placed parallel and close to the surface of a plane mirror (or black glass) and the reflected source B_1 interferes with the original light waves in the shaded reegion S.

179. Interference with Coherent Sources

The mutual coherence of the interfering sources described in the last section arises from the dependence of each source on a single original source. Although the precise phase and structure of the wave front from the real source may be changing many times a second, the dependent sources are changing also. When we have a phase difference between the two sources it means that the precise parts of the waves which are interfering left the original source at different times. But the original source is changing many times a second and so if the time difference becomes too large we find that the two waves are no longer mutually coherent as the source changed between the times of their emission. The allowed time difference is called the **coherence time** of the source and when multiplied by the velocity of light gives a **coherence length** which is a more useful parameter.

It is found that this is related to how monochromatic the source is as if the source contains wave-lengths λ_1 to λ_2 then as k of equation 14.01 (§ 180) increases: x can equal $k\lambda_1$ and $(k + \frac{1}{2})\lambda_2$ and the fringe pattern fades away, with λ_2 filling in the dark bands of the λ_1 pattern. With an incandescent source followed by a filter the coherence length may only be a few micrometres. With a spectral line source such as a mercury lamp a coherence length of 20 to 30 cm. may be achieved. With continuous wave lasers source coherence over many metres may be readily obtained and special stabilising methods can extend this to many kilometres.

Fig. 171.—Lloyds mirror

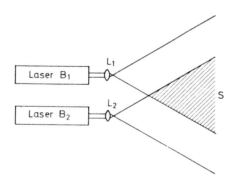

Fig. 172.—Interference with lasers — no
fringes are seen

The above discussion takes no account of the finite size of sources which effectively reduces the coherence length. With lasers the length of the resonant cavity has an effect on their coherence and fringes with highest visibility are generally obtained when the path difference is a multiple of the cavity length.

In Fig. 172, two similar lasers are arranged as interference sources B_1

and B_2. Their normally collimated output is caused to diverge by lenses L_1 and L_2, so that an interference pattern is formed at S on the screen With ordinary laboratory lasers no fringes will be seen. This is because the mutual coherence between the two lasers is limited by the random phase changes which limit their individual coherence length to a few kilometres at best. Although this is a considerable distance, the velocity of light is such that the **coherence time** is only a few microseconds. Thus the interference pattern is only stationary over this period of time and moves in a random manner. It is therefore not possible to see fringes in this way although special methods have shown they exist.

However, most laser systems use single lasers and the beam-splitting methods of § 173 to achieve measurements with long path differences or a multiplicity of interfering sources to give very fine fringes for high accuracy. These will be discussed in § 181 to § 185.

180. The Form of Interference Fringes

Referring again to Fig. 169 the presence of a bright or dark fringe at a given point on $A_1 - A_2$ depends on whether the waves arrive in or out of phase. If in Fig. 173 B_1 and B_2 represent two mutually coherent line sources emitting waves of equal amplitude, frequency and phase then the waves will arive in phase at point S on the screen as the distances $B_1 S$ and $B_2 S$ are equal if CS perpendicularly bisects $B_1 B_2$. S is therefore a bright fringe.

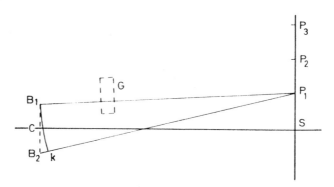

Fig. 173.—Width of interference fringes

For a point P_1 to one side of S the paths $B_1 P_1$ and $B_2 P_1$ will be unequal and the waves will arrive out of phase. If the phase difference is equivalent to one-half wave-length a dark fringe will result as crests and troughs are arriving together. A bright fringe will result at P_2 if the phase differences by then is equivalent to one whole wave-length. At P_3 the difference may have increased to one and a half wave-lengths and so a dark fringe results. Therefore on either side of the bright fringe at S there will be a series of bright and dark fringes as in Plate 1.

The bright fringes occur when the path difference is $k\lambda$ and the dark fringes for $(k + \frac{1}{2})\lambda$, where k is any integer.

The path difference for P_1 is found by constructing a circle centre P_1 through B_1. Then as $B_1 P_1$ and KP_1 are equal, the inequality is due to $B_2 K$. Normally $B_1 B_2$ is small compared with the distance to the screen if there is to be any apprciable separation of the fringes. Thus, $B_1 K$ approximates to a straight line perpendicular to $P_1 C$ and $B_2 K$. Representing $B_1 B_2$ by b, SP_1 by x and CS, which is approximately equal to CP_1, by d, we have by similar triangles:

for the centre of a dark fringe $\quad x = \dfrac{d}{b}(k + \frac{1}{2})\lambda$ $\quad \ldots \quad \ldots \quad \ldots (14.01)$

for the centre of a bright fringe $\quad x = \dfrac{d}{b}(k)\lambda$ $\quad \ldots \quad \ldots \quad (14.01\text{A})$

and the distances between centres of consecutive bright or dark bands will be

$$y = \frac{d}{b}\lambda \quad \ldots \quad \ldots \quad \ldots \quad \ldots (14.02)$$

It is thus seen that the spacing depends on the wave-length of the light used. If a source of more than one wave-length is used a number of interference bands will be produced which will mix in and out of step because of their different fringe spacings. The central band at S will remain white as all wave-lengths have zero phase difference.

If in Fig. 173, a block of glass is inserted into one beam only, a displacement of the fringe pattern will result. It is assumed that the frequency of any light vibration remains constant whatever medium it is travelling in. As the velocity changes, by the definition of refractive index, (see § 15) the wave-length inside a medium of index n will be $\lambda = \dfrac{\lambda_0}{n}$ where λ_0 is the wave-length *in vacuo* or in air. The calculation of phase difference due to a path difference in glass must take account of the different wave-length. Thus equation (12.06) becomes

$$\text{Phase difference } \alpha = \frac{2\pi}{\lambda} \times \text{Path difference}$$

$$= \frac{2\pi n}{\lambda_0} \times \text{Path difference} \ldots \quad \ldots (14.03)$$

Referring again to Fig. 173 the effect of a block of glass of index n and thickness (t) is to remove a path difference due to air and insert one due to glass. The resulting change in the phase difference is

$$\frac{2\pi n t}{\lambda} - \frac{2\pi t}{\lambda} = (n-1)\frac{2\pi t}{\lambda}$$

If the displacement of the central fringe due to the insertion of the glass is m fringes this represents a phase change of $2\pi m$ and so:

$$(n-1)t = m\lambda \dots \qquad \dots \qquad \dots (14.04)$$

The product nt is referred to as the **optical thickness*** of the glass block.

In the case of Lloyd's mirror (§ 178) it is found by experiment that the fringes produced are not as predicted by equations 14.01 and 14.01 A.

The fringe pattern is displaced by one-half fringe and this is due to the reflected source B_1 being π out of phase with the original source. This occurs due to a phase change of π occurring on reflection at the mirror and is considered further in § 183.

From expressions 14.01 and 14.02 the wave-length of the light can be determined by measuring the distances x or y, b and d in any of the methods described in § 178. Conversely, if the wave-length is known variations in b and d will show in the spacing of the fringes.

The interference fringes will not be sharply defined, because the resultant amplitude at different points on the screen gradually decreases as the waves get more and more out of step until a minimum is reached and then gradually increases to a maximum as the waves again come into step. The variation in intensity at different points on the screen may be found by applying the graphical method of § 136. If a_1 and a_2 (Fig. 174) are the amplitudes of the wave motion from b_1 and b_2 respectively arriving at a general point P with a phase difference of α, the resultant amplitude A at the screen is the vector addition of a_1 and a_2.

Analytically: $A \sin\left(\frac{2\pi t}{T} + \beta\right) = a_1 \sin\frac{2\pi t}{T} + a_2 \sin\left(\frac{2\pi t}{T} + \alpha\right)$ ... 14.05

where β is the phase of the resultant wave motion.

By vectors (Fig. 174): $A^2 = 2a^2 (1 + \cos\alpha) = 4a^2 \cos^2\frac{\alpha}{2}$ ... 14.05A

---

*Optical Thickness, nt relates to the difference in phase caused by an optical materials. (§ 15)

Apparent thickness, $\frac{t}{n}$ relates to imaging through an optical material (§ 33)

or: $A^2 = 4a^2 \cos^2\left(\dfrac{2\pi \text{ path difference}}{\lambda}\right)$ 14.05B

for the case where $a_1 = a_2 = a$.

Fig. 174.—Resultant amplitude

Fig. 175.—Intensity profile of two-beam fringes

It should be clear from 14.05 and Fig. 174 that A will be a maximum when the phase difference, α, is 0, 2π, 4π etc. and a minimum when α is π, 3π, 5π etc. The intensity of the fringes varies asas A^2 and so is proportional to the cosine squared of the path difference. These fringes are usually produced when two sources are used and are known as *cosine squared* fringes and have an intensity profile as in Fig. 175.

181. Multiple-Beam Interference

The cosine squared fringes of Plate 1 are the result of two interfering beams. When measuring their spacing it is difficult to locate accurately the centre of the fringe whether it is bright or dark. Using photo-electric methods it is possible to locate fringes to 0·01 times their spacing but better accuracies can be obtained if many more interfering beams are used. Although multiple-beam methods were employed in 1897 for accurate wave-length measurement the advent of the laser has brought them into common use.

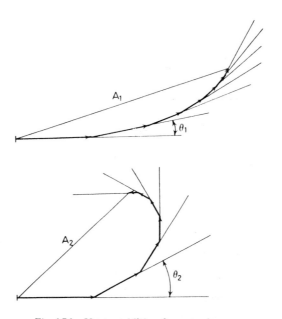

Fig. 176.—Vector addition for seven beams

In these methods the effective sources produce waves which are out of phase by multiples of the phase difference between the first two beams. This means that equation (14.05) becomes:

$$A \sin\left(\frac{2\pi t}{T} + \beta\right) = a_1 \sin\left(\frac{2\pi t}{T}\right) + a_2 \sin\left(\frac{2\pi t}{T} + \alpha\right) + a_3 \sin\left(\frac{2\pi t}{T} + 2\alpha\right)$$

$$+ a_4 \sin\left(\frac{2\pi t}{T} + 3\alpha\right) + a_5 \sin\left(\frac{2\pi t}{T} + 4\alpha\right) \quad \ldots\ldots \text{etc.}$$

The amplitudes a_1, a_2, a_3 etc. are generally reducing and the graphical calculation of the resultant appears as Fig. 176.

The analytical result gives the intensity of the fringes as proportional to:

$$\frac{F \sin^2 \dfrac{\alpha}{2}}{1 + F \sin^2 \dfrac{\alpha}{2}}$$

where α is the phase difference and F is a factor depending on the number of interfering beams and their amplitudes. Fig. 177 shows the profile of the fringes for different values of F.

If the interfering beams have the same amplitude then the fringe profile is still sharpened up but other fringes of lower intensity appear between them. (See § 209 Chap. XIII). The most common method of producing

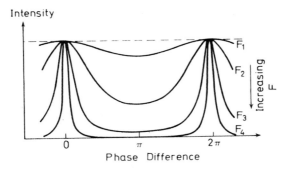

Fig. 177.—Intensity profile of multiple-beam fringes

multiple-beam fringes uses partial reflection (§ 183 ff) and this gives the fringe profile of Fig. 177 in the transmitted beam. The reflected beam normally shows the inverse of this, that is, sharp dark fringes on a bright background. A photograph of this type is shown in Plate 7.

182. Interference by Partial Reflection

In Fig. 169 of §176 the interference effects of two sources were shown as generating bright and dark regions on screens at $A_1 - A_2$ and $C_1 - C_2$. The methods of interference considered so far have formed apparent sources side-by-side to the screen as indicated by $A_1 - A_2$ of Fig. 169.

However, interference effects can also occur when the sources are effectively one behind the other as for screen $C_1 - C_2$. Apparent sources displaced along the line of viewing are readily generated by partial reflection*.

The best known phenomenon due to the interference of light is the brilliant colouring produced when white light is reflected from a thin film of a transparent substance. Examples of this are seen in the soap bubble, a patch of oil on a wet road, and a very thin film of air between two pieces of glass. Unlike the effects of interference previously described, which require the use of a small source, the interference colours of a thin film are best seen when the film is illuminated by an extended source such as the sky.

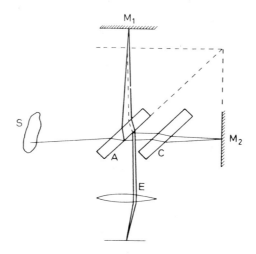

Fig. 178.—Michelson's interferometer

This action may be set up in the laboratory without the two surfaces being in close proximity. The Michelson interferometer is shown in Fig. 178. This uses the partially reflecting element A to divide the light from source S into two mutually coherent beams which are reflected by M_1 and M_2.

The partially reflecting element is often called a beam-splitter from this first action but it is also able to recombine the beams reflected by M_1 and

*The distinction "Division of Wavefront" for the methods of § 178 and "Division of Amplitude" for those of § 183 and following is commonly used.

M_2. Although some light is lost back to the source the appearance to the eye at E is that the two reflecting surfaces appear to be very close together or even superimposed at M_1 and M_2.

The light to M_1 passes twice through the glass support for the partial reflecting surface. Chromatic errors due to this may be cancelled out by inserting an identical block of glass to compensate at C. When the positions of M_1 and M_2 are such that the light paths are identical the two effective sources superimpose. With a small angle between the mirrors interference fringes can be seen even with white light. The different wavelengths produce fringe patterns with spacing proportional to λ (14.02) but with a central fringe coincident for all colours where the phase difference is zero for all colours. On either side a few highly coloured fringes occurs which rapidly merge to an even illumination as the fringe patterns of the different wave-lengths overlap. This action with white light can be very useful in positively identifying the point of effective coincidence of the mirrors and was used by Michelson to relate the wave-length of various monochromatic light sources to the unit of length, the standard metre.

A considerable number of interferometers are based on partial reflection techniques and some are described in § 187 and § 188.

183. Thin Films

In Fig. 179 light from an extended monochromatic source is incident upon a thin transparent film. A proportion of this light is reflected at each surface and the eye superimposes those coming from the same point on the film.

The actual ray bundles to the eye may be separated slightly before the action of the film and eye superimpose them on the retina but when the film is very thin this separation is very small and the bundles are largely mutually coherent. As they interfere the eye sees coloured fringes apparently localised in the plane of the film. The phase difference causing the fringes comes from the path difference between the light reflected from the top surface of the film and that which enters the film and is reflected at the lower surface. The colours arise as different wave-lengths generate different phase differences from the same path difference. Thus one area of the film may generate a bright fringe for red-orange light but a dark fringe for blue violet.

If a soap film is caught across a wire it may be supported vertically in a glass container to protect it from draughts. The film gradually becomes wedge-shaped and if illuminated with white light the reflection from it is seen to be crossed by broad horizontal bands of colour. After a time the upper region will get very tine and a perfectly black band will be formed at the upper edge. The same effect can be seen when two carefully cleaned glass plates are pressed into contact; where the air film between them is very thin compared with the wave-length of light a black patch will appear and the plates are said to be in **optical contact** at that place.

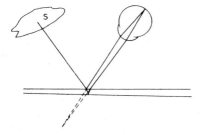

Fig. 179.–Interference with thin film

This dark region shows there is a phase difference between the two reflected waves for all colours. This cannot be due to the path difference which is very small and would produce various phase difference depending on wave-length. From electromagnetic theory it can be shown that a phase change of π occurs when a wave is reflected at a surface between a rarer and a denser medium. If the reflection is from a boundary between a denser and a rarer medium there is no change of phase. The two beams being reflected from the glass plates in optical contact come into each of these categories.

If crown glass is used for one plate and flint glass for the other the intervening space may be filled with a liquid of refractive index between the two glass types such as oil of sassafras. Under similar conditions of optical contact a bright patch is now seen as the light is being reflected from the same sort of boundary at each surface.

The path difference due to one beam traversing the film before reflection while the other is reflected from the first surface is required to be known so that the phase difference for a particular wave-length may be calculated. In Fig. 180 the difference in path length for the two beams which are subsequently superimposed by the eye is found by constructing DN perpendicular to the incident beams. There is then no path difference before that line and none after the point E where they combine. The difference then appears to be between DGE and NE. However, the path NE does not lie in a medium of the same refractive index as DGE. The difference can most easily be accounted for by constructing PE perpendicular to DG. Then as was shown in Chap. II the time taken to travel along NE is the same as for DP.

Thus the waves are still in phase at P and E. The distance PGE can then be shown, for a parallel film of thickness t and index n to be given by $2t \cos i'$ using the construction to H. The phase difference due to this is given

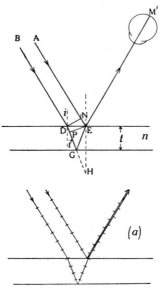

Fig. 180.–Reflection from a thin film.
Conditions for a light or dark band

by (12.06) to be $\dfrac{4\pi n t \cos i'}{\lambda}$ Remembering the phase difference due to reflection it is seen that for films bound by similar media a *dark fringe* will occur when:

$$\frac{4\pi n t \cos i'}{\lambda} = 2\pi k$$

$$\text{or } 2nt \cos i' = k\lambda \quad \dots \quad \dots \quad \dots \quad \dots \quad \dots \quad 14.07$$

and a *bright fringe* when

$$\frac{4\pi n t \cos i'}{\lambda} = 2\pi (k + \tfrac{1}{2})$$

$$\text{or } 2nt \cos i' = (k + \tfrac{1}{2})\lambda \quad \dots \quad \dots \quad \dots \quad \dots 14.07A$$

297

The dark fringes so calculated will not be completely dark unless the reflection from the two surfaces produce interfering beams of equal amplitude. The fraction of incident light reflected at normal incidence at the boundary between media of index n_1 and n_2 is given by Fresnel's expression $\left(\dfrac{n_1 - n_2}{n_1 + n_2}\right)^2$.

For a thin soap film in air or a thin film of air bounded by glass of identical type the intensity of the lower beam cannot equal that of the upper in Fig. 180 as this suffers two refractions as well as a reflection. However, there are other beams due to further reflections which provide interfering beams at larger phase differences. These give a multiple beam effect and provide for zero intensity in the dark bands. If the reflectivity of the two surfaces is increased a larger number of interfering beams occur giving the sharp dark fringes shown in Fig. 177.

184. Newton's Rings and Fizeau Fringes

Newton examined the effects produced on reflection from a thin film of air, by placing a convex lens of shallow spherical curve in contact with a glass plate (Fig. 181). The enclosed air film will then be extremely thin at the point of contact and will gradually increase in thickness with increased distance from this point, lines of equal thickness being circles with the point of contact as centre. Illuminated with monochromatic light, the film, as seen by reflection, will therefore have a black spot at the point of contact, surrounded by concentric alternate light and dark rings. These are known as **Newton's rings** and a photograph of these shown in Plate 2. If the film is illuminated with white light the rings will of course be coloured with colours which blur together 7 or 8 rings from the centre.

Up to that ring the colours appear in the following order from the centre outwards.

1st	order	Black, blue, white, yellow, red
2nd	"	Violet, blue, green, yellow, red
3rd	"	Purple, blue, green, yellow, red
4th	"	Green, red
5th	"	Greenish-blue, red
6th	"	Greenish-blue, pale red
7th	"	Greenish-blue, reddish-white.

This is known as **Newton's scale of colours** and is shown in the Colour Plate 4B.

From expression (14.07) the thickness t of the air film for any dark ring is

$$t = \frac{k\lambda}{2 \cos i'}$$

Fig. 181.—Newton's rings

FIG. 181.—NEWTON'S RINGS.

where k is the number of the ring (k = 0 for the central dark patch). If r is the radius of curvature of the lens surface and y the radius of the kth dark ring, then as t is small

$$r = \frac{y^2}{2t} \qquad \ldots \qquad \ldots \qquad \ldots \qquad \text{from (5.01)}$$

hence $$r = \frac{y^2}{k\lambda} \qquad \ldots \qquad \ldots \qquad \ldots \qquad \ldots \text{(14.08)}$$

for normally incident light when $\cos i' = 1$.

If the diameters of different rings are measured using a travelling microscope then it is possible to assess the curvature of the lens. This may be in error if the lens and plate are not in contact so a difference method is preferred using the expression

$$r = \frac{y_1{}^2 - y_2{}^2}{(k_1 - k_2)}$$

where y_1 and y_2 are the half-diameters of the $k_1{}^{th}$ and $k_2{}^{th}$ ring respectively.

If, instead of a lens, we use two flat glass plates with a thin piece of thin paper between them at one edge, we then have a thin wedge of air which produces **Fizeau fringes.** Lines of equal thickness of the air film are generally parallel to the edges of the glass plates which are in contact. Thus when the arrangement is illuminated with monochromatic light a series of dark fringes will be seen. If both plates have plano surfaces these fringes will be straight and equally spaced. With normal illumination the spacing is given by x where

$$x = \frac{\lambda}{2\alpha}$$

where α, the angle between the plates is assumed to be very small. This expression may be calculated in a similar manner as (14.08).

In both cases the interference pattern is said to comprise **fringes of equal thickness.** Any variation from flatness of the plates or sphericity of the lens will change the thickness of the air gap and this will show in the fringe pattern. With extended source illumination the fringes are localised in the plane of the film. For thin films changes in the angle of viewing across the lens or plates only introduces a fraction of a finge error in the contouring action as $(1 - \cos \theta)$ is less than 2% up to $10°$ from normal. The method constitutes a test of lens surfaces and is further discussed in § 186.

185. Interference with Thick Plates

Interference colours will be seen with white light only when the film is quite thin, that is, of only a few wave-lengths thickness. From expression (14.04A) it is seen that with a film of air illuminated with light at normal incidence, a maximum for a particular wave-length will occur when the thickness of the film is any *odd* number of quarter wave-lengths. Thus wave-lengths of 4 times, $\frac{4}{3}$ times, $\frac{4}{5}$ times, etc., the thickness of the film will be in the same phase on reaching the eye.

Suppose a film to have a thickness $t = 0.0005$ mm., the wave-length of green light, then the following wave-lengths will be in the same phase on reflection from the front and back surfaces, and will reinforce.

$\lambda = 4t = 0.0020$ mm.   — infra red

$\lambda = \frac{4}{3}t = 0.00067$,,   — reddish orange

$\lambda = \frac{4}{5}t = 0.00040$,,   — violet

$\lambda = \frac{4}{7}t = 0.00029$,, etc. — ultra violet.

Of these, only the second and third are in the visible spectrum and the reflected light is a mixture of these two wave-lengths and therefore distinctly coloured. As the thickness of the film is increased more of the wave-lengths of the visible spectrum are reinforced. Thus if the thickness of the film was 0·005 mm., proceeding as above, it will be found that twelve different wave-lengths in the visible spectrum, from $\frac{4}{27}t = 0.00074$ mm. to $\frac{4}{49}t = 0.00041$ mm. will be reinforced, and the mixture of all these wave-lengths cannot be distinguished by the eye from white light.

If the light reflected from a thin film is received on the slit of a spectroscope, the spectrum will consist of bright bands separated by dark spaces; thus with a film 0·0005 mm. thick there will be two bright bands with their centres corresponding to wave-lengths of 0·00067 mm. and 0·0004 mm. respectively, while with the film of 0·005 mm. thickness there will be twelve bands. This method may be used in determining the thickness of films.

As the separation of the surfaces providing the interfering beams is increased the visibility of the fringes is reduced with an extended source even if this is very monochromatic. This occurs because the pupil of the eye collects rays of different angles which give fringe patterns which spread across each fringe. An important exception occurs with a parallel-sided plate or air film for then the fringes are localised at infinity and the eye subtends a small angle. Such fringes are called **fringes of equal inclination** as the pattern then *depends* on changes of the angle of view as the eye looks to different parts of the plate.

However, when the plate or film is a wedge, fringes can only be seen using a point source of collimated illumination. In both cases the unaided eye can only see a small area of the fringe pattern at a time, but with collimated light fringes of equal thickness are obtained and the collimating lens can be used for the reflected light as well. In this case the eye can be placed to receive all the fringe pattern. Fig. 182 shows a **Fizeau Interferometer** using collimated light.

It will be seen that the source size and pupil size do not allow large wedge-angles to be used. In fact, a common method of adjusting this interferometer is to bring the two images of the source at S^1 into coincidence before placing the eye there to observe the fringes.

The fringes seen are "cosine squared" fringes and the source must have a coherence length greater than the path difference used. Mercury isotope lamps provide sufficient coherence for several centimetres but the use of a laser source normally allows for many feet or, if the surfaces have high reflectance, for multiple beam fringes with a basic path difference of up to 2 or 3 feet. Best fringes are obtained when the path difference is equal to or a multiple of the resonant cavity length of the laser. With a laser source this interferometer is often called a **Laser-Fizeau interferometer**.

186. Contour and Index Mapping

In optical component fabrication it is important to know the precise curvature of any surface and also its figure, that is any variation in curvature over the suface. A traditional method, introduced by Fraunhofer uses **test plates** and is based on Newton's rings. Two glass plates are ground and polished to the exact curve required by the test surface, one being convex and the other concave. The accuracy of the curves is tested by means of an accurate spherometer and by placing the plates in contact and, observing the interference pattern from the enclosed air film, any departure from the spherical can be seen. If both surfaces are spherical and of equal curvature a uniform illumination is seen over the whole surfaces.

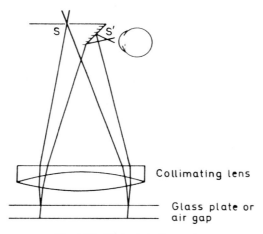

Fig. 182 – Fizeau interferometer

In testing a lens surface during manufacture the test plate of the opposite curvature is placed in contact with the lens surface, both surfaces being perfectly clean. If the lens is spherical but not of the correct curvature, circular fringes will be seen while any departure from spherical will show up in the shape of the rings. The rings do not immediately show whether the air film is thicker at the centre or at the edge, but in the latter case the lens can be rocked causing the fringe centre to move while in the former a slight pressure in the centre of the lens will cause the fringes to move outwards.

In spite of precautions it is all too easy to scratch the lens surface when a contact method such as this is used. Non-contact interferometers are discussed in § 187 and § 188.

For plano optical components Fizeau interference is commonly used. The fringes formed with an arrangement such as in Fig. 182 are fringes of equal thickness but more properly they constitute lines of equal *optical* thickness. Thus irregularities in the fringe pattern may be due to thickness variations or inhomogeneities of the material. In order to separate the two effects the interference fringes may be obtained from two known flats having between them the test component submerged in oil of equal refractive index. With such an arrangement irregular fringes must be due to inhomogeneities. Alternatively the fringes may be obtained with the component at a significant angle to the collimated beam. This will necessitate another collimating lens for the returning beam but if the effects at two angles are subtracted one from the other, that due to changes in refractive index may be found. This is normally done using automatic scanning techniques and a computer program to analyse the results.

When the plano optical component is intended to be an optical window its quality can be measured by inserting it between two plates forming the the air film of a Fizeau interferometer. The use of a laser source allows the size of the air film to be increased and the component can be conveniently placed in the gap.

187. Equal-Path Interferometers

The Michelson interferometer shown in Fig. 178 was modified by **Twyman and Green** to form a very useful testing instrument for optical shop use. When using collimated light from a monochromatic source the system can be considered to be a Fizeau interferometer when the reflected image of M_2 is close to M_1 but with the two beams separately available. The system as arranged in Fig. 183 can be used for testing prisms or optical windows. If mirror M_2 is replaced by a mirror under test that too may be interferometrically compared with M_1. For lenses the arrangement shown in b may be used where l is a lens producing a convergent beam. The test surface m is placed so that the beam is returned along its own path and therefore is re-collimated by l to interfere with the beam from M_1. A range of different radii of curvatures may be measured by this system and concave surfaces by replacing l with a divergent lens.

The results are seen as fringe patterns in which departure from circles or straight lines indicates a less-than-perfect component. The interpretation of the patterns can be difficult and interferometric testing is generally reserved for only the highest quality of optical component.

As has been indicated in § 186 variations in refractive index affect the fringe pattern and the **Mach-Zehnder** interferometer uses two partially reflecting elements (A and B) to provide the square arrangement in Fig. 184. This has been used extensively in wind-tunnels where the different refractive index of air at different pressures shows up in the fringe pattern the airflow and shock-waves occurring in the test area of the tunnel which is crossed by one of the interfering beams.

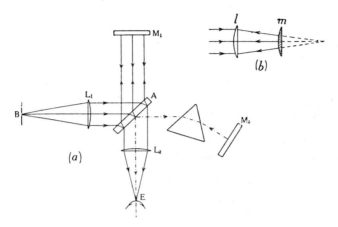

Fig. 183.–Interferometer method of testing prisms and lenses

188. Unequal-path Interferometers

Large path differences between interfering beams make heavier demands on the coherence of the light source particularly if the path difference is traversed many times as in multiple-beam interference. The availability of lasers as sources has stimulated considerable developments in this type although the **Fabry-Perot** was first conceived in the last century. This interferometer consists of nothing more than two glass or quartz plates with very plane surfaces held parallel a small distance apart by a hollow cylinder of invar or silica as in Fig. 185. When the separation is fixed as is normally the case the instrument is called **Fabry-Perot Etalon.** The inner surfaces of the plates are coated to give very high reflectivities and the plates themselves are made slightly prismatic to avoid any interference effects inside them.

The prime purpose of the device is as a form of spectrometer. When the light from a source is collimated and passes through the mirrors the multiple reflections result in very fine circular fringes, shaped as the sharpest in Fig. 177 so that the fine circles are bright against a dark background. Any other wave-lengths present in the source are easily seen even if their wave-lengths differ by considerably less than one Ångstrom unit.

Fig. 184.—Mach-Zehnder interferometer

Fig. 185.—Fabry-Perot etalon

The **Fizeau** interferometer vies with the Twyman-Green for the testing of optical components. The collimating lens of Fig. 182 may be replaced by a multi-element lens based on aplanatic theory (§ 132) as with a point source only spherical aberration has to be corrected. Fig. 186 shows the design of this system and different amounts of convergence of the beam may be obtained by taking out the component lenses starting with L_1 and working upwards. In the figure a convex surface is being compared with the exit surface of the lens with a path difference of a few centimetres. A

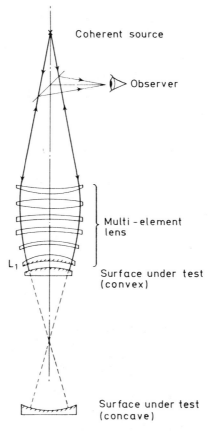

Coherent source

Observer

Multi-element lens

L_1

Surface under test (convex)

Surface under test (concave)

Fig. 186.—Fizeau interferometer for spherical surfaces

concave lens can be located at a position as shown for comparison with the same surface with a path difference of some feet. These path differences may be easily accommodated with a laser source which will even allow multiple-beam fringes.

A type of interferometer developed only since the advent of the laser is now in common use for measuring distances in workshops and laboratories (Fig. 187). It may be considered as a very unequal path version of the Michelson interferometer. The system uses a stabilised laser (L) as its source. Light from this encounters a beam-splitter (B) which reflects a small amount of light. The main beam then traverses some distance before being returned by a corner cube prism (P) to another section of the beam-splitter which reflects most of it on to a photo-detector (D). The internal beam also reaches the photo-detector and broad interference fringes can be caused to occur. As the corner cube is moved, the path difference is changed and these fringes move rapidly across the photo-detector, which therefore generates electrical pulses as each bright fringe moves across it. If

Fig. 187.—Distance measuring interferometer

these fringes are counted and divided by the appropriate factor the distance moved by the corner cube can be displayed to an accuracy of better than 0·000004 inch. It is necessary that the counting system be such that it counts negatively for movement of the corner cube in the opposite direction and that the refractive index of the atmosphere is known as variations in this requires adjustment of the factor used in the computation.

189. Single-Layer Anti-Reflection Coatings

The interference effects of thin films as typified by Newton's rings and Fizeau interference have a usefulness beyond that of optical shop measurement. While the particular shape, spacing and order of the fringes convey information on the film thickness so it is possible by controlling that thickness to obtain a particular level of light reflectance which is uniform over a lens or other optical component if the film thickness is uniform. This effect was first observed by Fraunhofer as reducing the reflection at glass surfaces when a thin film or tarnish developed on exposure to the atmosphere of the glass used in lenses. Artificial methods of obtaining the tarnish were based on chemical etching and developed by Dennis Taylor in the early years of the century. Modern methods are based on the evaporation of particular substances in a high vauum, so that no air molecules impede transfer to the optical surface.

For a single film to give zero reflection it is necessary to create the conditions for a dark fringe all over the surface. However, a further condition exists as the principle of superposition requires that the interfering beams must be of equal amplitude if they are to cancel out completely.

The fraction of the incident light reflected at normal incidence from a surface between two media of refractive indices, n and n' is given by Fresnel's expression as $\left(\dfrac{n - n'}{n + n'}\right)^2$. When a film is deposited on to glass, at the first surface of the film $n = 1$ (air) and if n_F is the refractive index of the film, the fraction of the incident light reflected will be $\left(\dfrac{1 - n_F}{1 - n_F}\right)^2$

At the second surface of the film in contact with the glass of refractive index n_G the fraction of the light reflected will be $\left(\dfrac{n_F - n_G}{n_F - n_G}\right)^2$. Hence if the intensities of the light reflected from each surface of the film are to be equal

$$\left(\frac{1 - n_F}{1 + n_F}\right)^2 = \left(\frac{n_F - n_G}{n_F + n_G}\right)^2$$

from which we have $n_F = \sqrt{n_G}$ or the refractive index of the film should be equal to the square root of the refractive index of the glass. As reflection takes place each time at a surface between a rarer and a denser medium there will be no difference in phase between the two beams due to reflection alone. Therefore, in order that there shall be extinction of a particular wave-length at normal incidence, the thickness of the film must be such that

$$nt = \left(k + \frac{1}{2} \, \frac{\lambda}{2}\right)$$

that is, nt the optical thickness, must be an odd number of quarter wavelengths. Usually only one is applied and this is known as a **quarter-wave coating.**

Not all substances evaporate easily or deposit on a clean glass surface to form a hard durable film. The following is a short list of commonly used materials with their refractive indices.

Magnesium fluoride	MgF_2	1·38	Silicon monoxide	SiO	2·0
Silicon dioxide	SiO_2	1·45	Zinc sulphide	ZnS	2·3
Lead fluoride	PbF_2	1·7	Titanium dioxide	TiO_2	2·35

For optical crown glass extinction requires a film of index $\sqrt{1\cdot52} = 1\cdot233$ but the lowest useful film index is magnesium fluoride which has a refractive index of $1\cdot38$. Using Fresnel's expression we have for the amplitude reflected at the glass/air surface $\dfrac{1 - 1\cdot52}{1 + 1\cdot52} = 0\cdot206$ giving an intensity reflected of $0\cdot2^2 = 0\cdot04$ or 4%. For the film/air interface the amplitude reflected is $\dfrac{1 - 1\cdot38}{1 + 1\cdot38} = 0\cdot159$ while for the film/glass interface it is $\dfrac{1\cdot38 - 1\cdot52}{1\cdot38 + 1\cdot52} = 0\cdot048$. These amplitudes are in anti-phase for a quarter-wave coating and so the resultant amplitude is given by $0\cdot159 - 0\cdot048 = 0\cdot111$. This gives a reflected fraction of $0\cdot012$ or $1\cdot3\%$ which although not zero is a considerable improvement on the 4% of the untreated glass.

Although the index of magnesium fluoride is too low for spectacle crown glass, for higher index glass types the performance of the single-layer coating improves. With a glass index of $1\cdot9$ it is theoretically possible to obtain zero reflectance as $\sqrt{1\cdot9} = 1\cdot38$.

These figures have been calculated for normal incidence and are effective for the wave-length at which the refractive indices are measured and for which the coating has a quarter-wave optical thickness. For other wave-lengths the result is less good so it is general practice to choose a design wave-length near to the cnetre of the visual range such as 510 nm.

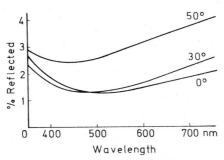

Fig. 188.—Single-layer anti-reflection coating

This means that most light is reflected in the red and blue extremes of the spectrum which gives rise to the purplish tinge on coated lenses and the old term, **bloomed lenses**. At angles other than normal incidence the optical thickness is affected by the $\cos i$ term and the Fresnel expressions change. Yet again any departure from the design angle gives a poorer result overall. Fig. 188 shows the reflectance values obtained at different angles of incidence for different wave-lengths.

The value of anti-reflection coatings is two-fold. In the first instance the reduction in reflection gives an equivalent increase in transmission as the absorption of the film is negligible. This may not seem large for a single surface but for a lens having six separate components there are 12 surfaces at which light is lost. If each surface transmits only 96% the total transmission of the lens is (0.996^{12}), 60%. For coated surfaces transmitting 98.8%, the total transmission is 80%.

The second advantage of coated surfaces is that the 40% portion of the light reflected in the example above is reflected about inside the lens and can end up in the image plane as stray light or flare spots. With coated lenses these flare spots are much reduced and are typically purple in colour.

190. Vacuum Coating Methods

The formation of a very uniform layer of material with a precise thickness in the range 0·1 μm to 1·0 μm on to an optical component is not a simple process. The now traditional method comprises electrically heating the material in a boat of tantalum or molybdenum so that hot molecules of the material are ejected, and holding the object to be coated in the path of these ejected molecules which lose their heat as they strike the cold object and adhere to its surface. If this were done in an ordinary room most substances would react with the atmosphere and even those molecules ejected would be impeded by the molecules of air in their path. This means that the process msut be done in a vacuum and values about 10^{-5} torr are commonly used.

As some materials liquify in the boat before evaporation the arrangement inside the vacuum chamber usually suspends the articles to be coated, commonly called **substrates**, over the point of evaporation so that they lie on the surface of a sphere. The density of the vapour stream follows the inverse square law and so it is necessary to have all the substrates to be coated at the same distance. If a large flat substrate is being coated or special accuracy is needed a rotating plate of metal may be positioned in the stream with a shape such that the stream is interrupted longer in the centre than at the edge and the propensity to deposit a greater thickness in the centre of the substrate is eliminated.

The material is evaporated at a rate depending on the temperature of the boat but this is too crude a relationship to give good control. If it is required to lay down a film with a quarter-wave thickness then the increasing thickness must be monitored during evaporation and the vapour stream interrupted as soon as the value is correct. This is done by continuously measuring the value of reflectivity from a particular substrate near the centre of the stream and having an iris diaphragm which can be closed when the reflectivity is at a minimum.

191. Front-Reflection Mirrors

One of the easiest materials to evaporate is aluminium and when deposited on to a polished glass surface forms a very good mirror both for light incident first on the coating and for light which passes first through the glass. Reflecting components for use in the former fashion are called **front-reflection mirrors**. Their advantage lies in having no chromatic effects as the glass is not used optically and also that wave-lengths in the ultra-violet can be used which would otherwise be absorbed in the glass. (Reflectance values of 92% are obtainable with freshly deposited aluminium.)

Disadvantages stem from the tarnishing of the aluminium and that it is easily scratched. To avoid this an *overcoat* of silicon monoxide is evaporated on top of the aluminium although this causes a loss in reflectivity, particularly for blue light.

In spectrophotometers the light beams are generally focused by mirrors as they require no adjustment for variation of wave-length. If a number of mirrors are used in sequence, however, a reflectance of only 90% on each can rapidly reduce the total amount of light finally reaching the detectors in the same way that transmission values less than 100% for air-glass surfaces reduce the overall transmission of a lens. Increased reflectivity up to 99% can be obtained with pairs of quarter-wave layers on top of the reflection and the refractive indices of the layers are chosen to be far apart as practicable. If MgF_2 (1·38) is deposited next to the aluminium and then ZnS (2·35), the total reflectivity increases from 92% to 97%. A further pair of layers gives 99%. This increase, as is typical with multilayer systems, is only achieved over limited wave-length ranges.

192. Multi-Layer Anti-Reflection Coatings

The disadvantage of single-layer coatings lies in the non-availability of materials of suitable refractive index for glass types of index lower than 1·9. A greater degree of freedom is found in design if two layers are used. Although a second coating material index may be even more incorrect, it now becomes possible to arrange for the larger amplitudes reflected to cancel to a more nearly zero value thereby obtaining a lower reflected intensity.

If, for example, a layer of refractive index 1·7 such as lead fluoride is interposed between the glass (1·52) and the magnesium fluoride (1·38) the three amplitudes of reflection are:

Air/MgF$_2$	0·17
MgF$_2$/PbF$_2$	0·105
PbF$_2$/Glass	0·06

If we arrange for the last two to be in anti-phase with the first the result is 0·005 giving a negligible reflectance value at the design wave-length. It should be remembered that the reflectance between air/MgF$_2$ and MgG$_2$/PbF$_2$ will undergo a change of phase of π but not that between PbF$_2$ glass so that two quarter-wave layers as in Fig. 189 will give the required phases.

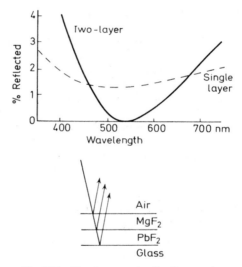

Fig. 189.—Two-layer anti-reflection coating

The result for various wave-lengths (Fig. 189) shows a complete extinction of reflection at the design wave-length but the reflectivity rises more sharply than for a single layer coating. It is possible with two layers to design not for zero reflection at the design wave-length but to obtain extinction at two other wave-lengths where the vector diagram forms a triangle but still gives a zero result.

If three-layer systems are considered the number of degrees of freedom becomes still greater. This allows other constraints to be invoked. The

312

durability of the coating may be considered important and the material chosen with this in mind. There is also the possibility that the coating can be "tuned" for different substrate refractive indices by changing only the thicknesses of the layers and not the materials used. This means that the optical coating department of an optical company may meet most requirements without having to handle numerous coating materials. The actual materials and thicknesses used in these top-quality coatings tend to be proprietary information to the companies employing them.

193. Multi-Layer Interference Filters

The narrower spectral response of the two-layer anti-reflection coating and the steeper sides of the three-layer types are related to the fine fringes found with multiple-beam interferometry. In general, the larger number of interfering beams, the sharper can be the effects on reflected or transmitted intensities for varying thickness or wave-length. This means that narrow spectral filters can be constructed with 10, 20 or even 100 layers of alternating materials so that the reflected light beams are alternately out of phase at a particular wave-length. This means that the transmission of the system is quite high (in the region of 60–70%) for that wave-length. However, for wave-lengths on either side of this the reflecting beams no longer cancel and high(99%) reflectance values can be obtained. The width of the transmitted light at the half-intensity points may be 5

Fig. 190.—Band pass filter (blue) for colour
television camera tube

nm or less with 20 layers. These systems are known as **band-pass** filters. Other transmission bands may exist at other wave-lengths but these can be cut out using a blocking filter either multi-layer or absorbing

Fig. 190 shows a band-pass filter designed for use in colour television cameras for selecting the blue channel. The red transmission is cut out using an absorbing filter.

Other multi-layer designs give high transmission for wave-lengths longer than a certain value and high reflectance for those shorter or vice versa. Such systems are known as **edge filters.** Both types are widely used in spectrophotometric instruments to assist in providing monochromatic beams, and in pollution-monitoring equipment where fixed wave-length beams are required.

The edge filter has found another application in the design of illumination systems in controlling the heat output of incandescent lamps. The heat is largely given out in the near infra-red part of the spectrum and would severely damage film material in high-powered projectors. Between the lamp and the film is placed an edge filter known as a **hot-mirror** which reflects the infra-red wave-lengths while allowing the visible light to pass through. Often lamps in these systems have reflectors behind them to concentrate their output on to the film. These reflectors often have edge filters deposited on them which reflect the visible light and allow the infra-red to pass through. They are naturally known as **cold mirrors** but the deeply curved shape of the reflector often makes a uniform coating difficult to achieve.

All multi-layer design are very sensitive to angle of inclination and to temperature as the latter causes the layers to change in thickness. Corresponding changes in refractive index rarely compensate for this. Filters can be designed to work at non-normal angles of incidence but a problem arises in that the reflectivities of the interfaces vary with the polarisation of the light. (See Chap. XVI).

194. Optical Coatings and Interference — Pertinent Questions

In the case of a single-layer anti-reflection coating the explanation of the its action may well be as follows. The light wave approaches the first surface and due to the difference in refractive index across the interface a small fraction is reflected. On reaching the second interface a further fraction is reflected which interferes with the first reflected wave giving zero (or reduced) reflectivity.

Many students have asked the question, how does the light on reaching the first surface know of the presence of the second and not be reflected? The answer to this may be given in various degrees of erudition and mathematical rigour which may or may not be of value to the student. At the level of this textbook it is not possible to provide a complete solution to a problem which has exercised the minds of eminent physicists during this century. However, the student is reminded to the points made in § 141 (developed further in § 206). The wave train builds up at the start in a gradual manner so that many waves crests have passed before full amplitude is reached. Thus between each half-wave (the phase difference of the "reflected" portions) there is very little difference even during the early waves. Thus it is not a case that the reflection at the first surface has to be annulled in some way but rather that it is not allowed to form as the wave train reaches and crosses the interface system.

This whole question becomes less the hypothetical when the modern theory of light is adopted. This states that the wave motion associated with light is no more than a useful description of the fact that any given photon need not react with matter in a standard manner. It has a range of reactions available to it and the wave motion calculations are in fact determining the probability of particular events. Over fifty years ago at the University of Cambridge the Young's slits experiment was carried out with such a weak light source that most of the time only one photon was on its way from the source to the photographic plate which was the screen. Yet after an exposure of some months the interference pattern was found to be on the plate. Recent experiments with a laser source and sensitive photon detectors have shown that the expected pattern slowly built up from the apparently random arrival of each photon.

Thus the wave-motion is associated with each photon and not with the beam as a whole. In other words we can say that photons are components of light whereas waves are a description of it.

CHAPTER XIV   *EXERCISES*

1. Explain the conditions necessary for interference to occur between two beams of light, and describe two methods of producing these conditions.

2. The effect at a given point of two beams of mutually coherent light depends upon the difference in phase between the two disturbances when they reach the point. Show that the difference in phase can be expressed in terms of the path difference by the relation

$$\text{phase difference} = \frac{2\pi}{\lambda} \times \text{path difference}$$

3. Two point sources emitting disturbances of 6 mm. wave-length in a horizontal plane are 2 cm. apart. Show in a diagram the lines along which there is no difference in phase and along which there are differences in phase of 1, 2 and 3 wave-lengths.

4. Describe a laboratory method of obtaining interference fringes in a manner based on Young's experiment.
Explain what measurement should be made and the calculation necessary to enable the wave-length of the light to be determined from the experiment.

5. Explain clearly why, to exhibit interference effects with incoherent sources, some arrangement such as Fresnel's bi-prism or mirrors is necessary.

6. A narrow slit illuminated with sodium light ($\lambda = 589$ mμ) is placed 15 cm. from a bi-prism of angle 179° ($n = 1.5$); find the separation of the dark interference bands on a screen 1 metre from the prism.

7. A Fresnel bi-prism ($n = 1.5$) with angles of ½° at its two edges is used to produce interference. It is placed 10 cm. in front of a narrow illuminated slit and the interference fringes on a sereen 1 metre from the prism are found to have a separation of 0.8 mm. What is the wavelength of light used? Explain what would happen if a very thin sheet of glass were placed in the path of the light from one half of the prism.

8. Describe carefully an interference method of determining the wave-length of light. State clearly the measurements made and precautions taken and the manner in which the calculation is made.

9. Light from a narrow slit passes through two parallel slits 0.2 mm. apart. The interference bands on a screen 100 cm. away are found to be 3.29 mm. apart. What is the wave-length of the light used? Describe a method other than the above for determining the wave-length of light.

10. Bi-prism fringes are produced with sodium light ($\lambda = 589$ mμ). A soap film ($n = 1.33$) is placed in the path of one of the interfering beams and the central bright band moves to the position previously occupied by the third. What is the thickness of the film?

11. In a Newton's rings experiment a curved glass surface rests in optical contact on a plano glass surface. The diameter of the 5th dark ring viewed in reflection is 9.96 mm. for light of wavelength 628 mμ. The surfaces are then separated, coated to give a reflectivity of 90% and replaced in contact as before. Describe the new fringe pattern and derive the diameter of the 5th fringe. (Assume that the change of phase on reflection remains the same.)

12. Explain carefully the reason for the colours seen in thin films.
A convex lens is placed on a plane glass surface and illuminated with sodium light incident normally; if the diameter of the tenth black ring is 1.5 cm., find the curvature of the surface (wave-length = 589×10^{-7} cm.).

13. Explain the appearance of coloured bands when light is reflected from a thin film of variable thickness. Describe a practical application of this phenomenon.

14. Newton's rings are formed between a plane surface of glass and a lens. The diameter of the third black ring is 1 cm. when sodium light ($\lambda = 589 \times 10^{-7}$ cm.) is used at such an angle that the light passes through the air film at an angle of 30° to the normal. Find the radius of curvature of the lens.

15. Circular interference fringes are formed by reflection at the air film enclosed between a plane and a spherical surface (Newton's rings). Show that the radii of the dark rings are proportional to the square roots of the natural numbers and the radii of the bright rings to the square roots of the odd numbers.

16. Explain with the aid of a diagram the formation of Newton's rings between two surfaces of approximately equal curvature.
A plano-convex lens ($n = 1.523$) of exactly one-eigth dioptre surface power is placed convex surface downwards on an optical flat. The rings are observed by means of a travelling microscope provided with a vertical illuminator, sodium light being used. Calculate the radii of the first and the tenth dark rings observed.

17. Newton's rings are formed in the case of an air film between an upper surface

of radius *r* and a plane surface, normal incidence. Calculate the positions of the first six bright and dark rings for light of the three wave-lengths, viz.:

$$\lambda = 680 \text{ m}\mu, \ 589 \text{ m}\mu, \ 450 \text{ m}\mu.$$

Plot the results by smooth (sine) curves showing the variation of light intensity from the centre outwards and hence deduce approximately the first series of Newton's colour scale.

18. Explain the production of Newton's rings by the light *transmitted* by a thin transparent film. Why are these rings of complementary colour to those formed by reflection from the same thickness of film?

199. Explain the difficulty in obtaining interference fringes by reflection at a film of appreciable thickness when the source emits white light. How is the difficulty affected by using (a) Light from a sodium lamp and (b) Light from a He-Ne Laser?

20. Two plane plates of glass are in contact at one edge and are separated at a point 20 cm. from that edge by a wire 0·05 mm. diameter. What will be the width between the dark interference fringes formed when light of wave-length 589 mμ falls normally on the air film enclosed between the plates?

21. Give a brief account, with an explanatory diagram, of the optical arrangement of a Michelson interferometer. Good fringes were observed with such an instrument with monochromatic light; when the movable mirror is shifted 0·015 mm., a shift of 50 fringes is observed. What is the wave-length of the light used?

22. A Michelson interferometer is adjusted to give the brightest possible fringes with a source of sodium light, which is emitting light of two wave-lengths $\lambda = 589\cdot0$ mμ and $\lambda = 589\cdot6$ mμ. On moving one of the mirrors a position is found where the fringes disappear, the maximum of one set falling on the minimum of the other. How far has the mirror been moved from its original position?

23. Whate are mutually coherent sources of light? Why are they needed in order to exhibit visible interference effects? How may they be produced from incoherent sources?

Interference fringes having been produced on a screen, a thin plate of transparent material is introduced into the path of the light from one of the sources. Explain briefly, with a diagram, what happens to the fringes, and why. (S.M.C.)

24. Interference fringes are formed by an arrangement such as a Fresnel bi-prism. When a thin flake of glass is introduced into one of the two interfering beams, the fringes are laterally displaced. Explain, with a diagram, why this is so and in which direction the displacement occurs.

If the refractive index of the glass flake is 1·5, the wave-length of the light used 6 $\times$ 10^{-4} mm and the central bright band moves to the position previously occupied by the fifth, calculate the thickness of the glass. (S.M.C.)

25. Anti-reflection coatings are applied to the surfaces of lenses and optical parts in order to reduce the amount of reflected light and increase the transmitted light. Explain in detail how this effect is brought about, making clear the role of the thickness and refractive index of the film. Give a diagram. (S.M.C.)

26. Explain how interferometry may be used to test the figure of a convex lens. Give diagrams of two methods.

27. Explain what is meant by a front-reflection mirror. How can a reflectivity of 98% be obtained?

28. Give an account of the method by which a very thin layer of material can be put on to a piece of glass.

29. A soap film illuminated by white light gradually becomes thinner as the liquid drains away. It is placed in front of the slit of a spectroscope so that the spectroscope receives the transmitted light. Describe what will be seen in the spectroscope eyepiece. (City University)

30. A simple interference filter is designed, using an air film bounded by parallel reflecting surfaces, to have a peak transmission at 546 nm. Calculate the thickness of film required for normal incident light and determine the shift in the wavelength of maximum transmission when light is incident at an angle off:

(a) 10°   and   (b) 30°   on the filter. (City University)

31. Magnesium fluoride of refractive index 1·38 is coated on to heavy flint glass of refractive index 1·7 to produce a non-reflecting surface for a wavelength of 500 nm. What thickness of coating is required? Why will magnesium fluoride not produce such an effective non-reflecting surface if it is coated on to glass of refractive index 1·5? (City University)

32. A film index 1·4 is coated on to a glass of index 1·6 to give minimum reflection for a wavelength of 500 nm at normal incidence. Calculate the thickness of the film and the effective reflection coefficients for wavelengths of 500, 400 and 600 nm. Deduce the probable colour appearance of normally incident light reflected from the surface, assuming no change in index with wave-length. (City University)

XV – Diffraction and Holography

195. Introduction

In chapters XII and XIV has been developed the concept that although the energy of light is propagated as photons the action of these can be most easily understood in terms of a wave motion. From an ordinary light source this wave motion is extremely complex as the motion comprises many waves in a random arrangement. The laser is a special source where the motion approaches that of a simple single wave. With wave motions such as water waves and sound it is clear that the waves bend round obstacles such as breakwaters and buildings and the discovery that the same is true of light held the wave theory of light in a position of pre-eminence for over a century.

This bending or **diffraction** of light has an important bearing on how precisely we can use light to measure size as the edges of shadows and images through lenses are found to have non-sharp forms with fringes. These effects will be described for incoherent sources and later for coherent sources where the fringing results in less precision. When these more pronounced effects can be utilised fully, the promise of **holography** can be realised as described in the final section of this chapter.

Previously we have made use of Huygen's principle in which every point on a wave front is assumed to be the centre of a system of secondary waves or wavelets and the new wave front is the common tangent of the secondary waves. Fresnel, by considering the mutual interference that takes place between these wavelets, succeeded in giving an explanation of diffraction effects. A more general theory was developed by Kirchoff and applied to optical images by Fraunhofer and Lord Rayleigh. Abbe applied it to microscope images where the different parts of the object being illuminated by the same source can become partially coherent although a full understanding of imagery by coherent and partially coherent sources has only been developed within the last twenty years, notably by Born and Wolf. The simplified theory presented in the following sections assumes that the illuminating source is monochromatic. With white-light sources the diffraction effects are coloured.

196. Half-Period Zones – Spherical Waves

Fresnel's treatment of diffraction was based on the idea of secondary wavelets emanating from all parts of the wave front. Fig. 191 shows a wave front arising from a point source at S. Fresnel maintained that the effect of P could be found either by allowing this wave front to proceed until it

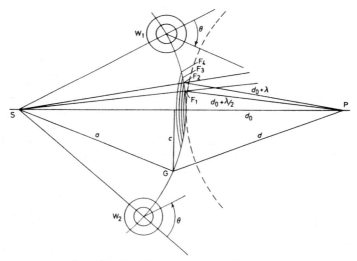

Fig. 191.–Wavelets and half-period zones

reaches P or by dividing the wave front into small areas so that each area can be assumed to generate a wavelet, as W_1 and W_2. Each of these wavelets then produces an effect at P which can be summed according to the principle of superposition to give the total effect of the wave front at P. Obviously these two methods should yield identical results but the second will allow us to calculate the effect at P when parts of the wave front are obscured by edges or apertures.

In his analysis Fresnel assumed that the *amplitude* of the effect at P of any wavelet is:

(1) Proportional to the amplitude of that part of the wave front.

(2) Proportional to the area of the element of the wave front generating the wavelet.

(3) Inversely proportional to the distance of the wavelet centre to P. (This is the inverse square law when intensities are considered.)

(4) Subject to an obliquity factor related to the angle θ which is zero when θ is 90° and rises to a maximum when θ is 0°. There is no wavelet in the backwards direction.

A precise equation for this obliquity factor was not required in Fresnel's analysis although Kirchoff later determined it to be

$$f(\theta) = \frac{-i}{2\lambda} (1 + \cos\theta)$$

where $i = \sqrt{-1}$ and λ the wave-length of the light.

This simple theory encounters some difficulties of interpretation and justification particularly in the last assumption but the major effect in the addition of the effects at P is the differences in relative phase of the disturbance due to each wavelet. This is correctly dealt with by this theory and remains an important contribution of Fresnel to the science of optics.

Referring to Fig. 191 we consider the shortest distance from the wave front to P to be d_0 (= OP). Other parts of the wave front will cause disturbances at P which are out of phase with that produced by the nearest part (at O) due to the longer routes giving a path difference. This may be most easily analysed by constructing a sphere centre P with a radius $d_0 + \lambda/2$. This sphere cuts the wave front in a circle centred on SP passing through F_1 on the wave front.

Within this circle the phase of the distances arriving at P varies between 0 and $\pi/2$. Another circle can be constructed using a sphere centre P of radius $d_0 + \lambda$. This circle is also centred on SP and passes through F_2. The disturbances arriving at P from the annulus between these circles have phases varying between $\pi/2$ and π.

Further annuli or zones can be constructed, each having an effect at P varying by 180° of phase. These hypothetical zones are known as **half-period zones** or **Fresnel zones**. It should be realised that the size of these zones will clearly depend on the wave-length of the light and the distance of the point P.

The reader may easily prove for himself that, when the wave front is spherical or plane and the wave-length is small compared with the distance d_0, the areas of the zones are all very nearly equal to $\pi d_0 \lambda$, and the radii of their boundaries are proportional to the square roots of the natural numbers. With the short wave-length of light, these half-period zones will be very small. If d_o is 50 cm. and the light of wave-length 6×10^{-5} cm. the area of each zone is 0·00942 sq. cm. and the radius of the first zone is 0·548 mm. assuming a plane wave front.

As the areas of these zones are equal they may be assumed to be sending out the same number of wavelets from equal elemental areas, but as the distance of P and also the obliquity are increasing as we go out from the central zone, the amplitude at P due to the zones is gradually diminishing from the central zone outwards. From the way in which the zones have been constructed it follows that the effect at P from any one zone is exactly opposite in phase from that of an adjacent zone. It is necessary therefore to find the resultant amplitude arising from a number of superimposed effects of very gradually decreasing amplitude and between every consecutive two of which there is a phase difference of π.

Let the amplitude at P due to successive zones be denoted by a_1, a_2, a_3, etc. As the average phase of the vibration due to adjacent zones differs by π, alternate amplitudes, a_2, a_4 etc. can be given negative values to show that the displacement is in an opposite direction to that of a_1, a_3, etc. Then the total amplitude at P will be

$$A = a_1 - a_2 + a_3 - a_4 + a_5$$

As the amplitudes due to consecutive zones, although gradually decreasing, are nearly equal, we may say that

$$a_2 = \frac{a_1 + a_3}{2}, a_4 = \frac{a_3 + a_5}{2}, \text{etc.}$$

Then if we write A in the form

$$A = \frac{a_1}{2} + \left(\frac{a_1}{2} - a_2 + \frac{a_3}{2}\right) + \left(\frac{a_3}{2} - a_4 + \frac{a_5}{2}\right) +$$

since each term in the brackets is equal to zero, the expression, if carried sufficiently far that the amplitude due to the outer zones is negligible, becomes

$$A = \frac{a_1}{2}$$

Hence, when a sufficiently large number of zones is considered, the total amplitude at a point, such as P, due to the wavelets from all points on a wave front is equal to half the amplitude due to the wavelets from the central half-period zone alone.

We have determined this result by dealing with the amplitude and

322

phase of the effect due to each half-period zone taken as a whole. Actually, there is a continuous change in phase from O to π across the zone. This can be most easily analysed by using the graphical method of § 136. For this purpose each zone may be sub-divided into smaller rings having equal areas but differing slightly in phase. The amplitude of these zones will therefore decrease only gradually due to distance and obliquity factors.

The graphical summation is shown in Fig. 192, first for finite zones having equal phase differences and then in the limit as these zones become infinitely small. In this case the first half-period zone is represented by the arc ABC which is very nearly a semi-circle and the resultant amplitude is given by the length AC.

In the same way the second half-period zone is given by CDE and the resultant amplitude due to both zones is the very small value AE. If the construction is continued for a large number of zones the curve is seen to be a spiral and approaches O. Thus the resultant for all the zones from the wave front is seen to equal AO which is closely equal to AC/2 as shown before.

Referring to any general point G on the wave front the path difference of the effect at P is given approximately as

$$\text{P.D} = \frac{c^2}{2a} + \frac{c^2}{2d} = c^2 \frac{a+d}{2ad}$$

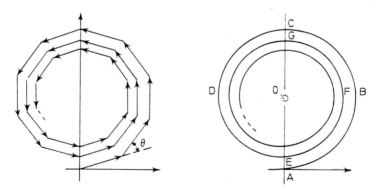

Fig. 192.—Vector addition of amplitudes – spherical wavefront

using the sag formula in Fig. 191. This gives a phase difference of $\delta = \pi \dfrac{a+d}{ad\lambda} c^2$ which is sufficiently accurate as c is usually very small compared with a and d.

197. Approximate Rectilinear Propagation of Light

It has been shown that in the case of light, the half-period zones will be very small, and a large number will be contained in quite a small area about the central point O. The effective portion of the wave front may therefore be considered as confined to this small area, as the effect of the outer zones will be negligibly small as compared with that of a large number at the centre.

An object ordinarily considered small has nevertheless dimensions large as compared with the wave-length of light and will cover a considerable number of zones; the resultant amplitude of the vibrations at a point P (Fig. 191) behind it from the remaining unscreened zones of the wave front will consequently be inappreciable, the effect at P being *as if* the light travelled from the source *along the straight line OP*. Thus, except that there will be diffraction effects around the edge of the shadow, of a nature indicated in the first paragraph and to be investigated below, the statement that light travels in straight lines is seen to be approximately true; and it is quite legitimate in practice to accept the rectilinear propagation of light as a law upon which to base the actions of optical instruments and to employ therefore the idea of "rays" of light in geometrical optics. (See § § 210 and 229).

198. Simple Diffraction Effects

The theoretical results obtained by considering the wave front as in § 196 are confirmed by the results obtained by experiment.

(1) *Small circular aperture.*—Light from a distant point source is passed through a perfectly circular aperture of 1 or 2 mm. diameter, and the light is received on a screen, or better by means of an eyepiece. As the area of the half-period zones will depend on the distance d of the screen or eyepiece, the number of zones contained in the aperture can be varied by varying the distance d. At a certain considerable distance the aperture contains only one zone, and the illumination at the centre of the light patch is a maximum, since $A = a_1$. As the eyepiece is moved towards the aperture until two zones are included, the centre becomes dark, as now $A = a_1 + a_2 = 0$ (approx.). Moving still closer until three zones are included, the centre again becomes bright, since $A = a_1 + a_2 + a_3 = a_1$ (approx.). In this way a series of alternate light and dark centres is found as the eyepiece is moved towards the aperture.

(2) *Small circular obstacle.*—One of the chief objections to the theory of Fresnel at the time of its publication was that put forward by Poisson, who showed that, according to the theory, there should be little loss of illumination at the centre of the shadow of a *small* circular object.

As the areas of the zones are approximately equal, a small circular obstacle intercepting a few central zones should have little effect on the total disturbance reaching the point P from the whole wave front, and there should be a bright spot at the centre of the shadow. Arago (1786–1853) showed by experiment that this was actually the case.

The experiment may be carried out as follows:

A circular opaque object, such as a smooth edged coin or a polished steel ball about 1 cm. diameter, is suspended in the path of the light from a pinhole aperture 2 to 3 metres away. An eyepiece is mounted in the centre of the shadow at about an equal distance from the object, and a small bright spot of light will be seen in the field.* (Plate 5, Fig. 1.) Removing the object will make little difference in the brightness of the spot. It is important that the object shall be perfectly circular and have smooth edges, as otherwise irregular portions of a number of zones will be exposed and the appearance confused.

If the eye, placed in the centre of the shadow, views the object without the eyepiece, the edge of the object is seen as a brilliant luminous ring, showing that the light entering the shadow is travelling as if originating at the edge of the obstacle. An interesting case of the same kind can often be seen in mountainous districts. If one is just within the shadow of a fairly near mountain or hill before the sun has risen over the edge or just after it is set, trees on the sky line appear to be lit with an intense brilliance, while birds and even flies, far too small to be seen at such a distance in the ordinary way, appear as brilliant points of light.

(3) *The Zone Plate.*—An interesting confirmation of Fresnel's theory is provided by the device known as a **zone plate.** It follows from our consideration of the effect of the various half-period zones (§ 196) that, if the disturbance from alternate zones can be prevented from reaching the point P, that from the remaining zones, since it all arrives in the same phase, will add up and produce a greatly increased illumination at this point. To construct a zone plate a series of concentric circles with radii proportional to the square roots of the natural numbers are drawn on white paper, and the alternate rings blackened. The diagram is then photographed, considerably reduced, on to a glass plate so that the rings are alternately transparent and opaque. The rings on such a plate will correspond to half-period zones for a certain distance of the point P, depending of course, on their size; and if the zone plate is set up at the correct distance in front of a screen, there will be a concentration of light on the screen corresponding to each point on the object. The zone plate thus behaves in much the same way as a positive lens. Zone plates

---

* About twenty half-period zones are covered in this case.

of any "focal length" can, of course, be made by reproducing the diagram to the required size.

A considerable increase in the concentration of the light could be effected if, instead of stopping out alternate zones, the phases of the wavelets from these zones were changed by π. The disturbances from *all* zones would then arrive at P with the same phase.

This can be done using photographic plates of bichromated gelatin. This material can be dissolved away when unexposed but the exposed areas remain to provide a transparent area giving a path difference from the clear areas. With a path difference of $\pi/2$ a **phase zone plate** is produced. These devices may be considered to be simple holograms. (§ 210.)

199. Half-Period Zones — Cylindrical Waves

Many of the diffraction phenomena are best seen when the source is in the form of a very narrow slit. In such a case the wave fronts may be considered as cylindrical, and the resultant amplitude along a line through a point, such as P, and parallel to the slit can best be found by dividing the wave front into strips rather than zones. The construction of these half-period elements is similar to that of the half-period zones. If d is the distance of P from the nearest point of the wave front, the distances of P from the outer edges of successive half-period elements will be $d + \dfrac{\lambda}{2}, d + \dfrac{2\lambda}{2}, d + \dfrac{3\lambda}{2}$ and so on.

The length of these strips being equal, their areas, unlike those of the half-period zones, decrease rapidly at first but more slowly as the distance from the centre increases, the more outer strips being practically equal in area. The amplitudes due to these outer strips, being approximately equal but of alternate phase, annul one another, and the effect of the whole wave front is due to a few central elements. The effects of these central strips are not however equal, as was the case with zones, which had equal areas.

200. Cornu's Spiral, Fresnel's Equations

Because, for a cylindrical wave front, the areas of the half-period strips are very unequal near to the axis we find that when the graphical method of § 136 is applied the spiral effect of § 196 still occurs but with a shift of centre. If narrow strips of equal phase difference are plotted as in Fig. 193 the amplitudes of the first few strips are large enough to give an offset before the spiral effect occurs. This can be reduced to a smooth curve when infinitely small strips are considered and the effects of those strips on the other side of the central strip show as a spiral to the opposite quadrant. The total curve known as **Cornu's Spiral** is given in Fig. 194.

In this curve, the amplitudes emitted by each of the elemental strips are laid end to end but as the lengths of the strips are equal, distance along the spiral is related to distance across the wave front. The resultant amplitude at the point P may be found from a line joining the ends of the available part of the spiral, the length of the line being proportional to the amplitude. Any part of this spiral has a slope given by the phase of the particular strip of the wave front.

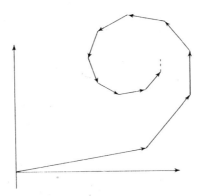

Fig. 193.–Vector addition of amplitudes – cylindrical wavefront

Thus from § 196 the phase, $\delta = \dfrac{\pi}{\lambda} \, \dfrac{a+d}{ad\lambda} \, c^2 = \dfrac{\pi}{2} v^2$ where v has been

defined as $c \sqrt{\dfrac{2(a+d)}{ad\lambda}}$. It can be shown that this is the distance from the origin along the curve to the point having a phase difference δ. To obtain the equation of the curve it is seen that for a short element dv of it the

the slope $\dfrac{dx}{dv} = \cos \delta$ and the slope $\dfrac{dy}{dv} = \sin \delta$.

These give
$$dx = \cos \delta \, dv = \cos \dfrac{\pi v^2}{2} dv$$
$$dy = \sin \delta \, dv = \sin \dfrac{\pi y^2}{2} \, dv$$

The co-ordinates of a particular point are then given by

$$\text{and } x = \int_O^y \cos \frac{\pi v^2}{2} \, dv$$

$$y = \int_O^y \sin \frac{\pi v^2}{2} \, dv$$

These are known as **Fresnell's equations** and they cannot be solved as they stand, but yield infinite series which can be evaluated for specific values to give a table of the co-ordinates of the curve. In reality the equations of Fresnel were formulated first from a more rigorous analysis of his theory. The curve was developed later by Cornu in 1874 as an elegant description of the mathematics.

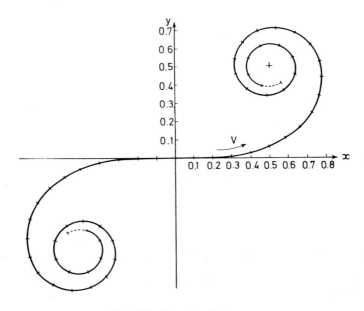

Fig. 194.—Cornu's spiral

201. Fresnel Diffraction Effects

The Cornu spiral (Fig. 194) describes the action of a cylindrical wave front at a given point P. Each part of the spiral relates to a part of the wave front. The part at the origin of the graph is given by that part of the wave front nearest to P which is the part intersected by the line SP at W (Fig. 195). If all the wave front has an effect at P the total amplitude of this effect is given by the line connecting the centres of the spirals. If some of the wave front is obscured the total effect at P is given by a straight line joining the ends of that part of the curve related to the unobscured portion. Thus when a straight edge, X – X, (Fig. 195a) parallel to the

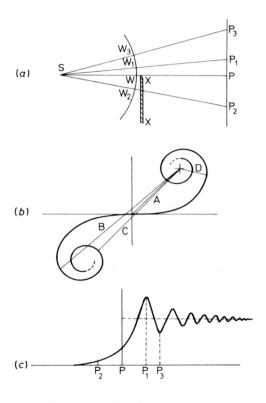

Fig. 195.—Diffraction at straight edge

source of the cylindrical wave front obscures part of that wave front, the effect at P is given by the line A on the Cornu spiral of Fig. 195b. When squared, this gives the intensity for the position P as shown in Fig. 195c.

For other points such as P_1 the nearest part of the wave front becomes

W_1 so that the unobscured part of the spiral is larger giving the line B (Fig. 195b) as the resultant amplitude. This is longer than A and the greater intensity is plotted in Fig. 195c.

For P_3 an even greater length of the curve is used but as this includes nearly a full turn of the lower spiral the resultant C is shorter than B. As more points are considered the amount of the curve used travels round the lower spiral so that the straight-line resultants oscillate in length. This generates the wave-like structure to the graph of Fig. 195c which shows as fringes when photographed with sufficient magnification (Plate 3).

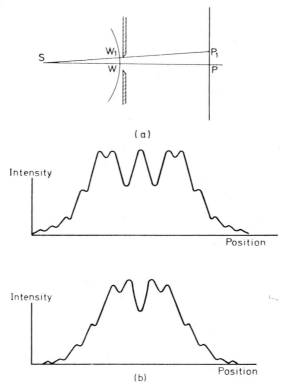

Fig. 196.—Fresnel diffraction effects

If points inside the geometrical shadow are considered (i.e. below P in Fig. 195a), the amount of the spiral used reduces. For example the point P_2 gives the resultant D which gives the intensity shown. Beyond P_2 the resultant straight line joins the centre of the upper spiral to points on the spiral giving a steadily reducing intensity, with no fringes, although the phase angle is varying rapidly.

Because the upper part of the wave front in Fig. 195a is completely unobscured the resultant straight lines all start at the centre of the upper spiral. If now a slit is considered as in Fig. 196a this fixed point no longer occurs. The width of the slit determines the length of the curve to be used to obtain the resultant amplitude but the location of this on the spiral is dependent on the position of the point on the screen. For example the point P on the screen uses a length of the spiral symmetrical about the origin while the point P_1 uses a length having one end at the origin. The resultant patterns show fringes but these are critically dependent on the apparent width of the slit, that is, the length of the spiral used. Two examples are shown in Fig. 196b for narrow and broader slits. If instead of a narrow slit a narrow wire is placed in front of the cylindrical wave front, again parallel to its source, a similar effect is obtained and this is shown in Plate 1, Fig. 2.

202. Fraunhofer Diffraction

The effect of apertures on wave fronts has been discussed so far with the resultant pattern being received as a shadow on a screen. When the source of the wave front is imaged by a lens system on to the screen we find that the image has a structure also due to the lens imaging the diffracted light. In this case there is no need to compare the spherical or cylindrical wave front with a spherical or cylindrical reference surface centred at each point on the screen. Any lens which images the source on to the screen converts the divergent wave front into a converging one centred on the image point. For locations on the screen near to this point the comparison is between spherical or cylindrical surfaces which are tilted with respect to each other. The comparison becomes simpler if we assume that the diffracting aperture operates in parallel light between two lenses as in Fig. 197. Any case where the screen is conjugate with the source is, however, known as **Fraunhofer diffraction**, and the images formed in optical instruments such as telescopes and the optical system of the eye are Fraunhofer diffraction patterns. The incident light does not need to be parallel but the instrument must be correctly focused.

203. Fraunhofer Diffraction Effects – Rectangular Apertures

In Fig. 197, collimated light from a narrow slit S passes through the aperture, CD, and is imaged by the lens L_2 on to a screen. To develop the one-dimensional case, the aperture will be considered as the cross-section

331

of a narrow slit parallel to the source slit S. According to geometrical optics a sharp image of the source should be formed at F′ in the focal plane of L_2, but close inspection in an experimental arrangement will show that a diffraction pattern is formed on the screen consisting of a central bright band bordered by fringes of rapidly reducing intensity. As with Fresnel diffraction these fringes are coloured if a white-light source is used.

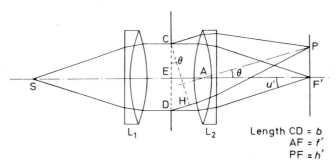

Fig. 197.—Fraunhofer diffraction. Single rectangular aperture

Considering first the point F′ we note that this is the focus formed by L_2 of the plane wave front emerging from L_1. If we consider the plane wave front at CED, then the wavelets arising from all points on it will produce effects over most of the forward direction. Those that travel parallel to the lens axis and are focused at F′ will all have the same path length as the lens creates the converging wave front by making all the ray paths equal. These effects, therefore, arrive at F′ with the same phase and this point, which is the centre of the pattern, is thus always a maximum and bright. In this important respect, the pattern will differ from that with Fresnel diffraction where the centre may be sometimes bright and sometimes dark.

For other points in the focal plane the lens will effectively scan the wavelets, taking into account their relative phases, over a plane tilted with respect to the axis. For the point P in Fig. 197 the ray paths CP and HP are equal as are the paths from all points on the line CH. With a single-slit source at S, the only wavelets available are in phase along the line CD. If they are compounded along the line CH there will be an increasing phase difference moving from C to H. If the aperture CD is divided into narrow strip elements of equal width parallel to the length of the slit, then the amplitudes due to each strip are equal and there will be a constant

332

difference in phase between successive elements. This means that the graphical summation of these effects gives Fig. 198a which in the limit reduces to the arc of a circle, Fig. 198b. As different points in the focal plane are considered, the rate at which the phase difference increases will change and will give different curvatures to the arc. As the same total area of wave front is being considered, the total amplitude and therefore the length of the arc, remains constant (neglecting the obliquity factor for small angles) but the resultant amplitude A, varies considerably, reducing to zero at times as shown in Fig. 198c.

The resultant amplitude may be calculated more easily than under Fresnel conditions. In the simplest case, Fig. 198b, the arc of the contributing vectors has an overall phase difference δ which is also the angle subtended at the centre, C, of the arc. If the length of the arc is l, the radius, r, is given by l/δ. The triangle gives the equation $A/2r = \sin \delta/2$ so that:

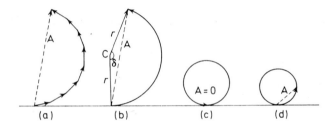

Fig. 198.—Vector addition for Fraunhofer diffraction

$$A = 2r \sin \delta/2 = \frac{2l \sin \delta/2}{\delta}$$

$$\text{or} \quad \frac{\sin \delta/2}{\delta/2} \quad \dots \quad \dots \quad \dots \quad \dots \quad 15.01$$

if the amplitude for zero $\delta (A = l)$ is taken as unity. The overall phase difference, δ, is generated by the path difference DH (Fig. 197) so that

$$\delta = \frac{2\pi DH}{\lambda} = \frac{2\pi b \sin \theta}{\lambda} = \frac{2\pi b \, h'}{\lambda f'} \quad \dots \quad \dots \quad 15.02$$

where the angles $D\hat{C}H$ and $P\hat{A}F'$ are equal for small values. Thus the equation for the amplitude becomes

$$A = \frac{\sin\left(\dfrac{2\pi b \sin\theta}{\lambda}\right)}{\left(\dfrac{\pi b \sin\theta}{\lambda}\right)} \quad \ldots \quad \ldots \quad \ldots \quad 15.03$$

while the intensity is the square of this.

The full circles in the graphical addition diagrmas occur when $\delta = 2\pi$, 4π, 6π, etc. and give zero amplitude. This is also shown in equation 15.03, by putting the numerator equal to zero which occurs when $\pi b \sin\theta = \dfrac{\delta}{2} = k\pi$. Thus the locations of P for zero intensity are given by $\sin\theta = \dfrac{k\lambda}{b}$

or
$$h' = k\lambda f' = \frac{k\lambda}{2\sin u'} \quad \ldots \quad \ldots \quad \ldots 15.04A$$

The first dark band occurs at $\dfrac{\lambda f'}{b}$ so that the width of the central bright fringe is seen to be $\dfrac{2\lambda f'}{b}$ which varies inversely with b.

The locations of the maxima are not so straightforward. The opposite phase value, $\delta/2 = (k + \frac{1}{2})\pi$ gives

$$\sin\theta = \frac{(k + \frac{1}{2})\lambda}{b} \quad \ldots \quad \ldots \quad \ldots 15.04B$$

but this only gives the *approximate* positions as the denominator of 15.03 has an effect. By differentiating 15.01 and setting to zero we have $\tan \delta/2 = \delta/2$ for the maxima. This gives $\delta/2 = 0, 1\cdot43\pi, 2\cdot46\pi, 3\cdot47\pi, \ldots$ which are different from $(k + \frac{1}{2})\pi$ although approaching it for the further locations. The amplitudes at these maxima may be calculated by assuming the central amplitude to be unity*.

The values of maximum amplitude and intensity for the first three bright fringes are given in the table, along with the phase differences, δ, across the slit which generate the maxima and minima.

---

*In ecuation 15.01, $\sin \delta/2 = 0$ and $\delta/2 = 0$ for the central maximum. As the sine term approaches zero more slowly than the angle, the value of unity may be assumed.

Phase difference(δ)	Amplitude	Intensity	Fringe No.
0	1	1	zero
2π	0	0	minimum
$2\cdot86\pi$	0·217	0·047	1
4π	0	0	minimum
$4\cdot92\pi$	0·125	0·017	2
6π	0	0	minimum
$6\cdot94\pi$	0·091	0·008	3
8π	0	0	minimum

It is thus seen that the central maximum is of much higher intensity than the fringes and contains most of the light. The curves for amplitude and intensity against phase difference are given in Fig. 199, and their appearance is shown in Plate 6A.

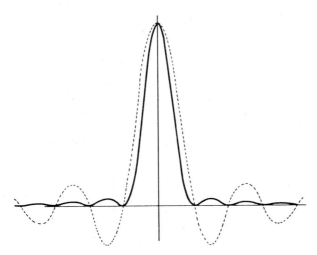

Fig. 199.—Amplitude and intensity curves for Fraunhofer diffraction from a slit

The above analysis has been restricted to one dimension, CD of the rectangular aperture between the lenses. If the other dimension, say CB, is considered it can be shown that this produces a fringe pattern orthogonal to the first where the locations of the maxima and minima are determined by the size of CB. The amplitude is given by

335

$$A(y) = \frac{\sin\left(\dfrac{\pi a \sin \varphi}{\lambda}\right)}{\left(\dfrac{\pi a \sin \varphi}{\lambda}\right)}$$

and the total pattern by

$$A(xy) = \frac{\left(\sin\dfrac{\pi b \sin \theta}{\lambda}\right)}{\left(\dfrac{\pi b \sin \theta}{\lambda}\right)} \times \frac{\left(\sin\dfrac{\pi a \sin \varphi}{\lambda}\right)}{\left(\dfrac{\pi a \sin \theta}{\lambda}\right)} \qquad \dots \ 15.05$$

Thus when one pattern gives zero amplitude, the total must also be zero so that the pattern has zero intensities along two sets of lines parallel to the sides of the rectangular aperture. Within the rectangles formed by these lines the intensity rises to a maximum.

204. Fraunhofer Diffraction Effects — Circular Apertures

The most common aperture in optical instruments has a circular rather than a rectangular shape. When this is divided into strips to calculate the effect at off-axis points, their lengths are no longer constant so that the graphical addition is no longer a circle. This makes the summation extremely difficult as a double integral is involved necessitating the use of Bessel functions. However, it may be noted immediately that the system has symmetry about the optical axis so that the pattern must be circular and indeed takes the form of a bright central disc surrounded by alternate light and dark rings (Plate 5, Fig. 2) which will be coloured when the source is white. As with the slit fringes, the diameters of these rings are inversely proportional to the diameter of the aperture and are unrelated to the size of the source although this must be a small pinhole for best fringe visibility.

With apertures usually found in lenses, the rings will be so small as to be visible only when the pattern is viewed with magnification. Also, unless the lens is of high quality, its aberrations will mask the diffraction effects. When a lens has residual aberrations which are negligible compared with the diffraction effects it is said to be **diffraction limited.** With poor lenses, the pattern may always be artificially enlarged by placing a small aperture over the lens.

The Bessel function mentioned above is usually indicated by the symbol* $J_1(x)$. With a circular aperture, equations 15.01 and 15.03 become

$$A = \frac{2 J_1 (\delta/2)}{\delta/2} = \frac{2 J_1 \left(\dfrac{\pi b \sin \theta}{\lambda} \right)}{\left(\dfrac{\pi b \sin \theta}{\lambda} \right)} \qquad \ldots \ 15.06$$

where b is the diameter of the circular aperture and θ the angle to a circle on the screen, centre F' and passing through P (as in Fig. 197). The function $J_1(\delta/2)$ gives zero values for particular values of $\delta/2$ showing that the graphical summation of the amplitudes from the strips still gives a closed circuit at particular phase differences. However, the location of these zeros and the maxima between them is no longer open to the simple analysis of 15.04A and 15.04B. The calculated values for the first three bright rings are given in the following table.

Phase difference (δ)	Amplitude	Intensity	Total fringe intensity	Fringe No.
0	1	1	1	Zero
2.44 π	0	0		Minimum
3·27 π	0·132	0·0175	0·084	1
4·47 π	0	0		Minimum
5·36 π	0·065	0·0042	0·033	2
6·48 π	0	0		Minimum
7·40 π	0·040	0·0016	0·018	3
8·48 π	0	0		Minimum

The fourth column shows the total intensity in each ring, and it may be deduced that about 87% of the available light appears in the central disc. The radius of this disc can be considered to be the radius, h', of the first dark ring. From equation 15.02 and the table above we have that

$$\delta = 2 \cdot 44 \ \pi = \frac{2 \ \pi b \sin \theta}{\lambda} = \frac{2 \ \pi b \ h'}{\lambda f'} \qquad \ldots \qquad \ldots \ 15.07$$

$$\text{so that } h' = \frac{1 \cdot 22 \ \lambda f'}{b} = \frac{0 \cdot 61 \ \lambda'}{\sin u'} \qquad \ldots \qquad \ldots \ 15.08$$

*The value of $J_1(x)$ for various values of x is available in tables in much the same way that $\sin \theta$ is given for various θ although the Bessel function is rather more complex.

where u' is the aperture angle (see § 222) on the image side and λ' the wavelength in the image space. This is usually, but not always, the same as that in the object space; it is not so in the eye or with the immersion microscope objective.

This disc of light may be considered as the image of a point as formed by the lens. It is known as the **Airy Disc,** after the astronomer Airy who first investigated in 1834 the distribution of light at the focus of a lens. The variation in intensity is shown in Fig. 200 and a photograph of the pattern is given in Plate 5 Fig. 2.

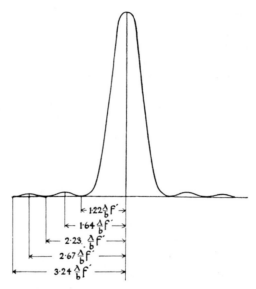

Fig. 200.—Distribution of light in the Airy Disc

It is thus seen that, whereas in geometrical optics we assume that a perfect lens will bring light from a luminous point to a *point* image, actually the image will be a disc of finite diameter depending largely on the wave-length of the light and the aperture of the lens. While the diameter of the Airy disc will be extremely small, it will affect the quality of images formed by lenses as described in § 205.

205. Limit of Resolution, Resolving Power

Suppose B and Q (Fig. 201) are two distant luminous points, such as two stars, subtending an angle w at a lens, then their images B' and Q' subtend the same angle at the second nodal point of the lens, and the separation of the images depends on this angle and on the focal length of the lens. Each image will consist of a diffraction pattern and when the separation of the images is large as compared with the diameter of the Airy discs the distribution of light is as shown in Fig. 201.

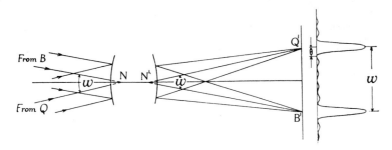

Fig. 201.—Images of two separated points

Under these conditions there will be no difficulty in seeing the two images as separate, that is, in "resolving" the objects. As the angle w is reduced the images come closer together, while the diameter of the discs remains the same, and when the angle is reduced below a certain value the two discs will overlap to such an extent that the eye sees them as a single patch and is no longer able to interpret the image as that of two object points. The smallest value of the angle w for which the images of two points can be detected as double is the **limit of resolution** or **resolving power** and is of very great importance in instruments, such as the eye, the telescope and the microscope, where objects with fine detail or point objects separated by small angles, such as double stars, are concerned.

In order that an expression may be obtained for the limit of resolution, it is necessary to adopt some standard for the separation of the Airy discs, such that the two object points shall be resolved. It is usually assumed that with a perfect optical system two points can be resolved if the centre of one diffraction pattern falls on the first dark ring of the other, that is, the angle w equals the angle θ, and the distribution curve is as shown in Fig. 202. Hence it follows from expression (15.07) that the smallest angle subtended by two point objects that can just be resolved — the limit of resolution is:

$$w = 1 \cdot 22 \frac{\lambda}{b} \quad \text{or} \quad h' = \frac{0 \cdot 61 \lambda'}{\sin u'} \qquad \ldots \qquad \ldots (15.09)$$

It is seen therefore that when an instrument, such as a telescope or microscope, is required to have a high resolving power its aperture must be as large as possible, and this is the chief reason for the very large apertures employed in telescopes for astronomy. The large refracting telescope at the Yerkes Observatory in America has an objective of 40 inches diameter, and it will be found from expression (15.09) that this will resolve two points subtending an angle as small as 6×10^{-7} radians or one-eighth of a second in the case of green light of $\lambda = 500$ nm. As the eye is unable to separate with comfort points subtending an angle of less than about 1½ minutes, the image formed by the objective will require to be magnified about 720 times by the eyepiece.

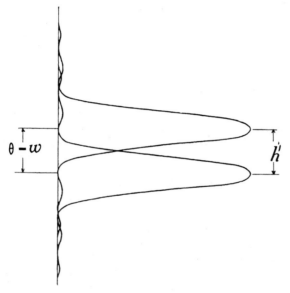

Fig. 202.—Intensity curve of two images just resolved

The affect of aperture on resolution may be easily demonstrated by the following simple experiment. A piece of fine wire gauze is held in front of a brightly illuminated background and slowly brought up towards the eye, until the position is found at which it is just possible to distinguish the separate wires. If now a card with a very small pin-hole is placed before the eye, the separate wires will no longer be visible, and it will be necessary to bring the gauze much closer, so that the spaces between the wires subtend a much greater angle at the eye, before the separate wires are again seen.

In the foregoing consideration of the nature of the diffraction pattern at a focus and hence the resolving power of an instrument, it has been

340

assumed that the optical lengths of all rays coming to the focus are equal, that is, that the emergent wave front is perfectly spherical. Such perfection cannot be attained in practice, but Lord Rayleigh (1842–1919) in 1878 showed that the conditions would be satisfied if the difference between the longest and shortest paths to the focus did not exceed one-quarter of a wave-length.

Example:

A telescope with an objective of 40 mm diameter is focused on an object 1,000 metres away; what is the smallest detail that could be visible with the instrument? If the eye can comfortably resolve two points subtending an angle of 100 seconds, what magnifying power is required for the telescope in order that the resolved detail shall be comfortably seen? (Take the wave-length of light as 500 nm.).

From (15.09), the angle subtended by smallest detail resolved

$$w = \frac{1 \cdot 22\lambda}{b} = \frac{1 \cdot 22 \times \cdot 00055}{40} = 0 \cdot 0000168 \text{ radians.}$$

As the angle is small, $w = \frac{h}{l}$ where h is the size of the object and l its distance.

$$\text{Then } \frac{h}{1000} = 0 \cdot 0000168$$

The image of this object formed by the telescope is required to subtend an angle of 100 seconds $= \dfrac{100}{206000}$ radians.

Therefore required magnifying power $= \dfrac{100}{206000 \times \cdot 0000168} = 29.$

In considering the resolving power of a microscope it will be convenient to consider the cones of light entering the objective from object points. From expression (15.08) the smallest distance between two image points that can be resolved, $h' = \dfrac{0 \cdot 61\lambda'}{\sin u'}$. Assuming that the objective satisfies the sine condition (see 18.07) i.e. $n \sin uh = n' \sin u'h'$, then the minimum resolvable interval in the object space

$$h = \frac{n'h' \sin u'}{n \sin u} = \frac{0 \cdot 61 n'\lambda'}{n \sin u} = \frac{0 \cdot 61\lambda}{n \sin u},$$

where λ is the wave-length in air.

As proposed by Abbe, $n \sin u$ is known as the **Numerical Aperture** of the objective and is represented by N.A.

In the foregoing discussion the resultant of the curves in Fig. 202 has been obtained by adding the intensities of the two diffraction patterns and it has therefore been assumed that the two luminous points are incoherent. This is by far the most usual case. If, however, the two sources are coherent or have a high degree of mutual coherence we must add the amplitudes of their respective patterns before squaring to obtain the intensity. When this is done it is found that at the just resolved position for incoherent sources, the pattern for coherent sources is not resolved. A comparison between the two patterns is given in Fig. 203 where the curve for two coherent sources also assumes that they are in phase.

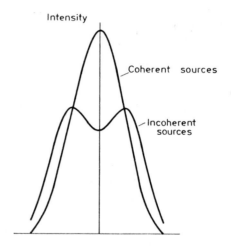

Fig. 203.—Resultant intensity curves for close objects

In the case where a telescope is being used to view two stars there is no question that the analysis of the diffraction patterns will be on an incoherent basis. The case of a microscope is more difficult, however, as the objects on the stage are being illuminated by a single light source which although incoherent can give rise to partially coherent conditions for objects close together.

206. Diffraction and Optical Processing

We have seen that the location of the first dark ring (or fringe) in a Fraunhofer diffraction pattern is inversely proportional to the diameter (or width) of the aperture used. The pattern is centred on the image of the source so that if the aperture is very small compared with the lens in an arrangement similar to Fig. 197, the pattern will be larger (and fainter) but still centred on the image of the source. If now a large number of equal apertures are irregularly distributed over the screen carrying the original aperture then each will form an identical pattern in the same place. Thus the intensity of the pattern will increase but except for the central disc where the effect is modified by the spacings of the apertures, the rings will have the same diameters for a large number of circular holes as for one circular hole.

It has been shown, first by Babinet (1774—1862), that if the circular apertures in the opaque screen are replaced by circular opaque particles of the same size on a transparent screen, the diffraction effect is exactly the same, and examples of this phenomenon can be easily produced. On look-ing at a white luminous point on a dark background through a plate of glass covered with lycopodium dust, which consists of minute spheres of equal size, brilliant coloured rings will be seen around a bright centre. Similar rings will be seen through a dried blood film, a steamy glass and also close round the sun and moon in hazy weather. In these last two cases the rings will not be so brilliant as with lycopodium and blood, as the particles of moisture causing the diffraction are not all of the same size. These rings are known as **coronas** or **halos,** although the latter term is also applied to the larger rings often seen round the sun and moon, and which are due to refraction.

If we can measure the angle subtended by a given ring for a particular colour, we can find the radius of the particles from expression (15.07).

Young devised a simple piece of apparatus, which he called an **erio-meter,** for this purpose. This consists of a metal plate having a small circular hole about ½ mm. diameter drilled in the centre and around this at a distance of about 10 mm. a circle of smaller holes. The plate is attached to a graduated bar and is illuminated from behind, preferably with mono-chromatic light. On looking at the central luminous point through a dif-fracting screen of small particles, the diffraction rings are seen, and by adjusting the distance between the screen and the plate a ring can be made to coincide with the circle of small holes. The distance is read on the scale; then if the angle is small, the distance will be approximately inversely proportional to the angle, and therefore proportional to the diameter of the particles. In this way Young compared the diameters of a great number of small particles, blood corpuscules, wool fibres, etc. The principle has been applied to the examination of the halos seen in glaucoma,* and to the measurement of fibre diameters in textiles.

*H. H. EMSLEY and E. F. FINCHAM. *Diffraction halos in normal and glauco-matous eyes.* Trans. Opt. Soc. 23, 1921—22.

Although very simple in use, Young's apparatus is, in fact, an example of what is now called **Optical processing,** and for an understanding of this and the remaining sections in this chapter it is necessary to consider the concepts of **Fourier transforms.** When the summation of simple harmonic motion was developed in § 136 it was seen (Fig. 140) that two sine waves of differing amplitude and frequency could generate a waveform of complex shape. Fourier demonstrated that any form, complex or simple, could be mathematically analysed in terms of sine and cosine wave forms of varying amplitudes and frequencies. This applies both to wave motions and to wave forms. A single perfect sine-wave motion can be described as a single frequency, as we have seen. If this wave motion is modified in any way other frequencies are required. This applies even to such simple modifications as starting and stopping the wave motion. Thus when light comprises short wave trains we find that many frequencies are required to define them and the light is described as "white". The longer the wave trains, the fewer frequencies are required and so we find that best interference effects occur with light having a long coherence length, which means highly monochromatic.

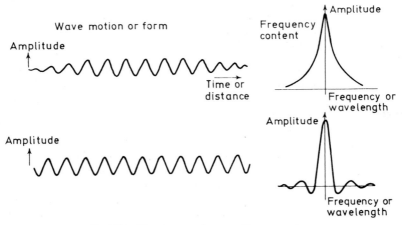

Fig. 204.–Fourier transforms – diagrammatic

Two particular examples of Fourier transforms are given in Fig. 204. The upper diagram shows that for wave motions that start and stop gradually the mathematically derived frequencies are spread about the main frequency. and this is the shape of most spectral lines. If the wave form is started and stopped abruptly the frequencies required are shown in Fig. 204b. Here it is seen that some frequencies are required to be negative or out of phase with the main frequency, and the shape bears a striking resemblance to the Fraunhofer diffraction patterns of § 203.

Considering Fig. 197 again it can be seen that for a wave front arriving at the aperture at an angle the wave form within the aperture is a sine wave abruptly terminated at each end. It is found that Fraunhofer diffraction patterns are related to the shapes of apertures in the same way as Fourier transforms, each a unique description of the other. From this, optical systems can be devised which use diffraction patterns to perform mathematical processes or pick out particular types of object from general scenes.

207. Fraunhofer Diffraction Effects — Multiple Apertures

If in place of the slit CD (§ 203) we have two equal narrow slits parallel to one another, each of these slits by itself would give rise to a diffraction pattern of the form described, the various maxima and minima due to one slit occupying the same position in the focal plane as those due to the other slit, if the incident light is parallel. There will, however, be interference between the light from the two slits. Since portions of the same wave front pass through each slit, we may consider the slits as collections of coherent sources in the same way as the virtual sources of the double mirror or biprism experiment (Chap. XIV).

If the width b of the slits is equal, the amplitude at the centre of the pattern will be twice, and the intensity therefore four times, that for a single slit. At any positions in the focal plane such that the difference in path $C_2 G_2$ (Fig. 205) of light from corresponding points on the two slits is an odd number of half wave-lengths, the intensity drops to zero, since the disturbance from any point in one slit annuls that from a corresponding point in another. Thus a series of dark interference bands are superimposed on the diffraction pattern. The separation of these bands will depend on the separation $(b + c)$ of corresponding edges of the two slits, c being the width of the opaque space, and a number may lie in the central maximum of the diffraction pattern. For example, when $(b + c)$ is equal to $2b$, the new minima are half the distance apart of those for a single slit of width b.

Between these minima there will be maxima at positions where the difference in path $C_2 G_2$ is an even number of half wave-lengths. The resultant amplitude at these positions is the sum of the amplitudes due to each slit along the corresponding directions, and the intensity is therefore four times that at the same position with a single slit. Thus the maxima will be absent when they fall in positions corresponding to the minima of the diffraction pattern of the single slit, for in these positions the intensity

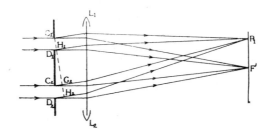

Fig. 205.—Fraunhofer diffraction. Two equal slits

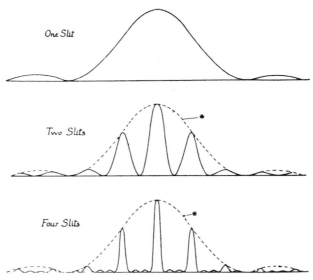

One Slit

Two Slits

Four Slits

Fig. 206.—Intensity curves of Fraunhofer diffraction from
one, two and four narrow slits. *The heights of the curves
for two and four slits are on a much smaller scale than those
of the single slit

due to each slit is zero. The intensity curve for two slits is shown in Fig.
206, and the appearance in Plate 6, Fig. B.

The curve of the interference is that given for two-beam interference in
§ 180 is $A^2 = 4a^2 \cos^2 \alpha/2$ (14.05A) where a is the amplitude of the
interfering beams and α is the phase difference between them. The curve
of the diffraction pattern is $A^2 = \dfrac{\sin^2 \delta/2}{(\delta/2)}$ (15.01) where δ is the phase

difference across each slit. The resultant curve is found from the product of these. However, both α and δ are dependent on the angle θ, subtended by P_1 (Fig. 205):

$$\alpha = \frac{2\pi (b + c) \sin \theta}{\lambda}$$

$$\delta = \frac{2\pi b \sin \theta}{\lambda}$$

so that the total intensity curve is given by

$$I = 4a^2 \frac{\sin^2 \left(\frac{2\pi b \sin \theta}{\lambda} \right)}{\left(\frac{2\pi b \sin \theta}{\lambda} \right)^2} \cos^2 \left(\frac{\pi(b + c) \sin \theta}{\lambda} \right) \qquad \ldots \; 15.10$$

which has a peak value four times that of a single slit.

As the first diffraction zero and the third interference zero coincide in the two-slit curve of Fig. 206 it can be shown from 15.10 that $c = 1\tfrac{1}{2}b$. If $c = 2b$ the zero of the diffraction pattern falls on the third interference maxima which is then virtually absent from the pattern although two very low intensity fringes mark the place.

When four slits are considered the interference becomes multiple-beam but without the reducing amplitudes assumed in § 181. Thus although the fringes become sharper their profile is more akin to the diffraction patterns of single wide slits which gave circular graphical summation curves in § 203. More simply, principal maxima occur in positions such that the difference in path between the light from corresponding edges of adjacent slits is an even number of half wave-lengths, as the amplitudes along these directions from each slit add up. The intensity of these maxima will therefore be sixteen times the intensity at the same positions due to a single slit, that is, N^2 times, where N is the number of slits. Thus, as before, the maxima are absent where the positions in which they should fall coincide with the positions of minima in the diffraction pattern from a single slit.

Dark bands are formed in positions such that the path difference between the light from corresponding edges of adjacent slits is an odd number of half wave-lengths, in which case the vibrations from any point in one slit are annulled by those from a corresponding point in an adjacent slit. There are, in addition, dark bands in directions such that the path difference from corresponding points on adjacent slits differs by an odd number of quarter wave-lengths, for in this case the path difference between alternate slits is an odd number of half wave-lengths, and the vibrations from the first slit cancel those from the third, and the vibrations from the second cancel those from the fourth. There are thus three minima between every two principal maxima. Subsidiary maxima occur between every two minima, but their intensity is very much less than that of the principal maxima. The intensity curve for four slits is of the form shown in Fig. 206, and a photograph of the pattern is shown in Plate 6, Fig. C.

In the same way it may be shown that with any number of equally placed slits there will always be $(N - 1)$ dark bands and therefore $(N - 2)$ subsidiary maxima between every two principal maxima. When the number of slits is very large the intensity of the subsidiary maxima becomes so small as compared with that of the principal maxima that they will not be seen. The diffraction pattern then consists of the principal maxima in the form of very narrow bright lines which may be considered as images of the slit source, these being separated by comparatively broad dark spaces. These principal maxima will still, of course, be absent where their positions would correspond to the minima of the pattern for a single slit.

208. The Diffraction Grating

As with multiple-beam interference, the sharper fringes associated with the four-slit pattern of the previous section means that the measurement of their position can be made more precisely. In general terms, the more slits the sharper the fringes, which leads to one of the most important applications of diffraction, the **diffraction grating**. This consists, in its simplest form, of a large number of extremely narrow equal parallel slits separated by equal opaque spaces that are usually of the same width as the slits.

The first diffraction gratings, made by Fraunhofer, about 1820, were formed of very fine wires closely and equally spaced. Later Rowland (1848–1901) produced gratings by ruling fine lines close together on the surface of glass with a fine diamond point. In these the diamond scratch may be considered as the opaque space between the transparent slits. Rowland succeeded in this way in making very accurate gratings containing as many as 20,000 lines to the inch. Owing to the very great difficulty of ruling such fine gratings, they would be too expensive for ordinary use, but replica gratings can be produced by a process invented by Thorp. A very thin cast is taken in polyester resin of the surface of a grating ruled on glass or metal, and the cast is mounted on a plane plate of glass. Such a replica gives effects very little inferior to the original ruling. The gratings in common use have rulings of approximately 10,000, 15,000 and 20,000 lines to the inch.

From what has been said in the preceding section it will be clear that when parallel monochromatic light from a distant slit or from the slit of a collimator passes through a diffraction grating and is focused by a lens on a screen or by a telescope, a central bright image of the slit is formed together with fainter bands on either side of it separated by dark spaces. As the light is passing through such a great number of slits and as these are so extremely narrow and close together, the various maxima will be sharply defined and widely separated.

The positions of the maxima, except the central one, are dependent upon the wave-length, and therefore if the light is white or any other compound light, the maxima for different wave-lengths occupy different positions, and a spectrum is formed. The different spectra corresponding to the various maxima from the centre outwards are termed the first, second, third, etc., order spectra respectively.*

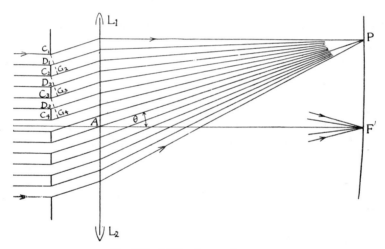

Fig. 207.—Diffraction grating

Fig. 207 represents the section of a grating on which parallel light is incident normally. The diffracted light is received by lens L and imaged on to its focal plane. The intensity at point P is governed by two effects: the diffraction of the light from each individual slit which would, if the light

*From the previous section it will be clear that certain order spectra will be absent, this being the case when their positions would correspond to the minima of the diffraction pattern for the single slit.

from the slits were incoherent, give rise to a single diffraction pattern in the focal plane; and the multiple interference effects between light from all the slits if this is mutually coherent. When illuminated by a parallel beam from a collimated slit parallel to the slits of the grating this mutual coherence is normally very high, but the overall intensity of the interference pattern is given by the diffraction from each slit as in the two- and four-slit cases of § 207.

If the width of the slits is given by b then the diffracted amplitude from each slit at P is

$$a = \frac{\sin \delta/2}{\delta/2} \qquad \ldots \qquad \ldots \qquad \ldots \qquad \ldots \quad 15.11$$

where δ is the phase difference and the amplitude at F' is assumed to be unity (§ 203). δ is given by the path difference between the sides of each slit, $D_1 E_1$, $D_2 E_2$, etc. (Fig. 207) which are identical, and so $\delta = \frac{2\pi b \sin \theta}{\lambda y}$.

The interference effect of all the slits can be calculated in a similar manner to that used in § 203 but with some important differences. The phase difference, α, between each slit is given by the path differences $C_2 G_2$, $C_3 G_3$, etc. which are equal and so $\alpha = \frac{2\pi(b + c) \sin \theta}{\lambda}$ where c is the opaque space between the slits. The distance $b + c$, that is, the width of one slit and one opaque space, is known as the **grating interval**.

Although the phase of the light from each slit varies across the slit, the phase difference between respective parts of slits is a constant and so we can show the phase of each slit as a single value in the summation. The length of each slit in a rectangular grating is also constant and so the amplitudes are equal. Thus the graphical summation is as shown in Fig. 208 where α and the length, a, of each vector are constants. This is very similar to Fig. 198, but the presence of the opaque strips means that polygon never becomes a circle. The calculation of the resultant amplitude follows follows from constructing bisectors of the angles between the vectors. As the polygon is regular, these all intersect at C. The angle subtended by each line at C is α while the length of each bisector may be designated by r. The resultant A subtends the angle $N\alpha$ at C where N is the number of slits. Then from the figure:

$$A = 2r \sin N\alpha/2$$

$$a = 2r \sin \alpha/2$$

Hence

$$A = \frac{a \sin N\alpha/2}{\sin \alpha/2} \qquad \ldots \qquad \ldots \qquad \ldots \qquad \ldots \quad 15.12$$

Where 15.11 is inserted for a, the equation becomes

$$A = \frac{\sin \delta/2}{\delta/2} \quad \frac{\sin N \alpha/2}{\sin \alpha/2} \quad \dots \qquad \dots \qquad \dots \quad 15.13$$

where the amplitude at F' is taken as unity.

The expression for intensity is, of course, the square of this and writing in the full terms for δ and α we have

$$I = \left[\frac{\sin \dfrac{\pi b \sin \theta}{\lambda}}{\dfrac{\pi b \sin \theta}{\lambda}} \right]^2 \left[\frac{\sin N \pi (b + c) \sin \theta}{\dfrac{\lambda}{\sin \pi (b + c) \sin \theta}{\lambda}} \right]^2 \quad 15.14$$

Putting $N = 1, 2$ and 4 gives the curves of Fig. 206, where the overall intensity is controlled by the $\dfrac{\sin \delta/2}{\delta/2}$ term and the sharpening of the fringes by the $\dfrac{\sin N\alpha/2}{\sin \alpha/2}$ term. This latter expression is a maximum when $\alpha/2 = k\pi$ at which points its value equals N* (k any integer). This means that for the maxima

$$\frac{\pi (b + c) \sin \theta}{\lambda} = k \pi$$

or $\qquad\qquad (b + c) \sin \theta = k \lambda \qquad \dots \qquad \dots \qquad \dots \quad 15.15$

In words, this means that the path differences $C_2 G_2$, $C_3 G_3$ in Fig. 207 are equal to a whole number of wave-lengths. The first, second, third, etc., order maxima occur in positions such that $C_2 G_2$ is equal to one, two, three, etc., whole wave-lengths respectively.

In contrast with the spectrum formed by a prism, the spectrum formed by a diffraction grating has the different colours directed into maxima whose angle θ is proportional to the wave-length. Thus the spectrum formed by one grating will be identical to that formed by another except in scale which will depend on the grating interval. The diffraction spectrum, for this reason, is often termed a **normal spectrum**. For values of k greater than 2 the maxima from different orders may overlap for different wave-lengths as the spread of the spectrum is proportional to the order.

*Although both sin Nx and sin x becomes zero for $x = k\pi$ the sines become very nearly equal to the angles when x approaches $k\pi$ and it may be shown that the quotient equals N.

When attempting to separate two wave-lengths close together it is preferable to use as high an order as possible as the width of the maxima does not change markedly with the order. However, the intensity of light in the higher orders is extremely small due to the overall diffraction pattern from a naroow slit.

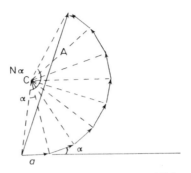

Fig. 208.–Grating – vector addition

It is possible to rule transmission gratings so that each slit is in fact a prism. This has the effect of tilting the overall diffraction pattern away from the zero order and aligning its maximum intensity with one of the higher orders to one side. Such gratings are known as **blazed gratings.** It is also possible to produce reflection gratings where the slits become long narrow mirrors which can be tilted to give the blazed effect. These can be made by vacuum coating the replica gratings described earlier. A reflection grating can also be made on a polished concave surface so that no lenses are required to give sharp spectra, the light being focused by the concave mirror. Such gratings were first made by Rowland in 1880 and are commonly used in commercial spectrophotometers operating in the U.V. or I.R. where achromatic or even transparent lenses are impossible to construct.

209. Grating Images and the Optical Transfer Function

The diffraction gratings described in the last section were considered as elements in a beam of light but we may also investigate the problem of imaging such very fine objects. This was first considered by Abbe (1904) when deriving his theory of the microscope. In this, he took a transmission grating illuminated by a parallel beam as an object for lens L (Fig. 209), to image on to a screen. The advantage of using a grating as an object lies in being able to apply the results of interference and diffraction previously considered. As this parallel illuminating beam is coherent the grating will give first, and higher order beams as shown in the figure.

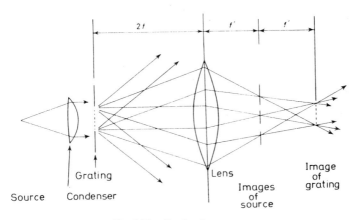

Fig. 209.—Grating images

If the imaging lens L is of limited aperture not all these diffracted beams will pass through it. Provided the first orders are accepted by this lens, it will then form three point images in its focal plane. The images act as coherent sources to provide interference effects in the final image plane which constitute the image of the grating. If the first order diffracted beams fail to pass through the lens no image of the grating is seen as the zero order "source" at F' has nothing with which to interfere. As the object grating is made finer the angle of the first order beam increases so that at some frequency of lines per inch the image of the grating is lost. With coherent illumination therefore it is found that the image of the grating quite quickly fades away as gratings of finer and finer structure are used as objects.

In the study of this fading it is normal to define the fineness of the grating in terms of its **spatial frequency**, that is the number of slits in a given length. This is commonly measured in cycles per millimetre or line-pairs per millimetre. Only if the grating has a sine-wave variation in trans-mission can it be said to have one frequency, according to Fourier's theory and even then the lack of *negative* transmission make this less than exact. However, it is usually a sufficient approximation to consider line/space gratings as having just one spatial frequency.

When such a grating is imaged by a lens it is found that diffraction and aberrations combine to reduce the clarity of the lines in the image. The ratio between the clarity of the image to the clarity of the object is generally called the **Optical transfer function (OTF)** or **Modulation** of the lens although both these terms have more precise definitions. This parameter is usually assumed to be unity at coarse grating frequencies and a graph of the variation in modulation against grating frequency for coherent illumination will look like Curve A of Fig. 210.

From the Abbe concept that the 1st orders must be imaged by the lens we have that $(b + c) \sin \delta = \lambda (15.15)$ where $(b + c)$ is the period of the grating and may be set equal to u (§ 222), the acceptance angle of the lens. So that with coherent light, we find that the maximum grating frequency $(f_c = \dfrac{1}{(b + c)})$ which can be resolved is given by:

$$\frac{1}{(b + c)} = f_c = \frac{\sin u}{\lambda} \qquad \dots \qquad \dots \qquad \dots \quad 15.16$$

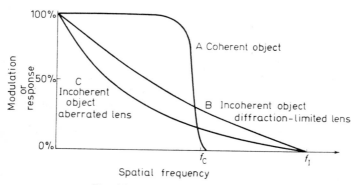

Fig. 210.—Image quality curves

The above has assumed that the light from the individual slits is mutually coherent. This is much the case with a microscope but if the grating is self-luminous or illuminated by having an extended source imaged on to it then the calculation becomes somewhat different and rather more complex. An idea can be gained from the minimum separation distance h' for just resolved points given by 15.09 and (for rectangular) apertures by 15.04A. Although not exactly correct, putting the grating frequency equal to $1/h'$ and the maximum resolved frequency, f_I, is then

$\dfrac{2 \sin u'}{\lambda}$ in the image plane or $\dfrac{2 \sin u}{\lambda}$ in the object plane. However, with incoherent light change of modulation with frequency is much more gradual and the curves B + C of Fig. 210 show typical results for a perfect diffraction-limited circular lens and a lens with aberrations respectively.

Various systems are now available for measuring the optical transfer function of lenses by presenting various gratings to the lens under test and measuring electronically the modulation of the image. When, as often happens in very sophisticated systems, the objects are not completely coherent or incoherent the calculation and measurement of the resolving action of lens is very difficult.

210. Holography

The basic concepts of holography were developed by Dr. Dennis Gaber in 1948, long before the laser was invented. Using mercury isotope lamps as his coherent sources he endeavoured to improve the resolution of the electron microscope and called his approach "*microscopy of reconstructed wavefronts.*" This description is the essence of the holographic process. Normal photography recreates an object in the light and dark tones of an emulsion, but Gabor appreciated that when an observer views a real object he receives only the wave front of the scattered light leaving that object and reaching his eyes. All the optical information must be in the phases, intensities and wave-lengths of that wave front which are extremely complex even for simple objects. To record this wave front for later reconstruction is normally impossible.

In holography the simplification of coherent monochromatic illumination is essential and even then it is not possible to record directly the phase values of the wave front as these are not seen by optical detectors such as photographic emulsions and image tubes. In order to record phase effects it is necessary to add a coherent reference beam into the system which will interfere with the scattered wave front so that differences in phase will show up as differences in intensity.

Leith and Upatnieks in 1961 introduced this reference beam at a fairly large angle to the scattered beam. This simple move creates very fine interference fringes in the plane of the photographic plate. When this is exposed and processed and returned to its original position the reference beam alone "reconstructs" the scattered beam from the object which can be seen by an observer located as in Fig. 211 (b) as an image having considerable clarity of detail and shading and in three dimensions. The angle of the reference beam means that the straight-through light and the "opposite" image are out of the field of view.

The three-dimensional aspect of the reconstruction is inherent in re-creating the wave front rather than a flat representation of the object. The eyes only use two small areas of the wave front at any one time. As the

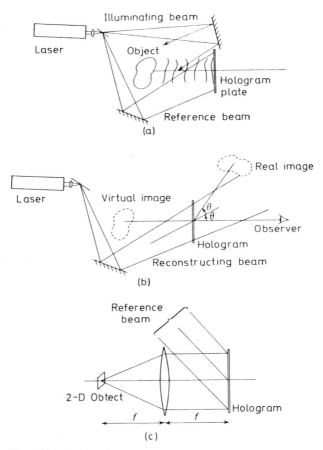

Fig. 211. (a), (b).—Fresnel holography; (c) Fourier-transform holography

head is moved other areas are used and the object is seen "in the round." The wave front is only reconstructed over the area of the interference fringes on the photographic plate so that the reconstructed object is only visible through the plate which acts as a window to the scene. The limited power of laser sources has restricted the size of hologram objects to a few feet although with special techniques holograms of rooms have been made. It is necessary to prevent incoherent illumination reaching the object or the plate.

Most holograms are made in darkrooms and it is generally necessary to mount the components on a vibration-free surface. If the exposed hologram is moved or tilted with respect to its original position in the reference beam the reconstruction distorts and detail is lost. If a reference beam of different wave-length is used the object may still be seen but at a magnification given by the ratio of the wave-lengths. Holograms recorded from radar signals or ultrasonic fields may be reconstructed using visible light and observed directly. These holograms are called **Fresnel holograms** or more accurately sideband fresnel holograms as fresnel diffraction effects are involved. As might be expected it is possible to produce **Fraunhofer holograms**. These are made when the photographic plate is distant from the object compared to its size either in actuality or because a lens is used to collimate the scattered light from the object. Fig. 211(c) shows a general arrangement using a lens. In the special case where the recording plane and the object are at the principal foci of the lens the hologram so formed is called a **Fourier-transform hologram**. This is because the intensity in the hologram at a given distance from the axis is found to be related to the amount of a given spatial frequency in the object (§ 207). This can be used to process transparencies so that particular spatial frequencies are enhanced and others rejected – another form of optical processing.

Another form of hologram can be made when the lens of Fig. 211(c) is positioned to form an image of the object close to the plane of the hologram. Such **focused holograms** are more tolerant on coherence especially in reconstruction and are used in display applications. If a thick emulsion is used on the photographic plate it is possible by virtue of interference effects within the emulsion to obtain a hologram which responds differently to different colours in the reconstruction beam thus allowing colour holograms to be made.

It is seen that the image results from the reconstruction of the wave front. If the object remains in place when the exposed hologram is returned to its original position, interference is possible between the actual object and the reconstructed image provided that both are illuminated from the same laser: thus if the object is moved slightly or deformed, two-beam interference fringes will be clearly seen. The techniques of **Interference holography** have become standard in mechanical testing and vibration analysis.

OPTICS

1. What must be the size of a circular opening in an opaque screen in order that it may transmit two Fresnel zones to a point 2 metres away? What will be the approximate intensity of light at this point? ($\lambda = 589$ mμ).

2. Describe, with a diagram, the Huygens-Fresnel principle of the approximate rectilinear propagation of light.

3. Write a short essay explaining how and why the geometrical laws of image formation are to be considered only as approximations from the standpoint of the wave theory of light.

4. Show that with a plane wave front, when the wave-length is small the areas of the half-period zones for a point at a distance d from the wave front are all very nearly equal to $\pi d\lambda$.

5. Find the diameters of the first four clear zones (the centre being opaque) of a zone plate that will "focus" parallel incident of $\lambda = 600$ mμ at 50 cm. from the plate.

6. The diffraction bands at the edge of the shadow of an opaque object decrease in width from the shadow outwards; explain why this is. When using white light these bands are coloured; state and explain the order of the colours.

7. Explain briefly the diffraction bands produced on a screen by light from a line source passing a straight edge parallel to it. Find the approximate intensity of illumination at the geometrical shadow edge and at the first bright band and first dark band; draw a graph to illustrate this variation of intensity.

8. When a thin wire is placed in the light from a slit source of monochromatic light its shadow on a screen is seen to be crossed with alternate light and dark bands parallel to the length of the wire; explain the formation of these bands. In an experiment the distance between the bright bands was 0·7 mm. when the screen was 1 metre from the wire. What was the thickness of the wire? ($\lambda = 650$ nm).

9. Derive the expression $\sin \theta = \dfrac{\lambda}{b}$ for the position of the first dark band in the case of diffraction by a rectangular aperture in front of a telescope objective.
Calculate the resolving power of a telescope objective of 1½ inches (circular) aperture, assuming $\lambda = 560$ mμ and that two stars can be resolved when the central maximum of one image falls upon the first dark ring of the other.

10. Explain how Cornu's spiral may be used to calculate diffraction patterns. To what type of pattern is it applicable?

11. Explain carefully the diffraction rings observed when a small light source is observed through a diffracting screen such as a glass plate covered with lycopodium powder.
In an experiment, the angular diameter of the rings was measured by using a plate containing a small central hole surrounded by a circular ring of smaller holes (Young's eriometer). Observed through lycopodium powder the distance from the diffracting screen to the hole was found to be 32·4 cm. when a certain diffraction ring coincided with the ring of holes. Observed through a screen of blood corpuscles the distance for the same ring was 8·92 cm. If the mean diameter of the lycopodium grain is 0·029 mm. find the diameter of the blood corpuscles.

12. Fraunhofer diffraction is observed using a square hole 2 mm. side. Explain qualitatively what differences can be seen when this is replaced by a circular hole 3 mm. in diameter?

13. Explain, giving a diagram, the formation of the diffraction pattern at the focus of, say, a telescope objective. What do the dimensions of this pattern depend upon and how are they associated with the resolving power of the instrument?

14. What is meant by the resolving power of a telescope? How may it be expressed and on what is it dependent?

A telescope with an objective of 2 inches diameter is focused on an object two thousand yards away: what would be the least size of the detail in the object which could be rendered visible with the instrument? ($\lambda = 570$ nm).

15. Assuming that two stars can be resolved when the central maximum of one image falls upon the first dark ring of the other, calculate the resolving power of a telescope objective, the aperture of which is 3 inches, and focal length 30 inches. Assume $\lambda = 560$ nm.

If the limit of resolution of the eye for comfortable working be taken as 100 seconds, find the focal length of the eyepiece necessary for clear resolution of the image when using the telescope.

16. Treating the eye as a perfect optical system of power 60D, aperture 3·5 mm. and refractive index of vitreous as 4/3, calculate the radius of the first dark ring in the Airy diffraction pattern formed of a distant point source. ($\lambda = 560$ nm).

Taking the diameter of the foveal cones as 0·0025 mm., comment upon the eye's known capability of resolving two stars or point sources separated by less than one minute of arc. (S.M.C. Hons.)

17. Explain how diffraction limits the detail that can be seen with an optical instrument. Why does a microscope using ultra-violet light offer a better theoretical performance?

18. The angular separation of two stars is 1½ seconds. Find the minimum aperture that must be given to the objective of a telescope in order that the stars shall be just resolved by the instrument. Assume $\lambda = 560$ nm.

19. The focal length of a telescope objective is 6 inches and the magnifying power of the whole instrument is eight. Taking the resolving power of the eye as 100 seconds for comfortable working, find the angular magnitude of the finest structure the eye and eyepiece are capable of resolving and hence the minimum aperture of objective that will suffice ($\lambda = 560$ nm.).

Why in practice would a larger aperture probably be used?

20. Draw carefully to scale the intensity curves of the diffraction patterns, as far as the third maximum, formed by two stars in the focal plane of a telescope objective of diameter 1½ inches and focal length 15 inches. Place the central (first) maximum of one curve over the first minimum of the other. Assume $\lambda = 560$ nm.

21. Calculate the radius of the first dark ring in the diffraction image produced by a telescope of 28 inch aperture, focal length 26 feet, $\lambda = 560$ nm.

On the assumption that two stars can be resolved when the centre of the image of one falls upon the first dark ring of the other, calculate the resolving power of this objective.

By how much would the image require to be magnified by the eyepiece in order that the eye may resolve it? The limit of resolutions of the eye may be taken as 1½ minutes for comfortable vision.

22. Explain the action of the diffraction grating. State the advantages and disadvantages of using a grating to produce a spectrum as compared with the use of a prism.

23. Explain the production of spectra by a diffraction grating. Describe an experiment for determining the wave-length of light by means of a grating, stating clearly what quantities are measured and how the calculation is made.

24. What will be the angular separation of the two sodium lines (λ = 589·0 nm., λ = 589·6 nm.) in the first order spectrum produced by a diffraction grating having 14,438 lines to the inch, the light being incident normally on the grating?

25. Parallel light from a mercury vapour lamp falls normally on a plane grating having 10,000 lines to the inch. The diffracted light is focused on a screen by a lens of 15 inches focal length. Find the distances in the first order spectrum between the lines corresponding to the wave-lengths.

579·1 nm   577·0 nm   546·0 nm   and   435·8 nm

26. A grating illuminated by imaging a source on to it is just resolved by a microscope. The condensing system is now changed to lens which collimates a point source. Explain why the grating is no longer resolved.

27. Explain briefly what is meant by the diffraction of light. Give an example, to be met with in everyday life, of diffraction in the case of (a) water waves, (b) sound waves, (c) light waves. In what respects does the spectrum produced by a diffraction grating differ from a prismatic spectrum? (S.M.C.)

28. If you look through a piece of fine gauze (40 wires to 1 cm.) at a narrow source, emitting light of wave-length 600 nm, placed 4 metres from the gauze, what will be the linear separation between the central and the first diffracted image?

Describe the advantages and disadvantages of using a diffraction grating to produce a spectrum as compared with the use of a prism. (S.M.C.)

29. Explain why it is possible to study much of the theory of lenses and of optical instruments generally on the basis of a ray theory rather than a wave theory of light propagation.

In what circumstances would the neglect of the wave theory introduce serious errors?

Explain why the image of a star viewed through a telescope appears smaller as the aperture of the telescope objective is increased. (S.M.C.)

30. Explain the term "Spatial Frequency". How can the performance of a lens be expressed using it?

XVI - Polarisation

211. Introductory

A vibratory disturbance may be propagated in one of two ways, (1) as a transverse vibration in which the vibration takes place perpendicularly to its direction of propagation; the simplest example of this type is seen in ripples on water, where the particles of the water vibrate vertically up and down, and successive vibrations are transmitted over a horizontal surface; and (2) as a longitudinal vibration in which the vibration takes place in the direction of its propagation, as in sound, where the vibrations consist of alternate compressions and rarefications of the air taking place along the direction in which the sound is transmitted. The various phenomena already described could be satisfactorily explained on the supposition that light was either one of these forms of wave motion, and the effects of interference and diffraction are found to occur in the case of both water waves and sound. There is, however, a series of phenomena, due to what is known as the polarisation of light, that can only be satisfactorily accounted for on the assumption that light is a form of transverse wave motion. Some of these effects will be dealt with in this chapter.

The polarisation of light may be demonstrated in the following simple way. Two similar parallel plates are cut from a crystal of tourmaline parallel to a particular direction in the crystal known as its optic axis, which will be described more fully later. On passing a beam of light through either of the plates, the emergent beam is coloured due to the natural tint of the crystal, but differs in no other way, as far as can be seen, from the incident light. Placing the two plates together with their optic axes parallel the emergent light has the increased depth of colour that would be expected, but on rotating one of the plates with respect to the other it is found that the emergent light gradually fades until,

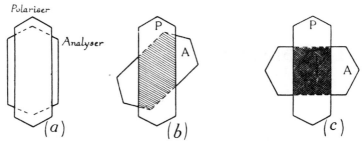

Fig. 212.—Polarisation by tourmaline plates

when the axes of the crystals are at right angles, the light is totally extinguished (Figs. 212a, b, c). On continuing the rotation the light gradually reappears until the original amount of emergent light is restored, when one plate has turned through 180° with respect to the other. We see therefore that the light on passing through the first crystal has undergone some change in its nature that prevents its being transmitted by the second crystal when this is in certain positions; that is, the light has acquired a definite one-sidedness, and it is difficult to conceive of this unless the vibrations are perpendicular to the direction of travel.

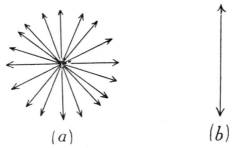

Fig. 213.—Ordinary and plane polarised vibrations

Some idea of what is taking place may be obtained from a simple analogy. A long cord is fixed at one end, and the other end is jerked about in all directions; then a series of transverse waves in all directions perpendicular to the length of the cord will travel along it. If now a board with a narrow slot in it is placed over the cord, the only vibrations getting past the board will be those taking place in a direction parallel to the slot. If a second similar board is placed over the cord, the vibrations passing the first slot can only pass the second when the two slots are parallel, and the vibrations are completely destroyed when the slots are at right angles. Thus we may imagine that the two tourmaline crystals are behaving in a similar way with regard to the light vibrations. We must not, however, carry this analogy too far, since it is not possible to polarise light by passing it through a narrow slit.

As a result of experiments, such as the simple one with the tourmaline crystals, it is concluded that in ordinary light the vibrations are taking place in continually and rapidly changing directions perpendicular to the direction of propagation (Fig. 213*a*). After passing the light through the tourmaline plate, however, the vibrations are all confined to one definite plane (Fig. 213*b*), and the light is said to be plane polarised. It will be seen later that under certain conditions the vibrations may be taking place in elliptical or circular paths, and the light is then said to be elliptically or circularly polarised. Any instrumental means by which light is polarised is known as a **polariser.** The eye alone cannot detect that light is polarised and some further instrument is required to detect this and also the direction of the vibrations. This is known as an **analyser,** and any device that will polarise light may be used as an analyser. In the case of the two tourmaline plates, the first is acting as polariser and the second as analyser. A polariser and an analyser together constitute the chief parts of the instrument known as a **polariscope.**

212. Double Refraction

In 1670 Bartholinus (1625–1698) showed that when light passed through the crystal calcite or Iceland spar it was refracted as two separate beams.

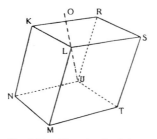

Fig. 214.– Rhomb of calcite

This property of producing **double refraction** is possessed by many crystals and a few other transparent substances. It is especially marked in the case of calcite, which is a transparent crystalline form of calcium carbonate that cleaves readily into the form of rhombohedra or rhombs of the form shown in Fig. 214. At the two corners L and U the three edges make angles of 102° with each other and at the other corners two of the angles are 78° and the third 102°.

When a *near* object, such as a small hole in a card, is viewed through a rhomb of calcite, two images are seen, and on rotating the rhomb about the direction of view one image is seen to remain stationary while the other moves round it. The line joining the two images will always be parallel to a line, such as OL, passing through one of the obtuse corners and making equal angles with the three edges. Any direction parallel to this line is the **optic axis** of the crystal, and if the corners L and U are cut off by plane faces, so that it is possible to view the object along this direction, it will be found that there is then only a single image, and therefore no double refraction of the light travelling along the optic axis.

A beam of light falling normally on one face of a rhomb of spar is doubly refracted as shown in Fig. 215. One beam, travelling in the direction AC_0D_0, corresponds to the beam which would be transmitted if the substance were an isotropic medium, such as glass; this is called the **ordinary beam.** The other AC_ED_E behaves in quite a different way; for example, its direction is changed on entering the crystal, although the incident light is normal to the surface; this beam is called the **extraordinary beam.**

By measuring the angles of refraction of the ordinary and extra-ordinary beams for different angles of incidence and in different planes, it is found that the value of $\dfrac{\sin i}{\sin i'}$ is a constant for the ordinary beam, but varies with the angle of incidence in the case of the extraordinary beam. Hence we conclude that the velocity of the light and therefore the

Fig. 215.—Double refraction by calcite

refractive index of the crystal are constant in all directions for the ordinary beam, but differ in different directions for the extraordinary. Only along the optic axis of the crystal will the two velocities be equal. In a number of doubly refracting crystals, notably calcite, quartz, tourmaline and ice, there is only one direction along which the velocities of the ordinary and extraordinary beams are equal; these crystals are said to be **uniaxial.** In other doubly refracting crystals, such as mica, selenite, topaz and aragonite, there are two directions along which the two velocities are equal; these are called **biaxial** crystals. A doubly refracting crystal may be considered as having an "optical grain" analogous to the grain of wood, the direction of the grain being parallel to the optic axis. In light travelling through the crystal in a direction perpendicular to the axis the vibrations of the extraordinary light are taking place along the grain and those of the ordinary across the grain. When the light is travelling along the axis all the vibrations are taking place across the grain, and thus they are all propagated with equal velocity.

The explanation of double refraction was first given by Huygens by applying the principle of wavelets to the case of a crystal. From any particle which has been disturbed in a uniaxial crystal two wavelets are propagated. One, corresponding to the ordinary beam, will be spherical as in an isotropic medium, since the velocity is uniform in all directions. The other, corresponding to the extraordinary beam, as the velocity in this case differs in different directions, will be an ellipsoid of revolution, that is, the surface generated by the rotation of an ellipse about its major or minor axis. As the velocities are equal along the optic axis, the two wavelets touch each other at the points where they are intersected by the optic axis.

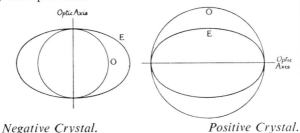

Negative Crystal.   *Positive Crystal.*

	Ordinary n_O	Extra-ordinary n_E
Calcite -	1·6585	1·4864
Tourmaline -	1·669	1·638
Sodium nitrate	1·5874	1·5361

	Ordinary n_O	Extra-ordinary n_E
Quartz -	1·5442	1·5533
Ice -	1·309	1·313

Fig. 216.—Forms of Huygens' wavelets in uniaxial crystals

365

In some uniaxial crystals, such as calcite and tourmaline, the spherical wavelet is entirely enclosed in the ellipsoidal wavelet, and hence the velocity of the extraordinary is greater, or the refractive index less, than that of the ordinary. Such crystals are termed *negative*. In the others, such as quartz and ice, the ellipsoidal wavelet is entirely within the spherical wavelet and hence the velocity of the extraordinary is less or the refractive index greater than that of the ordinary. These are termed *positive* crystals. Sections of the wavelets in uniaxial crystals are shown in Fig. 203. The forms of the wavelets in biaxial crystals are naturally much more complicated and will not be dealt wtih here.

In Fig. 217 parallel light is shown incident normally on one face of a crystal of calcite, the optic axis of which is in the plane of the paper. When the plane wave front meets the surface AD, wavelets of the form described above travel into the crystal from each point on the surface; those originating at A and D are shown in the figure. The refracted wave fronts of the ordinary light M_0N_0 and P_0Q_0 are tangential to the spheres, while those of the extraordinary light M_EN_E and P_EQ_E are tangential to the ellipsoids. It is seen that, in this case, the two sets of wave fronts are

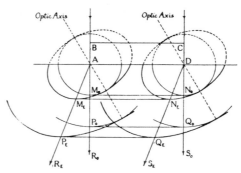

Fig. 217.—Huygens' construction for ordinary and extraordinary wave fronts and ray directions

parallel to one another. The rays of the extraordinary beam are not, however, perpendicular to the wave fronts, and may not always lie in the plane of incidence; the extraordinary beam is therefore deviated from the normal. The rays of the ordinary beam are perpendicular to the wave fronts as in an isotropic medium. Other examples with the incident light reaching the surface at various angles should be solved by similar graphical constructions. It will be obvious that when the light incident in the direction of the optic axis the ordinary and extraordinary beams coincide, since the spheres and ellipsoids are in contact along this direction.

An important case occurs when the optic axis is perpendicular to the plane of incidence and parallel to the face of the crystal. In the graphical construction of this (Fig. 218), as the optic axis is perpendicular to the plane of the paper, the sections of both ordinary and extraordinary wavelets will be circles. Therefore in this plane the velocity of the extraordinary light is the same in all directions, and the value $\dfrac{\sin i}{\sin i'}$ is a constant. For this reason the refractive index n_E of a crystal for the extraordinary beam is defined as *the ratio of the velocity of light in air to the velocity of the extraordinary beam in a plane perpendicular to the optic axis.* The values of the ordinary and extraordinary refractive indices for the D line of a few of the more important uniaxial crystals are given in Fig. 216.

When the two images seen through a rhomb of calcite are viewed through a plate of tourmaline, as described in § 211, or any other form of analyser, it is found that when the optic axis of the tourmaline is in a certain position only one of the two images is visible. On rotating the tourmaline this image commences to fade and the other to become visible, and when the tourmaline has turned through 45° the two images are equally bright. Continuing the rotation, the first image seen gradually

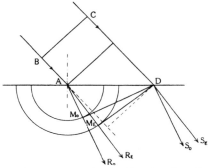

Fig. 218.—Huygens' construction for double refraction. Optic axis perpendicular to plane of incidence

fades while the second gets brighter, and when the tourmaline has turned through 90° only the second image is visible. Both images are again equally bright when the tourmaline has turned through 135°. We must therefore conclude that the two beams from the calcite are plane polarised, and have their vibrations taking place in planes at right angles to each other. From experiments it is found that the vibrations of the extraordinary beam in a doubly refracting crystal take place in a plane containing the transmitted ray and the optic axis, sometimes called a

367

principal plane, while the vibrations of the ordinary beam are perpendicular to this plane.

213. The Production of Polarised Light. Double Refraction Methods

(1) *The Tourmaline Plate.*—Tourmaline, a negative uniaxial crystal, occurs in various colours, and of these, red, brown and green specimens are generally used for polarisation purposes. If a parallel plate of tourmaline is cut with the optic axis in the plane of the plate, double refraction takes place, and with a suitable thickness of plate the ordinary beam is completely absorbed, while the extraordinary is transmitted. This effect is known as **dichroism** and such crystals are said to be **dichroic.** Owing to the small size in which suitable tourmaline crystals are found and the small amount of light transmitted even in the extraordinary beam, the tourmaline plate is of little use for general polarisation work. A simple polariscope—the tourmaline clip—consisting of two tourmalines mounted in a wire spring handle, was used for the purpose of sorting glass from quartz or "pebble" spectacle lenses and for detecting whether the latter were "axis" or "non-axis" cut.

(2) *The Nicol Prism.*—An efficient form of polariser or analyser is the **Nicol prism** or one of its modified forms. The original form of prism, as

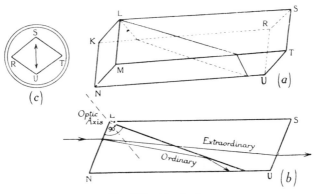

Fig. 219.—Nicol prism

invented by Nicol (1766–1851) about 1833, is constructed from a rhomb of calcite about three times as long as it is broad. The two short end faces of the rhomb, which naturally make an angle of about 71° with the long side of the rhomb, are ground and polished to an angle of 68°;

the rhomb is then cut in two diagonally through one of the obtuse angles, so that the cut face is at right angles to the end face. The two cut surfaces are polished and then cemented together in their original position with Canada balsam. The form of the prism is shown in Fig. 219a, and a section through the prism in Fig. 219b. Light entering one of the end faces of the prism is doubly refracted and divided into an ordinary and an extraordinary beam. On reaching the film of balsam, which has a refractive index of 1·526, the ordinary beam, for which the calcite has a refractive index of 1·6585, is totally reflected and passes through the side of the prism where it is absorbed by the mount. The extraordinary beam, for which the calcite has a *minimum* refractive index of 1·4864, is not totally reflected on reaching the balsam, since it is travelling from a rarer to a denser medium, but passes through the prism and emerges parallel to its original direction. In this way one of the two beams is eliminated, and a single beam of plane polarised light is obtained. The vibrations of the transmitted extraordinary beam will be taking place in a principal plane of the prism, that is, parallel to the shorter diagonal of the end face. (Fig. 219c).

If two Nicol prisms are placed end to end with their principal planes parallel, light transmitted by one is also transmitted by the other. On rotating one of the prisms the transmitted light is gradually reduced until when the principal planes are perpendicular no light is transmitted. The Nicols in this position are said to be *crossed*. If I is the intensity of the beam transmitted when the principal planes of the Nicols are parallel and I_θ the intensity when the principal planes are inclined at an angle₃, it can be shown that

$$I_\theta = I \cos^2 \theta \qquad \dots \qquad \dots \qquad \dots \quad (16.01)$$

This law was first suggested by Malus (1775–1812) after whom it is named.

Two Nicols, one of which can be rotated with respect to the other, are sometimes used for reducing the intensity of a beam by known amounts, as for example in certain forms of the spectrophotometer.

The beam of polarised light transmitted by the ordinary Nicol is limited to an angle of about $24°$ as shown in Fig. 220. Beyond one edge of this beam an ordinary ray meets the balsam film at less than the critical angle and is transmitted, while beyond the other edge of the transmitted beam an extraordinary ray will be totally reflected, since the minimum refractive index for an extraordinary ray only applies when the ray is perpendicular to the optic axis and gradually increases as the direction of the ray approaches that of the axis. Many modifications of the Nicol prism have been suggested with the object both of increasing this field of view and also of economising the amount of spar required in the construction. Details of these modifications are given in a paper by S. P. Thompson,* from which Fig. 220 is taken.

Because of the great scarcity of Iceland spar, particularly in large pieces, and its consequent high cost together with the comparatively

Fig. 220.—Field of view of nicol prism

small angular cone of the transmitted beam the Nicol prism has been very largely replaced by the Polaroid filter in practical applications of polarisation.

(3) *The Double Image Prism.*—In some investigations with polarised light it is necessary to have the ordinary and extraordinary beams inclined at an angle to one another so that two separate images will be formed. It will be seen from Fig. 215 that with a distant object only a single image will be formed with the ordinary calcite rhomb as the two beams emerge parallel. When a beam of unpolarised light is incident normally on one face of a prism (Fig. 221a) cut from a doubly refracting crystal with its optic axis parallel to the face AC, the two polarised beams emerge at different angles from the second face, having been deviated by amounts depending on the two refractive indices of the crystal. The beams are, however, also dispersed and the two images are spectra. Various combinations of prisms termed **double image prisms** have been devised with the object of separating the two beams and at the same time neutsalising the dispersion.

A typical example of this kind is the **Wollaston double image prism.** Two prisms of equal angle made of a doubly refracting crystal, such as quartz or, if a wide separation of the two beams is required, calcite, are cemented together by their hypotenuse faces. One prism ACD (Fig. 221b) is cut so that the optic axis of the crystal is parallel to the surface AC and in the plane of the paper, while the other prism DEA is cut so as to have the optic axis parallel to the surface DE, but perpendicular to the plane of the paper. A beam of unpolarised light incident normally on AC becomes an ordinary and an extraordinary

* SILVANUS P. THOMPSON, F.R.S. *On the Nicol Prism and its Modern Varieties.* Trans. Opt. Convention, 1905.

beam on entering the prism and these, since they are perpendicular to the optic axis, travel in the same direction but at unequal speeds. As the

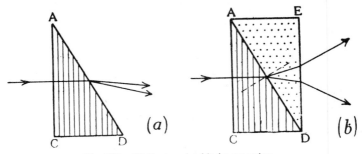

Fig. 221.—Wollaston double image prism

optic axes of the two prisms are perpendicular to one another, the ordinary beam in the first prism becomes the extraordinary beam on entering the second, because the vibrations taking place perpendicularly to the axis in the first prism are parallel to the axis in the second. Thus, in the case of quartz, this part of the light has passed from a medium of refractive index $1 \cdot 544$ (n_0) to one of refractive index $1 \cdot 553$ (n_E), and it is therefore refracted at the surface AD *towards* the normal to the surface, and is refracted again on leaving the prism. The vibrations taking place along the axis in the first prism are perpendicular to the axis in the second prism, that is, the extraordinary beam in the first prism becomes the ordinary in the second. This part of the light is therefore travelling from a medium of refractive index $1 \cdot 553$ (n_E) to one of refractive index $1 \cdot 544$ (n_O), and hence is refracted at the surface AD *away* from the normal, and undergoes a further refraction on leaving the prism.

In this way considerable angular separation of the two beams can be obtained, while the dispersion of one prism is almost neutralised by that of the other. Prisms of this kind are sometimes used for doubling the image in measuring instruments, where the fact that the light is polarised is of no importance, for example, the Javal-Schiotz Ophthalmometer or Keratometer.

(4) *Polaroid Screens.*—Herapathite, a compound of iodine and quinine sulphate crystallising in the form of thin flat plates, and first discovered by Herapath in 1851, possesses similar properties to tourmaline. These artificially produced crystals have been used for polariscope purposes, but like tourmaline are only of small dimensions besides being quickly affected by moisture. E. A. Land* succeeded in producing a very

* Brit. Patent No. 412,179. Dec. 16, 1932.

efficient polarising medium in the form of a sheet of nitro-cellulose packed with ultra microscopic crystals of Herapathite with their optic axes all parallel to one another. These sheets were mounted between plane glasses and behaved in the same way as a single extended crystal. Screens or filters of this form were manufactured under the trade name of Polaroid. Modern Polaroid is made by heating and stretching to several times the original length thin films of a plastic, polyvinyl alcohol. This has the effect of orientating parallel to one another the previously randomly arranged long-chain molecules. The film is then doubly refracting and by dyeing the film in a solution of iodine one of the two polarised beams is absorbed leaving the other plane polarised beam to be transmitted. The polarising film is then laminated between two sheets of a tough plastic or mounted between glasses. The simple form of these filters together with the large fields of view obtainable, due to the much larger apertures available than are possible in other types of polarisers, have considerably simplified many applications of polarisation, besides making possible others that were scarcely practicable with previously existing polarising devices.

214. Polarisation by Reflection and Refraction

Malus, in 1810, discovered that light reflected at a certain angle from a polished glass surface is polarised. If a beam of light reflected from a glass surface is received by a Nicol, it will be found that the intensity of the transmitted light varies as the Nicol is rotated, and has its minimum intensity when the principal plane of the Nicol is parallel to the plane of incidence on the glass surface. With the Nicol in this position, varying the angle of incidence will vary the intensity of the light transmitted by the Nicol, and for a certain angle of incidence almost the whole of the light is extinguished. Therefore, for this angle of incidence the reflected light is almost completely plane polarised, the vibrations taking place in a direction perpendicular to the plane of incidence. At the other angles of incidence the reflected light is only partially polarised and will consist of a mixture of plane polarised and unpolarised light. The angle of incidence at which the polarisation of the reflected light is most complete is termed the **polarising angle.** Brewster (1781–1868) by measuring the polarising angle for a large number of transparent substances, found that for any transparent substance

$$\tan i_P = n \quad \dots \quad \dots \quad \dots \quad \dots \quad (16.02)$$

where i_P is the polarising angle and n the refractive index of the substance. This is known as **Brewster's Law.** The polarising angle for crown glass of refractive index $1 \cdot 516$ is thus $56° \ 36'$. It will be seen from expression (16.02) and the law of refraction that the most complete polarisation occurs when the sum of the angles of incidence and refraction is 90°.

The vibrations of unpolarised light may be resolved into two equal vibrations taking place in directions at right angles to one another, and these directions may be regarded as being, one parallel to the plane of incidence, and the other perpendicular to this plane. If, for the incident light, I_1 is the intensity due to the vibrations parallel to the plane of incidence and I_2 the intensity due to the vibrations at right angles, and I'_1 and I'_2 are the corresponding intensities of the reflected light, then according to **Fresnel's law of reflection**

$$I'_1 = I_1 \frac{\tan^2 (i - i')}{\tan^2 (i + i')} \quad \dots \quad \dots \quad \dots \quad (16.03)$$

$$I'_2 = I_2 \frac{\sin^2 (i - i')}{\sin^2 (i + i')} \quad \dots \quad \dots \quad (16.03\text{A})$$

where i and i' are respectively angles of incidence and refraction. It will be seen that when $i + i' = 90°$, I'_1 will be zero, and only vibrations perpendicular to the plane of incidence are reflected. This is in agreement with Brewster's Law. This law does not apply to polished metallic surfaces, although there will be some difference between the intensities of the two sets of reflected vibrations in these cases.

The amount of polarised light reflected may be increased by using a number of plates of glass instead of one only. At the polarising angle, the reflected beam contains about 13% of the incident light for one plate 20% for two and 28% for four plates, rising to about 35% for a large number of plates.

A simple and useful form of polariscope, particularly for use when comparatively large specimens are to be examined in polarised light, may be constructed as illustrated in Fig. 222. Light diffused by a plate of ground or opal glass is reflected at an angle of 57° from the polished surface of a plate of black glass or of clear glass with the lower surface blackened. The polarised reflected light passes through the specimen to be examined and is received by an analyser, which may be a Nicol prism suitably mounted so that it can be rotated. As the eye is placed close to the Nicol, this can be quite small.

When light that has already been plane polarised is incident on a polished glass surface, the amount of reflected light will depend on the angle of incidence and on the direction of the incident vibrations with respect to the plane of incidence. At the polarising angle practically no light is reflected when the incident light contains only vibrations in the plane of incidence. If two plane glass plates are arranged as in Fig. 223 so that light is incident on each at the polarising angle, the first will act as polariser and the second as analyser and on rotating the upper plate about the axis CD, keeping the angle of incidence constant, the light which suffers two reflections will vary in intensity in the same way as with two Nicols.

When plane polarised light is incident on an unpolished surface, the diffusely reflected light is no longer polarised. The light is then said to be **depolarised.**

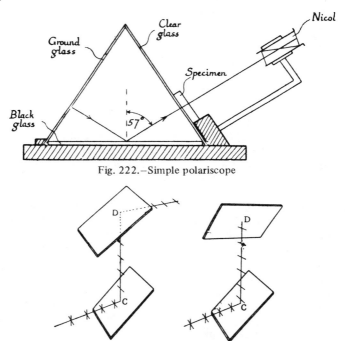

Fig. 222.—Simple polariscope

Fig. 223.—Polarisation by reflection

At the polarising angle, since a certain percentage of the vibrations perpendicular to the plane of incidence and none of the vibrations parallel to this plane are reflected, the transmitted beam will contain an excess of the latter vibrations over the former; that is, it is partially polarised. As the number of plates is increased, more and more of one set of vibrations is reflected, and the transmitted beam becomes more completely plane polarised. This concept can now be realised using optical thin films (Chap XIV). If a number of layers of alternately high and low refractive index are deposited, then for the correct angle a very efficient polariser is formed. This is particularly so at the design wave-length but as is usual with thin-film systems, the effect falls away for longer and shorter wave-lengths.

215. Polarisation by Scattering

If a beam of ordinary light is passing through a medium containing minute particles that scatter the light, and the light scattered in a direction perpendicular to the original direction of the beam is examined with a Nicol, it will be found that the intensity of the light changes from maximum to minimum every 90° as the Nicol is rotated. The scattered light is therefore plane polarised, and its vibrations are found to be taking place at right angles to the direction of the original beam. In the same way, if a beam of plane polarised light from a Nicol is passed through a scattering medium and the scattered light observed with the unaided eye, the intensity of the scattered light will be seen to vary as the Nicol is rotated. The light reflected from a blue sky is polarised in this way.

216. Interference of Polarised Light

Among the most important phenomena of polarisation are those due to interference of the polarised light. Fresnel and Arago have established the following laws governing the conditions under which interference can occur with polarised light.

(1) Two rays polarised in the same plane interfere with each other under the same conditions as for ordinary light.

(2) Two rays polarised at right angles do not interfere under these conditions.

(3) Two rays polarised at right angles, if obtained from unpolarised light, may be brought to the same plane of polarisation without acquiring the quality of being able to interfere with one another.

(4) Two rays polarised at right angles, and afterwards brought to the same plane of polarisation, interfere like ordinary light if they originally belonged to the same beam of polarised light.

Suppose a thin parallel plate, cut from a doubly refracting crystal with the optic axis of the crystal parallel to the surface of the plate, is placed between crossed Nicols between which parallel white light is passing, the surfaces of the plate being perpendicular to the direction of the light (Fig. 224a). On rotating the plate about an axis perpendicular to the direction of the light, it is found that in the four positions where the optic axis of the crystal is parallel to the principal section of either the polariser or analyser the plate produces no apparent effect, and no light passes through the analyser. In any other position of the plate light is transmitted by the analyser, and this light is coloured. The colour transmitted will vary with the thickness of the plate for a given crystal, and different crystals will give different colours for the same thickness. When either polariser or analyser is rotated through 90°, so that they are in their parallel positions, white light is transmitted with the crystal plate in the positions that gave darkness with the crossed Nicols; while in the other positions of the plate the colour transmitted is complementary to that transmitted with crossed Nicols.

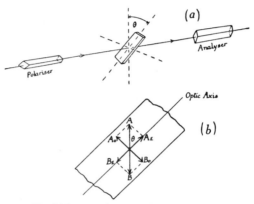

Fig. 224.—Interference of polarised light

When the vibrations of the plane polarised beam leaving the polariser are taking place in a direction parallel to or perpendicular to the optic axis of the crystal plate they are transmitted unchanged by the plate, except for a reduction in velocity while in the plate, and are therefore transmitted or extinguished by the analyser according to the direction of its principal plane. Thus the plate in these positions produces no apparent effect. When the plane polarised beam is incident on the plate with the vibrations taking place obliquely to its optic axis, the vibrations on entering the plate may be regarded as resolved into two components, one vibrating parallel to and the other perpendicular to the optic axis of the crystal. This is shown diagrammatically in Fig. 224b, where AB is the incident vibration, and this is resolved into the ordinary vibration A_0B_0 perpendicular to the optic axis and the extraordinary vibration A_EB_E parallel to the optic axis. The relative amplitudes of these two components will obviously depend on the angle θ that the incident vibrations make with the optic axis. These two sets of vibrations travel through the plate with different velocities, and therefore emerge with a difference in phase depending on the thickness of the plate and the difference in the two refractive indices. This difference of phase may be found as follows: The difference in optical path of the two beams

$$= (n_E - n_O)\,t$$

where t is the thickness of the plate, and therefore the difference in phase (see § 136, Chapter XII)

$$\alpha = \frac{2\pi}{\lambda}(n_E - n_O)t \quad \dots \quad \dots \quad \dots \quad (16.04)$$

These two beams emerging from the plate, although they differ in phase, cannot interfere as the vibrations are at right angles.

On entering the analyser each of the two sets of vibrations may again be regarded as resolved into two components, parallel to and perpendicular to the principal plane of the Nicol; the first of these will be

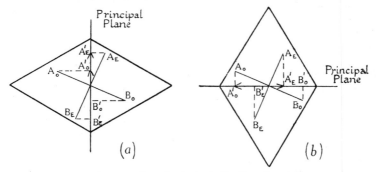

Fig. 225.—Resolution of polarised vibrations by analyser

transmitted while the second are totally reflected. Thus, portions of each of the two sets of vibrations are transmitted by the analyser and, since the disturbances are now taking place in the same plane, interference is possible. Fig. 225a and b show diagrammatically the resolution of the two sets of vibrations $A_O B_O$ and $A_E B_E$ by the analyser when it is in a direction parallel to the polariser (Fig. 225a) and in its crossed position (Fig. 225b). As only the vibrations in the principal plane of the analyser are transmitted, the components in this direction are alone represented in the figures, thus $A_O B_O$ is resolved in the principal plane into $A'_O B'_O$ and $A_E B_E$ into $A'_E B'_E$.

It will be seen that when the analyser is rotated from its parallel to its crossed position a change in phase of π between the two vibrations is introduced, $A'_O B'_O$ and $A'_E B'_E$ being in opposite directions in the second case. For a given wave-length therefore, the intensity of the light

377

leaving the analyser in its parallel position will be a maximum when the difference in optical path of the two beams through the plate is $k\lambda$ and a minimum when it is $(k + \frac{1}{2})\lambda$, k being an integer. With the polariser and analyser in the crossed position there will be maximum intensity when the plate causes a difference in path of $(k + \frac{1}{2})\lambda$ and minimum when the path difference is $k\lambda$. With a given thickness of plate a particular wavelength which is missing from the white light when the polariser and analyser are crossed will be reinforced when they are parallel, and so the colours transmitted in the two positions are complementary. That this is so may be seen by substituting a double image prism for the analyser. Both sets of components are then transmitted, and the two coloured beams can be received together on a screen. If the deviation of the prism is such that the two beams overlap, the overlapping portion will be white.

As the interference will be complete only when the amplitudes in the two beams are equal, the colours will be most intense when polariser and analyser are either crossed or parallel and the optic axis of the plate is at 45° to their principal planes.

If the parallel plate is replaced by a thin wedge cut from a doubly refracting crystal with its thin edge parallel to the optic axis of the crystal, and the wedge be placed between the crossed Nicols with its axis at 45° to the vibrations of the incident polarised light, a series of light and dark bands parallel to the edge of the wedge will be formed when the light is monochromatic. In white light a series of coloured bands is formed, similar to and in the same order as those formed with the thin film described in Chapter XIV.

Glass, which in the ordinary way is isotropic and therefore singly refracting, becomes doubly refracting when considerably strained. Strain may be produced by the glass being cooled too quickly from its molten state, or by mechanical pressure. Lenses or prisms constructed of strained glass will give bad definition of an image owing to the double refraction, particularly if the thickness is at all considerable, while thin lenses, such as spectacle lenses, are particularly apt to break at the position of the strain. When a piece of strained glass is placed between crossed Nicols, light and coloured bands are seen indicating the positions and extent of the strain. Simple polariscopes, such as the form shown in Fig. 222 and known as **strain-testers**, are used for detecting strain in glass.

The polariscope has been applied in engineering as a means of determining the distribution of stresses in various parts of a construction. A model of the part, for example, a girder, is constructed in glass or transparent celluloid and examined in a polariscope. From a study of the form and colour of the interference bands seen when proportional stresses are applied to the model, a determination can be made of the stresses to be expected in the actual structure.

217. Elliptically and Circularly Polarised Light

In the preceding paragraph it has been stated that a plane polarised vibration on passing through the thin parallel plate of crystal is resolved into two mutually perpendicular vibrations which travel at different velocities and therefore have changing differences in phase as they travel through the plate. These two components travel along the same path and the resultant vibration produced on an ether particle at any position in the plate is due to the combination of these two components. We have seen in Chapter XII, § 136, that when a particle is acted upon by two vibrations taking place at right angles to one another, the resultant vibration of the particle will be along an elliptical, circular, or straight line path, depending on the difference in phase and the relative amplitudes of the two components. While passing through the plate the vibrations have successively the paths shown in Fig. 226, and as the

Fig. 226.—Elliptical and circular polarisation

vibration leaving the plate persists in the space between the plate and analyser, the emergent light from the plate may be plane, elliptically or circularly polarised. For example, circularly polarised light may be produced by passing plane polarised light through a crystal plate with the optic axis of the crystal at 45° to the direction of the vibrations of the incident light, the thickness of the plate being such as to produce a difference in phase of $\frac{\pi}{2}$ i.e. $\frac{\lambda}{4}$. Such a plate is termed a **quarter-wave plate** and has several uses in polarisation work.

218. Interference with Convergent Polarised Light

Important information regarding the nature of a crystal can be obtained by passing a convergent beam of plane polarised light, the separate pencils of which are parallel, through a parallel plate of the crystal, cut with the optic axis perpendicular to its surfaces, and receiving this light with an analyser. The simplest way of doing this is by means of the tourmaline clip. If light from the sky is passing through a tourmaline clip held close to the eye, the separate pencils are parallel but are travelling in converging directions through any crystal plate placed between the tourmalines (Fig. 227). Suppose the plate be cut from a uniaxial crystal with the optic axis perpendicular to the surfaces (*axis cut*) and the tourmalines crossed. Then in one direction the eye will be looking along the optic axis, and as there is no double refraction in this direction no light gets through. At any angle with the axis there is double refraction and a phase difference and therefore interference between the two beams. This phase difference is the same for all rays making equal angles with the axis and so is the same along circles about the optic axis. On examining a uniaxial crystal in this way, a series of circular interference bands is seen, the centre of the circles corresponding to the axis direction. In addition to the rings a black cross will be

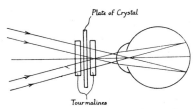

Fig. 227.—Examination of crystal in "convergent" light

seen superimposed on the rings, its arms being parallel and perpendicular to the principal sections of polariser and analyser. On rotating the analyser this cross becomes white. The reason for the cross will be understood from Fig. 228, which represents a front view of the crystal plate. In the figure the incident vibrations are taking place vertically and at the points B, C, D and E and at all points on lines through these points and OA, the vibration is taking place parallel or perpendicular to the principal section of the crystal. These vibrations are therefore transmitted unchanged and will be extinguished or transmitted by the analyser according to whether it is crossed or parallel to the polariser. In other positions, such as L, M, N and P, the vibrations are resolved into ordinary and extraordinary vibrations, and a difference in phase and therefore interference occurs. With a biaxial crystal the pattern is more complicated as there are two directions along which there is no double refraction.

When a greater angle of convergence and therefore a more extensive pattern is required a microscope is used. Polarised light is converged, by means of the substage condenser, through the specimen, which is placed on the stage, and the light is received by the objective. The interference pattern is formed in the upper focal plane of the objective and can be focused by a lens, known as a Bertrand lens, screwed into the lower end of the drawtube.

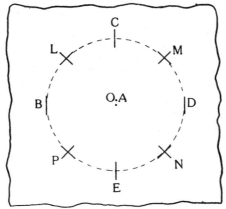

Fig. 228.—Vibrations in axis cut crystal in "convergent" light

219. Rotatory Polarisation. Optical Rotation

A parallel plate of calcite, cut with the optic axis perpendicular to its surfaces (axis cut), produces no apparent effect on polarised light passing through it normally, and when such a plate is placed in parallel light between crossed Nicols, no light leaves the analyser. When a similar plate of quartz is placed between crossed Nicols in monochromatic light, light is transmitted by the analyser, and on rotating the analyser through an angle that varies with the thickness of the plate, a position is found where the light is again extinguished. It follows, therefore, that the light leaving the quartz plate is plane polarised, but the plane in which the vibrations are taking place has been rotated.

With a given substance the angle through which the plane of the vibrations is rotated is proportional to the thickness of the plate, and with a given thickness of plate, this angle of rotation varies inversely with the wave-length, being approximately inversely proportional to the square of the wave-length. Hence, with white light there will be no position of the analyser at which all light is extinguished and, on rotating the analyser, different colours will be transmitted in turn as each wave-length is successively extinguished. This phenomenon, which was discovered by Arago in 1811 and fully investigated by Biot (1774–1862) is known as **optical rotation** and occurs not only with quartz but

381

also with some other crystals and with certain liquids, such as turpentine, solutions of sugar and tartaric acid. All such substances are said to be **optically active.**

Some substances will rotate the plane of polarisation to the right and others to the left, and plates of quartz cut from different crystals may rotate the plane in either direction. A substance is said to be right-handed (positive) or left-handed (negative) rotating, according to the direction of rotation to an observer looking towards the polariser. Thus with monochromatic light, if on looking through the analyser it is necessary to turn it in a clockwise direction to extinguish the light again, the substance is right-handed, and vice versa.

The rotation produced by two plates is the algebraic sum of the rotations due to the separate plates. Thus a plate of right-handed quartz just neutralises the rotation of the plate of left-handed quartz of the same thickness. The rotation produced by a plate of quartz 1 mm. thick for wave-lengths corresponding to the chief Fraunhofer lines is as follows:

A	B	C	D	E	F	G
12·67°	15·50°	17·22°	21·67°	27·46°	32·69°	42·37°

There is no double refraction or optical rotation in the case of fused quartz, the crystalline structure having been destroyed and this makes it possible for this material to be used in the construction of high-class microscope objectives for use in the ultra-violet, to which it is transparent.

Cane-sugar is right-handed rotating and grape-sugar left-handed.

In the case of an optically active solution the rotation for a given thickness of solution is very closely proportional to the weight of the active material dissolved in unit volume of the solution, and is expressed in terms of the **specific rotation** as follows:

$$\text{Specific rotation} = \frac{\text{angle of rotation for a thickness of 10 cm.}}{\text{density of solution in grams per c.c.}}$$

The following are the specific rotations of a few solutions for the D line for 10 cm. thickness at 20° C. Positive values are right-handed rotations and negative values left-handed.

Cane Sugar	+ 66°·67
Tartaric Acid	+ 15°·06
Rochelle Salt	+ 29°·73
Camphor (in alcohol)	+ 54°·4
Turpentine (pure)	− 37°
Nicotine (pure)	− 162°

The rotation given by a sugar solution provides a convenient means of ascertaining its strength and is largely used in industry. Instruments used for determining the optical rotation of a substance are known as **Polarimeters** and the special modified forms used in the case of sugar, as **Saccharimeters.** In its simplest form the polarimeter consists of two Nicols, the analyser being fitted with a divided circle on which its rotation can be read. Monochromatic light, usually from a sodium flame, is passed through the polariser and the analyser turned until the light is extinguished. The substance to be examined, such as an axis-cut plate of quartz, or a solution contained in a tube with plane glass ends, is placed between polariser and analyser, and the analyser is rotated until the light is again extinguished, the angle of rotation being read on the scale.

Difficulty is experienced in practice in determining the exact position at which the light is cut off, and a more accurate setting is obtained by the use of devices that divide the field into two halves, which have equal brightness at the correct setting, and are unequal at other positions of the analyser. One such device is the **Cornu-Jellet prism,** which replaces the ordinary Nicol as polariser. This prism is formed by cutting from a Nicol a thin wedge-shaped piece along its length as in Fig. 229 and cementing the two halves together.

The beams leaving the two halves of the prism have their vibrations inclined at a small angle to one another, and the analyser can be set so that one beam is completely cut off, in which case a small amount of light in the other will be transmitted, giving one half of the field dark and the other half faintly illuminated. Or it may be set with its principal plane perpendicular to the direction equally inclined to the two vibrations from the polariser, in which case the two halves of the field are equally bright. The vibrations in the two beams from the polariser are rotated through an equal angle by an optically active substance and

Fig. 229.—Cornu-jellet prism. End view

the correct setting of the analyser is given when the two halves of the field are equally bright, a position which can be found with considerable accuracy, because of the contrast between the two halves of the field in any other position of the analyser.

Descriptions of other forms of polarimeters and saccharimeters and fuller details of procedure will be found in modern books on polarimetry.

The explanation of optical rotation was given by Fresnel who showed that when a plane polarised vibration enters a plate of axis-cut quartz the vibration is resolved into two equal circular vibrations revolving in opposite directions. These two circular vibrations travel through the plate at unequal velocities, and therefore emerge with a difference in angular phase. Two circular vibrations of equal amplitude and period but rotating in opposite directions when acting together are equivalent to a rectilinear vibration, for, as will be seen from Fig. 230a, if the two vibrations pass one another at A and B, the displacements such as CD and CE perpendicular to AB are equal but opposite and neutralise, leaving a displacement only along AB. If the velocities of propagation of the two vibrations in the plate are unequal they will leave the plate with a difference in phase, so that while the clockwise vibration is making a full revolution, the counter-clockwise vibration is rotating only from A to A' (Fig. 230b). The resultant is then a rectilinear vibration along A''B'', and in this case the plane of vibration is rotated in a right-handed direction. Similarly it will be seen that the rotation of the plane of vibration will be left-handed when the counter-clockwise vibration has the greater velocity.

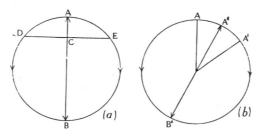

Fig. 230.—Optical rotation

220. Magneto-Optics and Electro-Optics

As light is an electromagnetic disturbance it may be expected that it should be influenced by electric and magnetic fields. This is particularly so when the field is acting on the source of the light. In this case the reaction is primarily between the field and the electron orbits of the source emitting the light. The **Zeeman** effect relates to magnetic fields and the **Stark** effect relates to electric fields. In both cases each spectral line of the source is observed to divide into a number of closely spaced lines.

Related to the Zeeman effect is the **Faraday** effect. If a magnetic field is generated along a piece of glass, a plane-polarised light beam directed parallel to the field has its plane of polarisation rotated. This effect is greater with high refractive index materials but occurs over a wide range of solids, liquids and gases. A related phenomenon, the **Voigt** effect, occurs when the light path is perpendicular to the magnetic field. Under these circumstances the material may become doubly refracting.

Related to the Stark effect is the effect in which a material placed in a strong electric field exhibits double-refraction to light beams perpendicular to the field. All the effects so far considered in this section relate inter-actions *within the atomic structure.*

Other, and usually considerably larger, effects can occur when the magnetic or electric field affects the alignment of the molecules within a material. The **Cotton-Mouton** effect occurs with liquids like nitrobenzene when anisotropic molecules are all lined up by a transverse magnetic field and bulk double-refraction results. Nitrobenzene exhibits a similar action with a transverse electric field which is known as the **Kerr** effect. This is used in the **Kerr cell** which uses the fast switching of an electric field to alter the polarising properties of the liquid. When placed between a polariser and analyser the system comprises an **electro-optical shutter** which may be used for chopping light very quickly. This has been utilised for velocity measurements, optical radar, television and other applications.

More recently, techniques of single crystal manufacture have been developed. A small crystal is dipped into a molten bath of the same material and very slowly withdrawn. Given the correct conditions the material solidifies into a sizeable volume with all the crystal structure in perfect alignment. These can react very strongly to electric fields and also to the presence to an electromagnetic field such as a light beam itself. As well as providing modulators and deflecting systems, these materials can produce light at twice the frequency of a beam incident upon them or if two beams are incident new light beams can be generated of frequencies equal to the sum and the difference of the incident beams. This is a similar phenomena to beats formed with sound waves and for it to occur the light beams must be coherent.

Some liquids having long thin molecules are known as liquid crystals as the molecules have a tendency to line up in a regular order. This can be disturbed by electric fields and currents and thereby change the appearance of the liquid between transparent to scattering or optically active. When placed between glass sheets carrying conductive transparent coatings these materials can be used in numerous display applications.

CHAPTER XVI   *EXERCISES*

1. Describe two experiments which show that light is propagated as a transverse wave motion. How is the law of rectilinear propagation explained on the wave theory?

2. Give a short account of the nature of light, describing experiments to prove the truth of your statements.

3. Explain what is meant by plane polarised light and describe two methods by which it can be produced.

4. Explain why two crossed tourmalines or Nicols allow no light to pass through, and why interposing a plate of crystal between them may restore the light (S.M.C.)

5. Describe the phenomenon of double refraction. The end face and the diagonal face of a Nicol prism are at right angles. Show on a careful diagram the approximate paths of the ordinary and extraordinary rays within the prism, the light being incident on the end face at an incident angle of 50°. The refractive indices of Iceland Spar are approximately 1·66 (ordinary) and 1·49 (extraordinary) and of Canada balsam 1·53. (S.M.C.)

6. Show that when $i + i' = 90°$, $\tan i = n$. What is the difference between the incident and reflected light for this particular angle of incidence?

7. Describe Brewster's law concerning the polarisation of light by reflection. Describe a form of double image prism, making quite clear, with the help of a diagram, its construction; and explain carefully the reason for the course taken through the prism by an incident ray. (S.M.C.)

8. If a picture is hung so that it is badly seen owing to light reflected from its surface it can often be clearly seen through a Nicol prism or other analyser. Explain this. (S.M.C.)

9. Using Fresnel's law of reflection plot the curves showing the variation of the intensity of the reflected light with variation of the angle of incidence when light is reflected from the surface of a medium of refractive index 1·6, for vibrations taking place (a) parallel (b) perpendicular to the plane of incidence. From the curves determine the polarising angle.

10. The angle of incidence of a parallel pencil of white light on the plane polished surface of a block of glass of refractive index 1·760 is 60° 24′. What will be the angle between the reflected beam and the beam refracted into the glass? Explain carefully any differences there will be in the nature of the incident, the reflected and the refracted light.

11. Explain the construction and action of a Nicol prism. Describe two practical applications of the use of polarised light obtained by the use of the Nicol prism.

12. A Wollaston double image prism is made of two 45° prisms of quartz, the refractive indices being 1·544 for the ordinary ray and 1·553 for the extraordinary. Find the angle between the two emergent beams if the incident beam of unpolarised light is normal to the first surface.

13. Describe and explain the appearances which are seen when a thin plate of selenite cut parallel to its axis is placed in a polariscope, when the Nicols are (a) parallel, (b) crossed.

14. Plane polarised light ($\lambda = 600$ nm) is incident on a plate of quartz 0.5 mm. thick cut with its faces parallel to the axis. Find the difference in phase in the two beams leaving the plate.

15. A thin plane plate of a doubly refracting crystal cut with the optic axis in the plane of the surfaces is placed between crossed Nicols. Explain carefully the appearances seen as the plate is rotated about a line joining the Nicols. When the plate of crystal is placed with its optic axis at 45° to the principal plane of the polariser, what will be the effect of rotating the analyser?

16. A beam of plane polarised sodium light is incident on a plane plate of quartz cut with the optic axis in the plane of the surfaces, the plane in which the vibrations of the incident light are taking place being at 45° to the optic axis. Find the least thickness of the plate so that the resultant vibration produced on an ether particle by the ordinary and extraordinary beams immediately on their emergence from the plate is (a) plane, (b) circularly polarised.

17. Two Nicol prisms are placed in line to form a polariscope; compare the intensities of the transmitted beam when the principal planes of the Nicols are inclined at (a) 0°, (b) 30°, (c) 45°, (d) 60°, (e) 90°.

18. Explain what is meant by a quarter wave plate. The refractive indices of mica being 1·561 (ordinary) and 1·594 (extraordinary), what must be the thickness of a film of mica to produce a quarter wave plate for sodium light? ($\lambda = 589$ nm).

19. What is meant by optical rotation? Describe a practical application of this phenomenon.

20. A solution of sugar in a tube 30 cm. long rotates the plane of polarisation through 35°. If the specific rotation of the sugar is 66·7°, how many grams of sugar are there per cubic centimetre of the solution?

21. Along what direction in a uniaxial crystal must light travel in order that the velocities of the ordinary and extraordinary rays shall be equal? Define the refractive index of a crystal for the extraordinary ray.

A ray of plane polarised light, the vibration direction being vertical, is incident normally on a Wollaston prism made of quartz, the axis direction of the first component prism being vertical. Show on a diagram how a ray will pass through and emerge from the prism. Give a brief explanation. (N.B.—In quartz the extraordinary index exceeds the ordinary.) (S.M.C.)

22. A double image prism consists of a 30° prism of calcite cemented to a glass prism. The calcite prism is cut with the optic axis of the crystal in the plane of the first surface and parallel to the base-apex line. If light falls normally on the first face of the calcite prism, find the angle of the glass prism required in order that the extraordinary beam shall emerge undeviated. What will be the deviation of the ordinary beam?

Refractive indices: Calcite, ordinary ... ... 1·659
extraordinary ... 1·486
Glass ... ... 1·510

23. A beam of plane polarised light of wave-length 540 nm falls normally on a thin plate of quartz cut so that the optic axis lies in the surface. If the ordinary and extraordinary refractive indices are respectively 1·544 and 1·553, find for what thickness of the plate the difference in phase between the ordinary and extraordinary beams is π radians on emergence from the plate. (S.M.C.)

24. Plane waves of light are incident obliquely on the surface of a uniaxial crystal such as calcite. Give a careful diagram with explanation showing the passage of the light within the crystal when the optic axis lies perpendicular to the plane of incidence and parallel to the surface of the crystal.

Hence define the refraction index of the crystal for the extraordinary ray.

In what important respect do the ordinary and extraordinary beams emerging from the crystal differ from one another? (S.M.C.)

XVII – The Effects of Limitation of Beams in Optical Systems

221. Introductory

In the earlier chapters on geometrical optics little attention has been paid to the actual extent of pencils and beams passing through any optical system or instrument, except that most of the theoretical results obtained are limited, in the case of simple lenses, to the paraxial region, and therefore the pencils and beams are very narrow. Thus in the graphical constructions of Chapter VI the actual paths of light by which the image is formed were not considered, and the rays used in the constructions are not usually directions of the light that actually passes

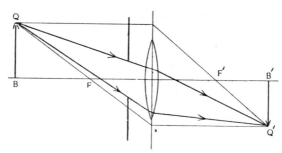

Fig. 231.–Graphical construction of image and actual path of rays from object point

through the lens or instrument. An example of this is shown in Fig. 231, where, in addition to the rays used in graphically constructing the image, the path taken by the light from one end of the object through the lens is shown.

Pencils and beams passing through any instrument are limited by apertures, which may be the actual apertures of the lenses, prisms, etc., or additional diaphragms or stops. These latter are used to cut off light not travelling in the required direction to form a perfect image, to cut off the outer parts of the image, where the definition is defective, that is, to limit the field of view, and to intercept stray light reflected from the interior of tubes, lens mounts, etc. The various apertures of a system will influence (1) the field of view, (2) the depth of focus, (3) the illumination of the image, and (4) the resolving power.

222. Entrance and Exit Pupils

One of the apertures of a system or instrument will be effective in limiting the width of the pencil of light from an object point on the axis, and this is known as the **aperture stop** of the system. This may be the mount of one of the lenses or a stop in front of, behind or within the system. When there are any lenses between the object and the aperture stop, the pencil in the object space will be travelling as though it passes through the image of the aperture stop as formed by any lenses in front of it, and in the image space the pencil is limited as if by the image of

Fig. 232.—Entrance and exit pupils

the aperture stop formed by any lenses behind it. In Fig. 232, S_1T_1 is the aperture stop of the system of two lenses at A_1 and A_2. It will be seen that the pencil from the object point B is incident as if passing through ST, the image of S_1T_1 formed by the first lens at A_1, while the pencil going to the image point B' emerges from the system as though it had come through S'T', the image of S_1T_1 formed by the second lens at A_2. Any rays which pass through an actual stop must pass through or towards corresponding points in the image of the stop, and any rays cut off by a stop will not pass through the opening in the image of the stop. In dealing with the limitation of pencils and beams in passing through a system it is very convenient to be able to consider all the apertures as being in the same medium, and this may be done by taking in the object space all the actual apertures in front of the system and the images of other apertures formed by any lenses in front of them. In the image space we have any actual apertures behind the system and the images of the other apertures formed by lenses behind them.

The stop or stop-image that subtends the smallest angle at the object point B is called, after Abbe, the **entrance pupil** of the system. When the aperture stop is in front of the system it will be the entrance pupil, but in other cases the entrance pupil will be the image of the aperture stop formed in the object space. The stop or stop-image subtending the smallest angle at the image point B' is called the **exit pupil** of the system, and will correspond to the aperture stop when this is behind the system or will be the image of the aperture stop formed in the image space. In Fig. 232, ST is the entrance pupil and S'T' the exit pupil. Since any ray in the object space passing through or towards a point in the entrance pupil, on leaving the system is travelling through or as though from a corresponding point in the exit pupil, the exit pupil is the image of the entrance pupil formed by the complete system, and the centres of the pupils R and R' are conjugate points with respect to the system. The angle subtended at B by the radius RS of the entrance pupil is known as the **aperture angle** of the system for that particular object.

223. Field of View

In addition to limiting the width of the pencil from an axial object point and therefore the amount of light going to the image, the apertures of a system will restrict the extent of object of which an image can be formed by the system. This extent of object is called the **field of view** of the system, and is usually expressed as the angle that the field subtends at some given point in the system. In the case of an instrument, such as a telescope, used with the eye, the field of view is the extent of object that can be seen through the instrument by the eye rotating about its centre of rotation. As a simple example of field of view we may consider the case of a person in a room looking through a window. Here the aperture stop of the system is the pupil of the person's eye, and the entrance pupil is the image of this stop formed by the cornea. The field of view is obviously dependent upon the size of the window opening and the position of the eye, and the extent of the field of view may be determined by drawing straight lines from the centre of rotation of the eye past the edges of the window. The window here acts as the **field stop.**

Fig. 233 shows the field of view given by a lens of circular aperture DE with a circular stop ST placed before it. ST in this case is both the aperture stop and the entrance pupil with respect to an object point at B. In Fig. 233a, the full pencil of light passing through ST from the object point B is transmitted by the lens and goes to form the image at B'. It will be seen that pencils of the same width are transmitted from all object points between B and Q_1 (Fig. 233b), and the image B'Q'$_1$ will have uniform illumination. If the two apertures are centrally situated with respect to the optical axis, there will be a circular field of radius B'Q'$_1$ in the image plane in which the illumination is uniform, corresponding to a circular field of radius BQ$_1$ in the object plane. With object points lying farther from the axis than Q_1, the whole of the pencil passing through the entrance pupil is not transmitted by the lens, and the illumination of the image is accordingly reduced; thus in Fig. 233c, only one-half the incident pencil from Q_2 is able to pass through the lens. The width of the transmitted pencil is continually reduced as the object point is taken farther from the axis until as shown in Fig. 233d, the incident pencil from Q_3 just misses the lens aperture entirely. Thus the complete field of view in this case will consist of a central portion over which the illumination is uniform, this is known as the **field of full illumination,** and an outer "vignetted" portion in which the illumination

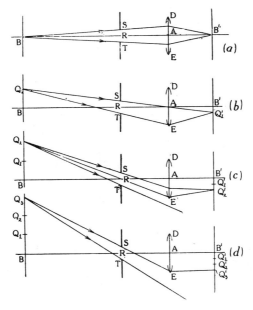

Fig. 233.—Field of view

gradually decreases to zero. From the figure it is seen that the field of full illumination is determined by drawing straight lines through corresponding points in the edges of the two apertures on the same side of the axis, such as the points E and T, D and S (Fig. 233b). The total field is determined by straight lines drawn through points on opposite sides of the axis, such as E and S, D and T (Fig. 233*d*). This case represents the field of view obtained with a photographic lens fitted with a diaphragm in front, the so-called landscape lens.

The stop or stop-image in the object space that with the entrance pupil limits the field of view is called the **entrance port,** and the actual stop or lens mount corresponding to this is the **field stop.** In Fig. 233 the lens aperture DE is both the field stop and the entrance port. The entrance port will be the stop or stop-image that subtends the smallest angle at the centre R of the entrance pupil. The image of the entrance port formed by the entire system is the **exit port** and limits the field of view in the image space.

In order to determine the field of view of any instrument, we must first find the images in either the object or image space of all stops that are likely to behave as entrance and exit pupils and as entrance and exit ports. Then if we consider the object side, the entance pupil is found as the actual stop or stop-image subtending the smallest angle at the axial point of the object and the entrance port as the actual stop or stop-

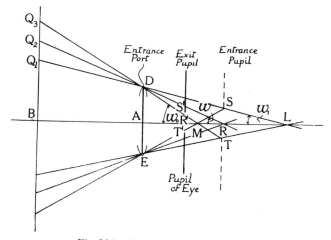

Fig. 234.—Field of view of magnifier

image subtending the smallest angle at the centre of the entrance pupil.

The fields of full and partial illumination on the object side can then be determined by drawing straight lines through corresponding and opposite edges of the entrance pupil and port, as in Fig. 233. When the object lies between the entrance pupil and entrance port, as in

§ 228, the field of full illumination is determined by lines joining opposite edges of the entrance pupil and port, and the total field by lines joining corresponding edges. Frequently the field of view of an instrument is defined as the field over which the illumination does not fall below half the illumination at the centre. This, as will be seen from Fig. 233c, is an angular field equal to the angle subtended by the entrance port at the centre of the entrance pupil, and is therefore independent of the size of the entrance pupil.

Similar constructions to the above, using the exit pupil and port, will determine the field of view on the image side, or this may be determined directly as the image of the object field. A few examples will illustrate the method.

Convex lens used as magnifying glass.—The aperture stop in this case is the opening in the iris of the eye, and the exit pupil $S'T'$ (Fig. 234) is the pupil of the eye seen through the cornea. The entrance pupil ST is the image of $S'T'$ formed by the magnifying lens, and if, as is usually the case, the eye is fairly close to the lens, that is, inside its principal focus, the entrance pupil is a virtual image behind the lens. The lens rim is both the field stop and entrance port in this case. Then the angle $BLQ_1 = w_1$ is the semi-angular field of full illumination, the angle

Fig. 235.—Field of view of Galilean Telescope

$BRQ_2 = w$ is the semi-angular field of half illumination, and the angle $BMQ_3 = w_0$ is the semi-angular total field.

Galilean Telescope (Fig. 235).—In this case it will be more convenient to consider the field of view referred to the image space. The aperture of the objective will be the entrance port and its image formed by the eyepiece the exit port, this will be a virtual image situated between the objective and eyepiece (see Chapter X). As the aperture of the eyepiece is always of much larger diameter than that of the eye pupil, the latter will act as exit pupil of the complete system consisting of telescope and eye. In Fig. 235 $D'_1E'_1$ is the image of the objective formed by the eyepiece—the virtual Ramsden circle—and $S'T'$ the eye pupil. Then

w'_1, in the image space, is the semi-angular field of full illumination, w' the semi-angular field of half illumination and w'_0 the semi-angular complete field.

From the figure

$$\tan w' = \frac{D'_1 A'_1}{A'_1 R'}$$

$$\tan w'_1 = \frac{D'_1 A'_1 - S' R'}{A'_1 R'}$$

$$\tan w'_0 = \frac{D'_1 A'_1 + S' R'}{A'_1 R'}$$

If y_1 is the semi-diameter of the objective, d_1 the separation between objective and eyepiece, p'_e the semi-diameter of the eye pupil, and d_2 its distance from the eyepiece, then the semi-diameter of the Ramsden circle is $\frac{y_1}{M}$, where M is the magnifying power of the telescope, and its distance from the eyepiece is $\frac{d_1}{M}$.

Hence

$$\tan w' = \frac{\dfrac{y_1}{M}}{\dfrac{d_1}{M} + d_2} \qquad \ldots \qquad \ldots \qquad \ldots \quad (17.01)$$

$$\tan w'_1 = \frac{\dfrac{y_1}{M} - p'_e}{\dfrac{d_1}{M} + d_2} \qquad \ldots \qquad \ldots \qquad \ldots \quad (17.02)$$

$$\tan w'_0 = \frac{\dfrac{y_1}{M} + p'_e}{\dfrac{d_1}{M} + d_2} \qquad \ldots \qquad \ldots \qquad \ldots \quad (17.03)$$

The field of view on the image side is called the **apparent field of view** of the telescope, and the field of view on the object side—the actual field of view—can obviously be determined by dividing the apparent field by the magnifying power of the instrument.

Astronomical Telescope.—In the astronomical telescope the Ramsden circle is a real image of the objective and in the ordinary way will coincide in position with the pupil of the eye. In this case the rim of one of the eyepiece lenses would be effective in limiting the field of view. In practice, however, a stop is usually placed in the common focal plane of objective and eyepiece, its diameter being such that the outer vignetted portion of the field is cut off. The field therefore has a well defined boundary, the image of the stop seen through the eyepiece. The apparent field will be the angle subtended at the eyepiece by the stop diameter, and the angle subtended at the objective by the stop diameter will be the actual field.

The aim in many instruments is to obtain as large a field of view as possible together with magnification. This entails a high value for the apparent field, and such a requirement demands the highest qualities in the design of the lenses. Thus in a binocular magnifying six times with a field view of 8°, the eyepiece is required to give good definition over a field of 48°. If the magnification is increased to eight times, for the same apparent field covered by the eyepiece, the actual field is reduced to 6°, or if the same actual field is required the eyepiece must be designed to give good definition over a field of 64°. Therefore an increase in the magnification tends to lead to a decrease in the field of view. The apparent field of view is a useful value in comparing the fields of telescopes of different magnifications and expresses what might be called the efficiency of an instrument.

224. Depth of Focus

The distance from a lens of any image point is dependent upon the distance of the corresponding object point, and there will be a solid image formed in space of any solid object. While the dimensions of the image in all directions perpendicular to the axis are the same fraction of the corresponding object dimensions, if aberrations are neglected, this is not so for dimensions along the axis, as was shown in Chapter VI. The image of a solid object formed in space is not, therefore, an exact scale reproduction of the object.

In practice, the image formed by an instrument is always received on a surface or screen, either directly, as in the case of photographic and projection apparatus, or on the retina of the eye in the case of the images formed by microscopes and telescopes. Only object points that lie in the same plane can, of course, have sharply defined image points on a receiving screen in a particular position, and object points in front of or behind this plane will be represented by patches of light that will be circular on the lens axis and elliptical in other positions. As these patches do not in the ordinary way depart far from a circular form they are usually referred to as **blur circles**. From Fig. 236 it is clear that the dimensions of these so-called blur circles depend on the position and diameter of the exit pupil and the distance of the true image position from the screen.

It is well known that in practice it is possible to produce apparently sharp images of objects at different distances from the lens simul-

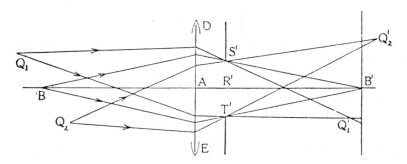

Fig. 236.—Depth of focus

taneously on a plane surface. Thus in a photograph of, say, a landscape, object planes that may be separated by considerable distances are all reproduced with apparent sharpness on the plate. The reason for this is that objects that subtend less than a certain angle, usually taken as one minute, at the eye appear to the eye as points, and if the blur circles are sufficiently small compared with the distance at which they are viewed, there will be no apparent difference between the portions of the image that are perfectly sharp and those that are blurred to this extent. A simple example of a similar kind occurs in the case of a picture printed in halftone. This is made up of light and dark dots, as can be seen with a magnifier, but when viewed at a suitable distance gives the effect of a sharp picture, the separate dots being indistinguishable.

The greatest distance the screen may be moved away from the theoretical image position without the image appearing unsharp is called the **depth of focus.** The corresponding distance on the object side, that is, between object planes that are apparently sharply imaged in one plane, is often called the **depth of field.** These distances are obviously largely dependent upon the standard of definition required, that is, the permissible diameter of the blur circles. In the case of an ordinary photograph to be viewed directly, $\frac{1}{100}$ inch is frequently taken as the limiting value of the blur circle diameter, but a better standard is a blur circle diameter of one-thousandth the focal length of the lens. This will allow for the enlargement usually required by the small pictures obtained with short focus lenses.

Fig. 237.– Depth of focus

Fig. 237 represents a system having entrance and exit pupils ST and S'T' respectively. Light from an axial object point focuses at B'_1 and from some other closer point at B'_2; then if a receiving screen is placed at the image point B'_1, the second image is represented by a blur circle of diameter z. From the figure

$$\frac{z}{x'} = \frac{2p'}{l'+x'} = \frac{2p'}{l'} \qquad \text{since } x' \text{ is usually very}$$

small as compared with l'. If B'_1 is at the principal focus of the system, l' will in most cases differ little from the equivalent focal length of the system, and we then have

$$\frac{z}{x'} = \frac{2p'}{f'} \quad \text{...} \qquad \text{...} \qquad \text{...} \qquad \text{... (17.04)}$$

$\frac{2p'}{f'}$ is the **aperture ratio** of the system, and is for most practical purposes the reciprocal of the f/No., as in most photographic objectives there is little difference in the diameters of entrance and exit pupils.

The distance $(f' + x')$ is the image distance of the nearest object sharply defined when the lens is correctly focused on a very distant object. This object distance is called the **hyperfocal distance,** and if the screen is placed at the correct focus for this distance, all objects from infinity to approximately half the hyperfocal distance have apparently sharp images on the screen.

225. The Telecentric Principle

If the aperture stop of a system is so placed that the entrance pupil falls in the first focal plane (Fig. 238a) the exit pupil is at infinity, and the rays through the centre of the entrance pupil from all points on the object are parallel to the axis in the image space. Similarly, if the exit pupil lies in the second focal plane (Fig. 238b) the entrance pupil is at

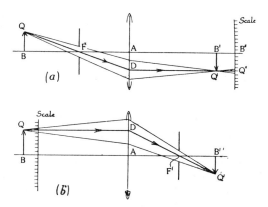

Fig. 238.—The telecentric principle

infinity, and the rays through the centre of the exit pupil are parallel to the axis in the object space. Such systems are said to be **telecentric** and have important applications in measuring instruments. Where the length of an object is to be measured by projecting its image on a scale, as, for example, in the various micrometer microscopes, errors frequently arise from the fact that, owing to depth of focus, the image may not be exactly in the plane of the scale, although both appear in focus together. An error may then be introduced as can be seen from Fig. 236, where if the scale is supposed to be in the position of the screen, the distance of Q_1 from the axis is measured by its image on the scale will be too large, while that of Q_2 will be too small. A similar error may occur when the scale is on the object side but cannot be placed in the object plane. This error can be avoided by the use of the telecentric system, for, as is shown in Fig. 238a, the length $B''Q''$ projected on the scale is the same as that of the focused image $B'Q'$. The telecentric principle is applied in the Wessely Keratometer, an instrument used in ophthalmic work for measuring the distance of the back vertex of a spectacle lens from the eye's cornea.

226. The Illumination of an Image

If the loss of light due to reflection and absorption of the light passing through a system is neglected, it can be shown that the illumination of an image is proportional to the luminance of the object and the area of the entrance pupil. Let h be the area of an object of luminance L at a distance l from the entrance pupil of area a, then if a and h are small compared with l, the luminous flux reaching the entrance pupil will be $\dfrac{Lha}{l^2}$. Neglecting reflection and absorption losses this flux is transmitted by the system and goes to the image. The area of the image h' will be $\dfrac{l'^2 h}{l^2}$, and since the flux is distributed over this area, the illumination of the image will be $\dfrac{La}{l'^2}$. In the case of the eye l' is constant, and the illumination of the retinal image is dependent only upon the luminance of the object and the pupil diameter. With a given pupil diameter, an object will therefore appear equally bright no matter what its distance, neglecting, of course, absorption due to the atmosphere.

When an instrument such as a telescope is used with the eye, with the exit pupil or Ramsden circle in the plane of the eye pupil, the luminous flux entering the eye is governed by either the Ramsden circle or the eye pupil, whichever is the smaller. The luminous flux entering the instrument is proportional to the area of the entrance pupil, that is, to p^2, where p is the radius of the entrance pupil, and after passing through the instrument this flux is concentrated in the area of the exit pupil of radius p', if reflection and absorption losses are neglected. Providing the eye pupil of radius p'_e is not smaller than the exit pupil of the instrument, so that the whole of the light can enter the eye, the ratio of the flux received by the eye with the instrument to that received from the same object by the unaided eye will be $\left(\dfrac{p}{p'_e}\right)^2$. The retinal image, however, is magnified M times, and the area over which the light is distributed is therefore increased M^2 times, hence the illumination on the retina of the image seen through the telescope will be $\left(\dfrac{p}{p'_e}\right)^2 \dfrac{1}{M^2} = \left(\dfrac{p'}{p'_e}\right)^2$ times that of the image received by the unaided eye. This factor cannot be greater than unity, for if the exit pupil is greater than the eye pupil the light from the outer parts are cut off by the iris, and the effective diameter of the exit pupil is then the diameter of the eye pupil.

Thus an extended object seen through a telescope can never appear brighter than when seen without the telescope, and will actually always be less bright through losses of light by reflection and absorption. To obtain the maximum illumination with a telescope in viewing an extended object, it is necessary that the diameter of the exit pupil should be equal to the diameter of the eye pupil and therefore the objective diameter should be M times the diameter of the eye pupil. Manufacturers of binoculars frequently specify the **light transmitting power** of an instrument as the square of the diameter of the exit pupil in millimetres. It will

399

be realised that this value has no significance when the diameter of the exit pupil exceeds the diameter of the eye pupil possible under any conditions.

When the object is a point, such as a star, the image is a disc, the size of which depends on the aperture of the objective, as explained in Chapter XV, and is independent of the magnification. Then, if the exit pupil of the instrument is smaller than the pupil of the eye, all light which passes through the entrance pupil—the objective aperture in this

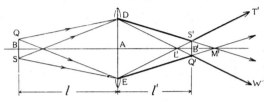

Fig. 239.—The nature of a beam

case—passes into the eye, and the luminosity of the image, as compared with the image seen by the unaided eye, is increased in the ratio of the area of the objective to the area of the pupil of the eye. When the diameter of the exit pupil exceeds the diameter of the eye pupil, the latter can be considered as the effective exit pupil and its image as formed by the instrument as the entrance pupil. The area of the effective entrance pupil is then M^2 times that of the eye pupil, and the luminosity of the image is increased in this ratio. Thus when the object is a point it is possible to increase the luminosity by means of a telescope, and in the case of large astronomical telescopes, where the diameter of the objective may be several feet, the increase in luminosity will be very considerable. Since the apparent brightness of a star can be increased in this way while the apparent brightness of the sky, this being an extended object, is unchanged or more often reduced, it is possible by means of a large telescope to see stars in full daylight.

227. The Nature of a Beam

Fig. 239 shows the beam from a source QS and the lens DE. The individual pencils converge to form an image of the source S'Q', and the edges of the beam are shown by the thick lines DS'T' and EQ'W'. This beam has its smallest cross-section at S'Q', and this position is termed the *waist* of the beam. Here the illumination will be greatest, since the flux is concentrated into the smallest space. Beyond the waist the beam is diverging from the point L'. The position of the waist is often important, for example, in the projection lantern the objective should occupy approximately this position in the beam from the condenser, in order that the whole beam shall be transmitted by an objective of reasonable aperture. When the image of the source is larger than the lens aperture or when it is virtual, the lens aperture will be the waist of the beam.

When an eye is situated in a beam between the positions L' and M' the whole lens is seen filled with light of uniform intensity as the out-of-focus image of the source subtends a greater angle than is subtended by the lens aperture. This appearance is known as **complete flashing** or the **Maxwellian view.** As the eye moves from this position, either towards or away from the lens, the area of the portion of the lens from which the light appears to be coming is reduced, a blurred image of the source being seen partially filling the lens aperture. The beam from a curved reflector may be considered in the same way, and the principle has important applications in dealing with motor headlights, search-lights and signalling lamps.

The principle of the Maxwellian view is sometimes useful in photo-metry, and is particularly applicable to the measurement of very small sources or extended sources of low intensity. A convergent lens forms an image of the source in the pupil of the eye, the eye being accommodated so that an image of the lens is formed on the retina. The eye then sees the lens as a field of uniform luminance. When the image of the source falls completely within the pupil of the eye, the apparent luminance L' of the field, neglecting any loss of light be reflection and absorption at the lens, is

$$L' = \frac{h\,L{l'}^2}{a_e\,l^2} = \frac{I\,{l'}^2}{a_e l^2} \quad \ldots \quad \ldots \quad \ldots \quad (17.04)$$

where h is the area, L the luminance and I the candle-power of the source, and a_e the area of the eye pupil. When the source is of such dimensions that its image more than covers the pupil, the apparent luminance of the field is equal to that of the source.

$$L' = L$$

again neglecting losses of light at the lens.

A simple application of the principle of flashing is used in the examina-tion of glass for veins or striae. Light from a bright point source B (Fig. 240) is rendered parallel by the lens L_1, the plate of glass to be examined is placed in the parallel beam, and the light is brought to a focus at B' by the lens L_2. The eye is placed at this focus, but in such a position that the focus falls just outside the pupil and, if the glass is homogeneous, no light enters the eye. Any veins of different refractive index will refract the light, and some of the light from them will enter the eye, which then sees the veins as bright lines on a black background.

A similar principle is used in the "knife-edge" test of lenses and curved mirrors and in retinoscopy.

Fig. 240.—Test for striae. Dotted line shows direction of light, refracted by vein, entering the eye

228. Object Illuminated by Transmitted Light. Condensers

In the foregoing considerations the object has been assumed as self-luminous or as one that reflects diffuse light in all directions. There are, however, a number of cases, particularly in optical projection and microscopy, where the object, usually in such cases called a *slide*, is more or less transparent and is illuminated by transmitted light from a source some distance behind.

Let us consider first the case of an axial object point B (Fig. 241), illuminated by a source NP which is supposed to be of uniform luminance and circular in form. ST is the entrance pupil of the instrument, in this case a single lens, that forms an image of B. It will be seen that when, as in the figure, the source subtends an angle at the object equal to or greater than that subtended by the entrance pupil, the cone of light transmitted by the system is the same as that from a self-luminous point

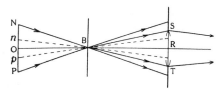

Fig. 241.—Point object illuminated by transmitted light

at B. When, however, the source, such as np, subtends a smaller angle, it governs the angular cone entering the system, and behaves as the entrance pupil.

With an extended object the extent of object from which full pencils can pass through the instrument is limited by the angle subtended by the source at the entrance pupil. This is shown in Fig. 242, where ST is the entrance pupil and DE the entrance port of the instrument forming an image of an extended object placed at B. If the object was self-luminous or diffusely reflecting, Q_1Q_2 would be the extent of object, of which the image would be evenly illuminated—the field of full illumination. If however, the object is transparent and illuminated by light from the source NP, the extent of object from which complete pencils can pass is limited to q_1q_2; this is determined by connecting the opposite edges of the source and entrance pupil. Thus for a point on the object, such as q_3, no light is passing from the source through the lower half of the entrance pupil, and the illumination of the image of this point is consequently reduced to half that of the axial point. Hence we see that the source may behave as the entrance port of the instrument, as it does in the case just considered.

It is necessary therefore that if the full aperture and field of view available in the instrument forming the image are to be used, the source must subtend a sufficiently large angle both at the object and at the entrance pupil. In a few cases where the required luminance of the source

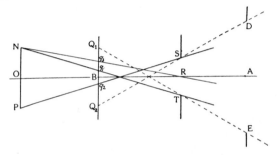

Fig. 242.—Object illuminated by transmitted light

is comparatively low, an evenly illuminated surface such as the white sky can be used as the source and will subtend a sufficiently large angle. Usually, however, because of the impossibility of producing an extended source of high luminance and also the necessary limitation of the distance at which the source can be placed owing to size of lamp and amount of heat radiated, the source cannot be made to subtend a sufficiently large angle directly. Under these circumstances we may make use of an image of the source formed by a positive lens or system, and this image may be made to subtend a sufficiently large angle at the object or at the entrance pupil. A lens used for this purpose is called a **condenser.** By making the condenser of large aperture as compared with its focal length, large angular cones of light may be collected from the source and passed through the system.

From the conditions to be satisfied in the illumination of a transparent object, which have been discussed above, it will be seen that the condenser may be used in two different ways according to the purpose of the instrument. When the instrument is required to form images of object points with pencils of large aperture, the angular extent of the object being small, the object should be situated at or close to the image of the source formed by the condenser, that is, in the waist of the beam. The condenser aperture should then be large enough to make use of the full angular aperture of the system. This is the arrangement used in the microscope with substage condenser and in the kinematograph. As an image of the source is formed at or close to the object and will therefore be more or less in focus in the final image it will be necessary that the source should be one that is uniform over its area, such as an arc lamp or flame. When the instrument is required to form an image of a large angular extent of object, as in the projection lantern, the image of the source should be formed at, or near, the entrance pupil. In this case the aperture of the condenser will act as the entrance port of the system, provided no other stop or stop image subtends a smaller angle at the entrance pupil, while the actual entrance pupil or the image of the source, whichever is the smaller, acts as entrance pupil. In the projection lantern the slide is placed close against the condenser and the slide mask is itself the field stop and entrance port (See Fig. 124).

CHAPTER XVII   *EXERCISES*

1. Explain the meaning of the terms: aperture stop, entrance and exit pupils, and field of view in relation to an optical instrument.

Find the position and magnification of the entrance pupil of the eye, taking the actual pupil to be 3·6 mm. behind the cornea, which is assumed to be a single surface of radius 7·8 mm. Assume the refractive index of the aqueous to be $\frac{4}{3}$.

2. The eye (assumed to be the 60D simple eye) with pupil diameter 6 mm. is placed behind a thin positive lens of 5 cm. focal length and 2 cm. aperture, the distance between lens and eye pupil being 8 cm. Find by construction (paraxial rays) the position and size of the entrance pupil of the system consisting of the lens and eye.

3. A system consists of two thin lenses having focal length + 10 cm. and + 15 cm. respectively separated 5 cm.; an aperture 2 cm. diameter is placed coaxially between the lenses and 3 cm. from the shorter focus lens which is turned towards the object. Find the positions and sizes of the entrance and exit pupils and ports.

4. If the lenses of the system in question 3 have diameters of 3 cm. and the object is situated 50 cm. from the first lens, find the field of view of full illumination and the total field.

5. An object 2 cm. long is placed 12 cm. from a positive lens of 6 cm. focal length and 6 cm. aperture. Make a careful diagram showing the pencils from each end of the object by which an eye with a pupil diameter of 8 mm. sees the aerial image, the eye being 25 cm. from the image.

6. An unaccommodated emmetropic eye having a pupil diameter of 4 mm. is placed 20 mm. behind a positive lens of 30 mm. focal length and 15 mm. aperture. An object is brought up towards the lens until its image is clearly seen. What will be the total extent of object that can be seen and over what extent of object seen will there be no diminution of illumination?

7. A telescope is made up of two thin positive lenses, the objective having a focal length of 6 inches and aperture 4 inches and the eyepiece a focal length at $1\frac{1}{2}$ inches and aperture 2 inches. There are no diaphragms in the instrument. Find by drawing a diagram carefully to scale the field of view and position of the exit pupil.

In what respects would the performance of such an instrument be defective?

8. A Galilean binocular magnifying four times has an objective of 160 mm. focal length and 40 mm. aperture. The pupil of the eye is 5 mm. diameter and situated 10 mm. from the eyepiece. Find (a) the apparent and actual field over which the illumination does not fall below half that at the centre, (b) the apparent and actual field of full illumination, (c) the apparent and actual total field.

9. Discuss the limitation of the field of view in a binocular of the Galilean type.

10. It is required to make up a telescope of the astronomical type to have a field of 10°, magnifying power of eight, aperture ratio of object glass $\frac{1}{8}$, using a one-inch Ramsden eyepiece.

Determine the dimensions of the parts and draw a careful full-size diagram (to scale) showing the passage through the instrument of a pencil from the limit of the field.

11. What is meant by "depth of focus"? On what will the depth of focus for any lens depend?

12. A $+ 6D$ lens, having an aperture equal to one-twelfth the focal length, is correctly focused on a very distant object. Find the position of the nearest object which would have its image apparently well defined in the same plane, assuming the image to be well defined when images of points do not exceed 0·1 mm. diameter.

13. A camera fitted with a lens of 6 inches focal length and aperture $f/8$ is correctly focused on an object 10 ft. away. What will be the range of apparent good focus on either side of the object position if the image is considered sharp when blur circles do not exceed 0·005 inch diameter?

14. With reference to the photographic camera explain (a) the purpose of the adjustable aperture or stop; (b) the meanings and significance of depth of focus.

If the correct exposure with an aperture of $f/8$ is $\frac{1}{30}$ second, what would be the exposure with the $f/16$ aperture?

If the depth of focus with the $f/8$ aperture is 0·016 inch, what would be the depth of focus with the $f/16$ aperture? (S.M.C.)

15. Explain carefully why the image of an extended object seen through an optical instrument can never appear brighter than when seen by the unaided eye. How may the apparent brightness differ in the case of a point object, such as a star?

16. A circular source of 5 mm. diameter is placed on the axis of, and 6·25 cm. from, a $+ 21D$ thin lens of 3 cm. aperture. Find the positions on the axis between which an eye, considered as a point, will see the whole lens filled with light—complete flashing.

Explain how this phenomenon can be utilised in examining glass for defects; give a diagram.

XVIII – Aberrations

229. Introduction

An optical system or instrument is required to produce, in a predetermined position and of a certain size, an image of a given object. In order that this image shall be a faithful and clearly defined copy of the object, each point of the latter must be imaged as a point, or at least as a small diffraction disc, in the image, and the totality of these image points must build up a geometrical figure similar to the object. We learn from Chapter XV that for the realisation of this it is necessary that the aperture of the system be appreciable; also that the emerging wave fronts corresponding to the several object points be spherical or that the "rays" of each emerging pencil, axial as well as oblique, shall reunite in common image point. We have previously seen (Chapters V, VI and VII) that when the aperture is appreciable, the whole of the light from a given object point does *not* reunite in a point after reflection or refraction at a spherical surface or refraction by a single lens; it does so only within the very restricted paraxial region. Hence in practice images suffer from defects or **aberrations** unless steps are taken—as by combining a number of lenses, probably made of different glasses and of various thicknesses—to eliminate or reduce such aberrations, in which case we have a "corrected" system.

Any complete study of aberrations and lens design is outside the scope of this book, but it is desirable to obtain an idea of the general nature of the more important aberrations and some of the principles involved in their correction. For a fuller account of aberrations and lens design the student should consult H. H. EMSLEY *Aberrations of Thin Lenses*. Constable. 1956.

406

230. Forms of Aberration

The primary or third* order aberrations may be classified as follows:

A. *Due principally to the material of the lens.*

Chromatic aberration, including chromatic difference of magnification.

B. *Due to the form of the lens.*
1. Spherical aberration.
2. Coma.
3. Astigmatism (radial or oblique).
4. Curvature of the image.
5. Distortion.

We may also classify them according to whether they affect axial or oblique pencils as follows:

C. *Affecting axial pencils.*
1. Chromatic aberration.
2. Spherical aberration.

D. *Affecting oblique pencils.*
1. Coma.
2. Astigmatism.
3. Curvature of the image.
4. Distortion.

When these third order aberrations have been corrected as far as possible, the image may still suffer from aberrations of higher orders, which have in their turn to be reduced in the case of high-class instruments, such as the microscope and photographic objectives.

The elimination of these various aberrations, particularly those of the higher orders, is extremely difficult and calls for the highest skill on the part of the designer. In any case it is impossible to correct then completely, and the final design will be a compromise according to the purpose which the instrument is to fulfil. Thus in the image formed by the objective of the microscope or telescope, which is to be considerably magnified by the eyepiece, particularly good correction is required. Since this high degree of correction cannot be obtained over a large field, in such cases the aberrations are fully corrected for image points close to the axis at the expense of the more peripheral portions of the

---

* The sine of an angle may be expressed in terms of the angle (in radians) as follows:

$$\sin a = a - \frac{a^3}{1.2.3} + \frac{a^5}{1.2.3.4.5} - \cdots\cdots$$

In the paraxial expressions the sines of the small angles involved are taken equal to the angles and the other terms neglected. On taking the second term into account we obtain the primary or *third order* aberrations and the higher orders with the further terms.

field. On the other hand, the photographic objective is required to give reasonably good definition over a large field, but as the subsequent magnification, if any, is comparatively small, in no part of the image will the same high quality of correction be required as in the former case. And the conditions of the problem in the design of spectacle lenses are different again. In these the aperture of the lens used is limited by the pupil of the eye, and consequently aberrations depending on the aperture, such as spherical aberration, are of little importance. The eye, however, rotates about its centre of rotation, looking through various portions of the lens and in different directions. In such cases the oblique aberrations, particularly astigmatism and curvature of the image will be important.

231. Chromatic Aberration

The nature of the chromatic aberration and chromatic difference of magnification and the method of correcting them have been dealt with in Chapter XIII and need not be further considered here. In considering the following aberrations, we shall suppose the light to be mono-chromatic, so that chromatic effects do not arise.

232. Spherical Aberration

The general nature of spherical aberration has been briefly discussed in Chapter V. It was seen that, when a wide pencil of light is refracted by a spherical surface, the axial intersection point of marginal rays is nearer to the surface than that of paraxial rays. The spherical aberration of a thin lens is shown in Fig. 243. The distance between B'_M, the focus

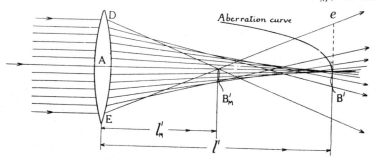

Fig. 243.—Spherical aberration

for the marginal rays, and B′, the paraxial focus, is called the **longi-tudinal spherical aberration.** The curve shows the variation of the longitudinal spherical aberration with aperture.

With a single convex lens forming a real image of a real object the aberration will always be of the form shown in the figure, the marginal rays focusing closer to the lens than the paraxial rays; this is known as undercorrection. By combining the lens with other lenses or by altering the form of the surfaces so that they are no longer spherical—"figuring" the surfaces—all the rays may be made to focus approximately at one position; or the paraxial rays may even be made to focus closer to the lens than the marginal ones; in the latter case the aberration is said to be overcorrected.

For rays meeting a single spherical refracting surface at any height above the axis we have the general expression

$$\frac{n'}{l'} = \frac{n}{l} + \frac{n'-n}{r} + 2\frac{n(l-r)}{rl}\ \frac{\sin\dfrac{i'-u}{2}\cos\dfrac{i-i'}{2}}{\cos\dfrac{i'-u'}{2}} \qquad \ldots (18.01)$$

where i and i' are the angles of incidence and refraction of a particular ray and u and u' the angles it makes with the axis in the object and image spaces. This general expression should be compared with the fundamental paraxial equation (5.02). The last term expresses the spherical aberration.

Tracing a number of rays, either graphically or by calculation, through thin lenses of different form, shows that the form of the lens will have a considerable influence on the amount of aberration. For a thin lens the longitudinal spherical aberration may be found from the following expressions:

$$l' - l'_M = \tfrac{1}{8}y^2\ l'^2F^3(a\,u^2 + b\,u\,v + c\,v^2 + d) \qquad \ldots (18.02)$$

where the form factor $u = \dfrac{r_2+r_1}{r_2-r_1} = \dfrac{R_1+R_2}{R_1-R_2}$ $\qquad \ldots \qquad \ldots (18.03)$

the conjugates or vergence factor $v = \dfrac{l'+l}{l'-l} = \dfrac{L+L'}{L-L'}$ $\qquad \ldots \qquad \ldots (18.04)$

and

$$a = \frac{n+2}{n(n-1)^2}, \quad b = \frac{4(n+1)}{n(n-1)}, \quad c = \frac{3n+2}{n}, \quad d = \frac{n^2}{(n-1)^2} \quad (18.05)$$

Thus the longitudinal spherical aberration of a thin lens of given form and focal length increases as the square of the aperture.

It will be found that for no form of single lens can the spherical aberration be completely eliminated. Minimum spherical aberration will be present when the form of the lens is given by

$$u = \frac{-2(n^2-1)}{n+2} \cdot v \quad \dots \quad \dots \quad \dots \quad (18.06)$$

This gives for an object at infinity ($v = -1$) and a refractive index of 1·5 a lens of the form $\frac{r_1}{r_2} = -\frac{1}{6}$. A lens of this form is often called a **crossed lens.**

The following table shows the way in which the aberration varies with the form of the lens for a distant object.

$$f' = +20 \text{ cm.} \quad y = 2 \text{ cm.} \quad n = 1·5$$

Form of Lens.	r_1	r_2	Spherical aberration $B'_M B'$
1st surface plane	∞	− 10 cm.	0·9 cm.
Equi-convex	+ 20 cm.	− 20 cm.	0·33 cm.
2nd surface plane	+ 10 cm.	∞	0·23 cm.
Best form (crossed)	+ 11·67 cm.	− 70 cm.	0·214 cm.
Meniscus $\frac{r_1}{r_2} = \frac{6}{11}$	+ 4·54 cm.	+ 8·33 cm.	1·896 cm.

The radius B'e (Fig. 243) of the patch of light at the position of the paraxial focus B' is known as the **lateral** or **transverse spherical aberration** and can easily be found when the longitudinal aberration is known. From the figure

$$B'e = \frac{AD \times B'_M B'}{l'_M}$$

and the lateral spherical aberration therefore increases as the cube of the aperture.

It will be seen from the figure that the beam has its smallest diameter and hence its best focus at a position between the paraxial and marginal foci and closer to the marginal focus. This is known as the circle of least confusion, the diameter of which also varies as the cube of the aperture.

From the figure given above for the spherical aberration of a thin lens, it is evident that the amount of aberration will depend largely on the directions at which the various rays meet the surfaces. For a single spherical surface there are three conditions under which all rays are refracted through a single point. These three cases are shown in Fig. 244. The first two are self-evident. In (a) the object point B coincides with the pole A of the surface; all rays, incident and refracted, pass through this point, so that there is no longitudinal aberration. In (b) the object point

B coincides with the centre of curvature C of the surface; all incident rays therefore meet the surface normally and all the refracted rays pass through the point C. In (c) it will be clear from Young's construction

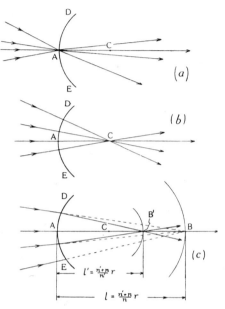

Fig. 244.—Aplanatic points of a spherical refracting surface

(Chapter V. § 50) that all incident rays travelling towards B pass after refraction through B', when B is in such a position that

$$\frac{l}{r} = \frac{n'+n}{n}$$

In none of these three cases is it possible to have a real image of a real object, but a combination of the three conditions is frequently used in the design of large aperture lenses to be free from spherical aberration, e.g., microscope objectives.

Since the amount of spherical aberration can be varied by altering the form of the lens, the focal length remaining constant, it is possible largely to correct the aberration by combining positive and negative lenses of suitable form, the aberration being in opposite directions in

Fig. 245.—Achromatic lens. Spherical aberration curve

the two cases. In the ordinary achromatic objective the light from two zones is brought to a focus at the same place and the residual aberration is small. The aberration curve is shown in Fig. 245.

The correction of spherical aberration in any lens system will only apply to one pair of conjugates. Thus in the case of the microscope objective, the image should be formed at the distance for which the objective was corrected; that is, the correct tube-length should be used. The telescope objective will usually be correct for an object at infinity.

233. The Oblique Aberration

The four aberrations, already mentioned, affecting the light passing obliquely through a lens, will in most cases all be present together in varying amounts, and it is difficult to separate their effects. It will, however, be convenient to describe each one separately, assuming that the others have been corrected.

234. Coma

This is very similar to spherical aberration, but affects the light coming from object points not on the axis. Light from an object point Q is refracted by a single lens of large aperture in the manner shown in Fig. 246. The image of Q is a pear-shaped or comet-like patch of light qr with the greatest concentration of light at q and becoming more diffused towards r. This one-sided flare may be either towards the axis as in Fig. 246 or away from it, depending on the form of the lens.

Fig. 246.—Coma

If the lens is free from all aberrations but coma, rays in the **meridian plane,** that is, the plane containing the object point Q and the optical axis, will reach the image plane B′Q′ in the way shown in Fig. 247. Pairs of rays from diametrically opposite points in any one zone of the lens meet the plane B′Q′ at points on a circle, those from outer zones meeting at greater distances from the chief ray than those from inner zones. Each zone of the lens gives rise to a circular patch of light and the centre of each circle will be at a distance equal to its own diameter from the point Q′ formed by the central zone. These overlapping circular patches produce the characteristic coma flare as shown enlarged in Fig. 247*a*. The length of the simple coma patch will vary as the square of the aperture and as the first power of the angular field.

Fig. 247.–Coma. Sine condition

Pure coma may be regarded as being due to the linear magnification of a small object being different for different zones of the lens. It can be shown that, in a system in which spherical aberration is corrected, coma for object points close to the axis will be eliminated when

$$\frac{\sin u}{\sin u'} = \frac{n'}{n} \cdot m \qquad \dots \qquad \dots \qquad \dots \quad (18.07)$$

is constant for every zone of the lens. This is the famous **optical sine condition** first stated by Abbe* in 1873 and proved by Helmholtz shortly afterwards. It will be seen that when u is small (paraxial region) the sine condition is the same as the Smith-Helmholtz equation (5.09)

$$n u h = n' u' h'$$

A system that is corrected for spherical aberration and satisfies the sine condition for a certain position of object is said to be **aplanatic** for the object in question. The object and its image are then called

---

* The condition for the elimination of coma has been called the Fraunhofer Condition, because it was found to be almost exactly satisfied by the Konigsberg Heliometer objective, constructed by Fraunhofer many years before the discovery of the condition, and which gave excellent definition.

aplanatic points. It will be easily seen that B and B′ in Fig. 244c are such points.

Coma is a particularly troublesome aberration because of the unsymmetrical form of the light patch, and every effort is made to eliminate it as far as possible in practical instruments. This is particularly the case in instruments to be used for measurement purposes owing to the difficulty experienced in deciding which part of the unsymmetrical patch shall be considered as the image point. A telescope objective with a large coma error will be particularly sensitive to being mounted out of square.

235. Radial or Oblique Astigmatism

If the oblique beam (Fig. 246) is limited by a stop with a small circular aperture placed, say at R, the comatic flare will be considerably reduced and the second oblique aberration will then be apparent. The narrow beam, instead of passing through a point, will pass through two short lines at right angles to one another, that is, the beam will be **astigmatic.** This aberration is known as **radial** or **oblique astigmatism.** It is due to the oblique incidence of the light at spherical surfaces and should not be confused with the astigmatic beam produced on refraction at normal incidence by a cylindrical or toric surface (Chapter VIII); there is, however, a general similarity in the form of the beam in the two cases.

We will consider the case of a narrow pencil incident obliquely on a spherical surface. In Fig. 248 ADEGH is a spherical surface between air on the left and glass on the right, having its centre of curvature at C, and CA is a radius lying in the plane of the paper. Q is an object point in the plane of the paper, and this plane is therefore the meridian plane (see § 234); it is also called the **tangential plane.** The chief ray, QA, that is the central ray of the pencil, is in this plane and remains in it when refracted to Q′. The plane perpendicular to the meridian plane and containing also the chief ray is the **sagittal** or **equatorial plane**; this plane will become tilted on refraction. In the figure the points Q A D E T S L M Q′ lie in the meridian plane and the points Q A G H T S N P in the sagittal plane.

Let us consider four limiting rays of the pencil, which is supposed to be very narrow, but is shown with its width much exaggerated in the figure. Rays QD and QE are in the plane of the paper—the meridian plane—while rays QG and QH are in the sagittal plane, QG being above, and QH below the plane of the paper. On refraction the rays QD and QE remain in the plane of the paper and meet the chief ray at T. The ray QG is refracted back towards the paper and the ray QH forwards towards the paper, and they meet on the chief ray at S. Rays in intermediate planes are refracted in such a way that they pass approximately through a short line at T and through a second perpendicular line at S. Of these two focal lines, that at T, called the **tangential** or **primary line focus,** is a short arc of a circle that has its centre on the axis of the surface; while that at S called the **radial, sagittal** or **secondary line focus,** is more exactly a narrow figure of eight (Fig. 248a), but may

Fig. 248.—Radial or oblique astigmatism

generally be considered as a short straight line perpendicular to the tangential circular arc. At a position roughly midway between the two lines the beam, if the aperture is circular, will have a circular cross-section; this is the circle of least confusion. The interval TS between the two focal lines is the **astigmatic difference.**

For a single refracting surface of radius r if i and i' are respectively the angles of incidence and refraction of the chief ray, t and s, the distances of the luminous point from the surface in the tangential and sagittal planes respectively (these will be equal for the original object point), and t' and s' the distances of the corresponding focal lines from the surface, it can be shown* that

$$\frac{n' \cos^2 i'}{t'} - \frac{n \cos^2 i}{t} = \frac{n' \cos i' - n \cos i}{r} \qquad \ldots (18.08)$$

and

$$\frac{n'}{s'} - \frac{n}{s} = \frac{n' \cos i' - n \cos i}{r} \qquad \ldots (18.08\text{A})$$

The astigmatic difference $t' - s'$ increases as the square of the distance of the object point Q from the axis or as the square of the angular field.

Astigmatic Image Surfaces.—If we consider the light from an illuminated plane surface perpendicular to the axis of the refracting surface, the pencil originating at each point on the illuminated surface will be of the form described above, and the focal lines corresponding to each object

---

* See H. H. EMSLEY, *Aberrations of Thin Lenses* or L. C. MARTIN, *An Introduction to Applied Optics*, Vol I.

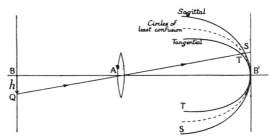

Fig. 249.—Sections through astigmatic image surfaces

point will lie on the chief ray from that point. For the object point on the axis there will be no astigmatism and the image will be a point. As the distance of the object point from the axis increases, both the astigmatic difference and also the distance of each focal line from a plane perpendicular to the axis and passing through the axial image point increases. Hence the various pencils from points on an illuminated plane surface pass, after refraction, through two curved surfaces containing, respectively, the primary and secondary focal lines (Fig. 249). These are known as the **astigmatic image surfaces.** Between these two surfaces is a third containing the circles of least confusion.

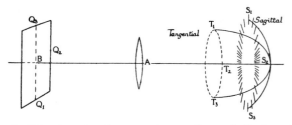

Fig. 250.—Radial astigmatism. Teacup and saucer diagram indicating astigmatic image surfaces

As the plane containing any object point and the axis of the refracting surface is the meridian plane, and the plane at right angles the sagittal plane of that particular pencil, it is easily seen that the primary or tangential focal lines for all object points at a given distance from the axis will lie on a circle perpendicular to the axis; while the secondary or sagittal lines will be short radial lines lying on a circular annular surface. Emsley and Swaine have likened the astigmatic image surfaces to the surfaces of a teacup and saucer, the teacup representing the tangential surface and the saucer the sagittal surface (Fig. 250).

Since the amount of astigmatism depends upon the obliquity at which the chief ray from any object point meets the surface, it will be largely influenced by the form of the surface and by the position of the stop. The astigmatic difference will be in opposite directions according as the refracting surface is convergent or divergent, and the relative positions of the astigmatic image surfaces will be different for convergent and divergent surfaces under otherwise equal conditions. The astigmatic differences for pencils at various angles for a convergent and divergent

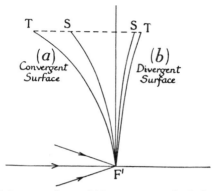

Fig. 251.—Astigmatic image surfaces of (*a*) convergent spherical surface, (*b*) divergent spherical surface

surface of equal power are shown plotted in Fig. 251. It is possible therefore by a suitable choice of surfaces, separations and position of stop to considerably reduce the astigmatism and in some cases to make the two astigmatic image surfaces meet in some zone in addition to meeting on the axis, the two surfaces nowhere being far apart.

236. Curvature of the Image or Curvature of Field

When astigmatism has been corrected, the two image surfaces coincide and, if the lens is free also from spherical aberration and coma, each object point will be represented by a point image. The various image points corresponding to points on a plane object surface may, however, lie on a curved surface. In many cases, for example, in photography, we require that the image of a plane object shall lie in a plane surface, and this further defect of curvature of the image must therefore be corrected as far as possible.

Petzval (1807–1891) showed in 1843 that when spherical aberration, coma and astigmatism have been corrected, the curvature of the image surface depends upon the refractive indices of the glasses and the radii of the lens surfaces. For a series of k refracting surfaces, the curvature, $\frac{1}{r'}$, of the image surface corresponding to a plane object is given by

$$\frac{1}{n'_k r'} = \Sigma_k \frac{1}{r_k} \left(\frac{1}{n'_k} - \frac{1}{n_k} \right) \dots \qquad \dots \qquad \dots \quad (18.09)$$

This must be made zero if the image surface is to be plane and this condition for flatness of field is known as **Petzval's condition.** The curved image surface corresponding to a plane object is called the **Petzval surface.** For a series of thin lenses in air, Petzval's condition becomes

$$\Sigma_k \frac{F_k}{n_k} = O \qquad \dots \qquad \dots \qquad \dots \quad (18.10)$$

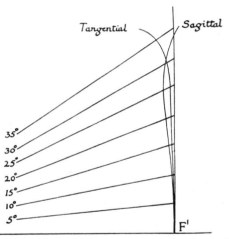

Fig. 252.—Astigmatic image surfaces with anastigmat lens. (Distances from the ideal image line magnified)

In order to satisfy this condition some of the positive lenses in the system must be made of higher refractive index glass than the negative lenses. Using the older types of crown and flint glass this requirement is opposed to the condition for achromatism, and it is only by the use of some of the newer types of glass having high refractive index and low dispersion, such as a dense barium crown (*see* Chapter XIII), that the conditions for achromatism and flatness of field can be satisfied.

In the modern photographic objective—the so-called **anastigmat**—a close approach to good definition over a flat field of considerable angular extent has been achieved to a remarkable degree. The form of the image surfaces in a typical objective is shown in Fig. 252.

237. Distortion

Having, by elimination of the previously described aberrations, obtained on a plane surface a sharply defined image of a plane object perpendicular to the lens axis, there may still remain distortion, which prevents the image being the same shape as the object. The shape of the image will depend upon the directions of the chief rays from all points on the object, and these must be made to meet the image plane in the correct positions, if this further defect is to be eliminated. In the pinhole camera (Fig. 6) it is clear that, as the chief rays pass undeviated through the centre of the aperture, object and image in parallel planes are geometrically similar figures. The centre of the aperture is the centre of projection and the image is said to be **orthoscopic** or true in projection.

Fig. 253.—Distortion. Positive lens with front stop

Suppose, as in Fig. 253, a stop is situated in front of the lens, then the chief rays from an object plane BQ all intersect the axis at the point R, the centre of the aperture. If the lens has spherical aberration for an object point R, the chief rays are refracted as shown in the figure and do not again pass through a point. It will be clear that in the image plane

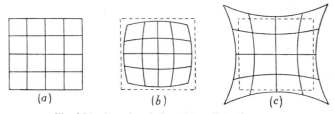

Fig. 254.—Barrel and pincushion distortion

B'Q' the outermost points are more crowded together than are those close to the axis, that is, the magnification decreases as we go away from

419

the axis. Thus the image of a network of straight lines at right angles as in Fig. 254*a*, will have the form shown in Fig. 254*b*; this is called **positive** or **barrel distortion.** The same effect occurs in the more usual case of the stop being inside the principal focus, its image then being virtual. When the stop is behind the lens, the magnification of the image increases towards the edge of the field, the image having the form shown in Fig. 254*c*. This is termed **negative** or **pincushion distortion.** The amount of distortion, measured by the distance of an image point from its ideal position, increases as the cube of the distance of the image point from the optical axis.

In order that images formed by a system shall be free from distortion the following two conditions must be satisfied:

(1) All rays passing through the centre of the entrance pupil must pass also through the centre of the exit pupil, that is, the system must be free from spherical aberration with respect to the entrance and exit pupils.

(2) The points such as D, d_1, d_2, and d_3 (Fig. 253) in which chief rays in the object and image spaces intersect when produced, must lie in a plane perpendicular to the lens axis. This is equivalent to fulfilling the **tangent condition** (due to Airy, 1827), that the ratio of the tangents of corresponding angles *w* and *w'* shall be constant for all inclinations of the rays.

A system in which these conditions are satisfied is said to be orthoscopic and the centres of entrance and exit pupils are called **orthoscopic points.**

238. Von Seidel's Theory

The five aberrations due to the form of the lens were mathematically deduced by von Seidel (1821–1896) and sometimes bear his name. This mathematician, in papers published in 1855–6, derived five expressions that must be introduced into the paraxial equations when dealing with rays outside the paraxial region. Each of these famous "Five Sums of von Seidel" corresponds to one of the above aberrations and must be made equal to zero if that aberration is to disappear. Calling these sums, S_1, S_2, S_3, S_4, S_5, spherical aberration is eliminated when $S_1 = O$, spherical aberration and coma are corrected when S_1 and $S_2 = O$, and in the same way for each of the other aberrations, in the order in which they have been considered above.

239. Trigonometrical Calculation of the Exact Path of a Ray Refracted at a Spherical Surface

In investigating aberrations, particularly those of the higher orders, and in designing lens systems, it is necessary to trace the exct path of rays incident at any angle at the surfaces and at any distance from the axis. This may be done graphically by the methods described in Chapter V (Young's construction and the ray plotter), but often we require the directions with greater accuracy than can be obtained by graphical methods. In such cases a trigonometrical calculation of the directions of the rays and the positions at which they meet the axis must be resorted to.

In Fig. 255, a ray QD is incident on a spherical surface DAE with centre of curvature at C, separating a medium of refractive index n from one of index n'. This ray is travelling towards B making an angle u with the axis and is refracted through B' at an angle u' with the axis.

In the triangle DCB

$$\sin i = \sin u \frac{l-r}{r} \qquad \ldots \qquad \ldots \qquad \ldots \quad (18.11)$$

also from the law of refraction

$$\sin i' = \frac{n}{n'} \sin i \qquad \ldots \qquad \ldots \qquad \ldots \quad (18.12)$$

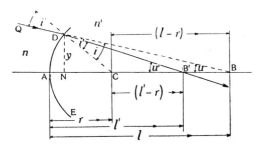

Fig. 255.–Spherical refracting surface. Trigonometrical ray tracing

The angle DCA is the exterior angle of both the triangles DCB and DCB', so that angle DCA $= u + i = u' + i'$ whence

$$u' = u + i - i' \qquad \ldots \qquad \ldots \qquad \ldots \quad (18.13)$$

In triangle DCB'

$$(l' - r) = r \frac{\sin i'}{\sin u'} \qquad \ldots \qquad \ldots \qquad \ldots \quad (18.14)$$

and

$$l' = (l' - r) + r \qquad \ldots \qquad \ldots \qquad \ldots \quad (18.15)$$

When the ray meets a second surface at a distance d from the first, the angle u' for the first surface becomes the angle u for the second,

421

that is $u_2 = u_1'$, and $l_2 = l_1' - d$. The calculation is then continued as for the first surface.

Where the incident light is parallel to the axis, the angle u is zero and l is infinitely great, then

$$\sin i = \frac{y}{r}$$

and the calculation is continued from equation (18.12).

At a plane surface

$$u = i$$

$$i' = u'$$

$$l' = l \frac{\tan u}{\tan u'} \qquad \dots \qquad \dots \qquad \dots \quad (18.16)$$

The calculation through a system of surfaces will be greatly facilitated by the adoption of a systematic method of writing down the work. The various steps in the calculation should be arranged in tabular form as in the examples given below.

The computation through the system for the paraxial rays may be conveniently carried out at the same time. As the angles are then very small, the radian measure may be substituted for the sines, and the equations become

$$i = u \frac{l - r}{r} \qquad \dots \qquad \dots \qquad \dots (18.11p)$$

$$i' = \frac{n\,i}{n'} \qquad \dots \qquad \dots \qquad \dots (18.12p)$$

$$u' = u + i - i' \qquad \dots \qquad \dots \qquad \dots (18.13p)$$

$$(l' - r) = r \frac{i'}{u'} \qquad \dots \qquad \dots \qquad \dots (18.14p)$$

$$l' = (l' - r) + r \qquad \dots \qquad \dots \qquad \dots (18.15p)$$

Theoretically, this paraxial computation should commence with an extremely small value of u, but it will be clear from the above paraxial equations that the size of u has no effect on the calculated value of l',

Example I.   Biconvex lens.

$r_1 = +1\cdot737$ in. $r_2 = -9\cdot739$ in. $d = 0\cdot40$ in. $n = 1\cdot5175$.

Ray to be calculated cuts the axis at $l_1 = -8\cdot75$ ins., $u_1 = -5°\ 15'\ 12''$

	FIRST SURFACE.		SECOND SURFACE.	
	Marginal	*Paraxial*	*Marginal*	*Paraxial*
l	$-8\cdot75$		$6\cdot5720$	$7\cdot8633$
r	$1\cdot737$		$-9\cdot739$	$-9\cdot739$
$l - r$	$-10\cdot487$		$16\cdot3110$	$17\cdot6023$
log sin u	$8\cdot96170$ n	$8\cdot96170$ n	$9\cdot08230$	$8\cdot98655$
log $(l - r)$	$1\cdot02065$ n	$1\cdot02065$ n	$1\cdot21248$	$1\cdot24557$
log sin u $(l - r)$	$9\cdot98235$	$9\cdot98235$	$\cdot29478$	$\cdot23212$
log r	$\cdot23980$	$\cdot23980$	$\cdot98851$ n	$\cdot98851$ n
log sin i	$9\cdot74255$	$9\cdot74255$	$9\cdot30623$ n	$9\cdot24361$ n
log $\dfrac{n'}{n}$	$\cdot18113$	$\cdot18113$	$\cdot18113$	$\cdot18113$
log sin i'	$9\cdot56142$	$9\cdot56142$	$9\cdot48740$ n	$9\cdot42774$ n
log r	$\cdot23980$	$\cdot23980$	$\cdot98851$ n	$\cdot98851$ n
log (sin $i' \times r$)	$9\cdot80122$	$9\cdot80122$	$\cdot47591$	$\cdot41325$
log sin u'	$9\cdot08230$	$8\cdot98655$	$9\cdot35707$	$9\cdot27330$
log $(l' - r)$	$\cdot71892$	$\cdot81467$	$1\cdot11884$	$1\cdot13995$
u	$-5°\ 15'\ 12''$	$-0\cdot091558$	$6\ \ 56\ \ 31$	$0\cdot096950$
i	$33\ \ 33\ \ 28$	$0\cdot552775$	$-11\ \ 40\ \ 44$	$-0\cdot175232$
$u + i$	$28\ \ 18\ \ 16$	$0\cdot461217$	$-4\ \ 44\ \ 13$	$-0\cdot078282$
i'	$21\ \ 21\ \ 45$	$0\cdot364267$	$-17\ \ 53\ \ 23$	$0\cdot265912$
u'	$6\ \ 56\ \ 31$	$0\cdot096950$	$13\ \ 9\ \ 10$	$0\cdot187630$
$l' - r$	$5\cdot2350$	$6\cdot5263$	$13\cdot1474$	$13\cdot8023$
r	$1\cdot737$	$1\cdot737$	$-9\cdot739$	$-9\cdot739$
l'	$6\cdot9720$	$8\cdot2633$	$3\cdot4084$	$4\cdot0633$
d	$\cdot400$	$\cdot400$		
l_2	$6\cdot5720$	$7\cdot8633$		

Note.—The letter n after a logarithm denotes the logarithm of a negative number. To avoid negative characteristics 10 has been added to the logarithms of numbers less than unity.

Difference between the paraxial and marginal intersection points gives the spherical aberration.

$$= 4\cdot0633 - 3\cdot4084 = 0\cdot6549 \text{ inches.}$$

Example II.—Objective for Prism Binocular × 6.

$r_1 = + 52.74$ mm. r_2 and $r_3 = + 29.31$ mm. (cemented) $r_4 = \infty$

$d_1 = 1.5$ mm. $d_2 = 4.0$ mm.

$n_{D1} = 1.6206$. $n_{D2} = 1.5107$.

Semi-aperture $y = 12$ mm. Parallel incident light.

	First Surface Marginal	First Surface Paraxial	Second Surface (r_2 and r_3) Marginal	Second Surface (r_2 and r_3) Paraxial
$\log y$	1·07918	1·07918		
$\log r$	1·72214	1·72214		
$\log \sin i$	9·35704	9·35704		
$\log \sin u$			8·94724	8·94017
$\log (l - r)$			2·02342	2·02902
$\log \dfrac{u(l - r)}{\log r}$			·97066 / 1·46702	·96919 / 1·46702
$\log \sin i$	9·35704	9·35704	9·50364	9·50217
$\log \dfrac{n'}{n}$	·20968	·20968	9·96950	9·96950
$\log \sin i'$ / $\log r$	9·14736 / 1·72214	9·14736 / 1·72214	9·53414 / 1·46702	9·53267 / 1·46702
$\log (\sin i' \times r)$	·86950	·86950	1·00116	·99969
$\log \sin u'$	8·94724	8·94017	8·80654	8·80625
$\log (l' - r)$	1·92226	1·92933	2·19462	2·19349
u	0° 0' 0"	0	5° 04' 51"	0·08713
i	13° 09' 06"	0·22753	18 35 45	0·31781
$u + i$			23 40 36	0·40494
i'	8 04 15	0·14040	20 00 15	0·34093
u'	5 04 51	0·08713	3 40 21	0·06401
$l' - r$	83·610	84·983	156·54	156·13
r	52·74	52·74	29·31	29·31
l'	136·350	137·723	185·85	185·44
d	1·5	1·5	4·0	4·0
l_2	134·850	136·223	181·85	181·44
r_2	29·31	29·31		
$l_2 - r_2$	105·54	106·913		

Third Surface. plane

	Third Surface Marginal	Third Surface Paraxial
$\log \sin u = \log \sin i$	8·80654	8·80624
$\log \dfrac{n'}{n}$	9·82082	9·82082
$\log \sin i' = \log \sin u'$	8·98572	8·98543
$\log l$	2·25971	
$\log \tan u$	8·80743	
$\log \tan u'$	1·06714	
$\log l'$	8·98776	
$\log l$	2·07938	2·25873
$\log \dfrac{n'}{n}$		9·82082
$\log l'$		2·07955
l'	120·06	120·10

and we may therefore choose any value for *u*. It is convenient to start the paraxial computation with *u* (in radians) = sin *u*, but it must be remembered that the angular values in this calculation are greatly magnified, and that the distances are true only for extremely small angles.

Only rays from points on the axis are considered above. For the methods of dealing with oblique rays and for further information on trigonometrical ray tracing and its application to the design of optical systems, the reader should consult A. E. CONRADY, *Applied Optics and Optical Design*, B. K. JOHNSON, *Optical Design and Lens Computing* and H. H. EMSLEY, *Aberrations of Thin Lenses*.

It will be seen from the results that the objective is practically corrected for spherical aberration, for the zone considered, for a distant object.

Obviously, such calculations are extremely laborious and do no more than indicate the extent to which spherical aberration, the simplest possible defect of the lens, is corrected. For off-axis cases the process is even more complex. Many fine lenses were designed using hand calculating machines even though each ray traced through the system took, on average, six minutes. Thus the analysis of a proposed design could take up to a day and various designs would need to be tested before a suitable balance of the aberrations could be found. Many ingenious short cuts based on approximate relationships were devised by the designers.

With the advent of the digital computer during the 1950s and its application to lens design, the time to trace a ray is now measured in fractions of a second and the analysis of a proposed design is available to the lens designer almost as soon as he has fed the data into the computer. The next stage was the development of auto-design programs which would examine the results of the analysis and alter the parameters of the design to improve the level of residual aberrations. The computer will therefore, by this iterative process, improve the design to the best possible *within a restricted range of parameters*. The understanding of the basic equations covering third and higher order aberrations is essential to the designer so that he can recognise what features are impeding the further improvement and modify the basic design accordingly.

CHAPTER XVIII   *EXERCISES*

1. Explain what is meant by the spherical aberration of a lens. An object is viewed over the edge of a card placed near to the eye. When the card is moved up and down, the object appears to move, the motion being *with* the card when the object is distant, and *against* the card when the object is near. What optical defects of the eye does this indicate? Illustrate your answer with diagram. (B.O.A.)

2. (*a*) Using the ray plotter, find graphically the paths of the refracted rays con-

jugate to rays parallel to the axis, incident in one plane at heights (y) of 1, 2, ...
9 cm. on a spherical surface of radius $+$ 10 cm. Medium of object space is air and for
image space $n = 1.5$.

(b) Find by calculation the Gauss focal point F' and hence the distances along the
axis from this point to each of the points where the refracted rays cut the axis.

(c) Tabulate these quantities and plot them against y^3.

(d) Draw the spherical aberration curve.

(e) Plot log $B'_M B'$ against log y and so derive the relationship between $B'_M B'$
and y.

3. Determine by means of the ray plotter the axial spherical aberration of a
spherical surface of radius $-$ 10 cm. separating air from glass of index 1.5, with
the incident ray parallel to the axis and the incidence height 7 cm.

4. A parallel pencil of light is incident on a spherical surface of radius $+$ 10 cm.
separating air from glass ($n = 1.5$). Draw accurately the refracted rays, show the
caustic curve and the curve of aberration. What will be the positions of the pair of
conjugate points at which there is no spherical aberration and what will be the
magnification at those points?

5. Explain and prove Young's construction for finding the exact path of a ray of
light refracted at a spherical surface.

Find the positions of the conjugate aberrationless (aplanatic) points of a positive
spherical refracting surface of radius 2 inches separating air from glass of index 1.6.

6. Prove that the distances (from the vertex) of the aplanatic points of a spherical
refracting surface are given by

$$r \left(1 + \frac{n'}{n}\right) \text{ and } r \left(1 + \frac{n}{n'}\right)$$

What is the magnification at these points

7. Explain from first principles (using the conceptions of wave curvature, or otherwise) why a thin positive lens traversed obliquely by a pencil of light from a star
does not yield a point image. Describe the appearances which could be found on a
focusing screen held at different distances in the convergent beam if the lens were
a $+5$D. "sphere", and the light incident at 45° to the axis. (S.M.C. Hons.)

8. Apply Young's construction to determine the spherical aberration of the cornea.
Assume the latter to be spherical and of radius 7.8 mm.; refractive index of aqueous
humour 1.33. Take a ray parallel to the visual axis at an incident height of 4 mm.

9. Give a general description, with the aid of a diagram of the defect of lenses
called coma. How would you show this experimentally?

10. By means of Young's construction, or by using the ray plotter, find the astigmatism produced when a narrow pencil from an infinitely distant object is incident
at an angle of 20° to the axis on a spherical refracting surface of radius $+$ 5 cm.
separating air from glass of refractive index 1.5.

(a) When the beam is central.

(b) When the stop is placed 9 cm. in front of the surface.

(c) When the apparent stop is placed 9 cm. behind the surface.

11. A narrow pencil of light from an infinitely distant object is incident on a
spherical refracting surface of $+$ 5 cm. radius separating air from glass of $n = 1.5$.
The angle of incidence is 50°. Calculate the distance from the point of incidence to
the primary and secondary focal lines. Check by graphical construction.

12. A spherical surface of radius + 8 inches separates two media of refractive indices 1·36 and 1·70 respectively.

Light from the left is converging in the first medium, towards an axial point 18 inches to the right of the surface vertex. Find the position of the image point and show that there will be no spherical aberration.

13. A narrow pencil is incident centrally on a + 5D spherical lens in a direction making 30° with the optical axis of the lens. Explain briefly the nature of the refracted pencil.

14. A narrow pencil of parallel rays in air is refracted at a spherical surface of radius r into a medium of refractive index $\sqrt{3}$. If the angle of incidence is 60°, show that the distance of the primary and secondary image points from the point of incidence are given by $\frac{3}{4} r \sqrt{3}$, and $r \sqrt{3}$ respectively.

15. Explain with the help of a diagram the following terms associated with the refraction by a lens of an oblique pencil: meridian plane, sagittal plane, astigmatic difference, place of least confusion.

16. What is meant by radial astigmatism in lenses? Give examples of the effect produced by it in obtaining an image of a star covered field. (S.M.C.)

17. A spectacle lens of power + 5D is made to form the image of a distant axial source of light on a ground glass screen; the lens is then slightly tilted about one diameter. Explain with numerical details the changes thus produced in the image and the further changes observed as the screen is moved nearer the lens. N.B. If F is the axial power, and F_S and F_T are respectively the sagittal and tangential powers, while θ is the angle of obliquity, then

$$F_S = F\left(\frac{1 + sin^2\theta}{3}\right) \text{ approximately}$$

$$F_T = F_S \, sec^2\theta \text{ approximately. (S.M.C. Hons.)}$$

18. Write a short account of the aberrations in the image produced of a flat object by an uncorrected optical system. Give diagrams.

Explain, giving reason, which aberrations are of most importance in the case of (a) a telescope objective, (b) a spectacle lens.

19. Explain the terms barrel and pincushion distortion and show how they are produced by a single convex lens and stop. (S.M.C.)

20. Draw the all-positive diagram for refraction at a spherical surface and from it deduce the five simple equations used in the trigonometrical computation of rays through such a surface.

Write down also the corresponding equations for computing a paraxial ray and from these latter equations derive the fundamental paraxial equation:

$$\frac{n'}{l'} - \frac{n}{l} = \frac{n' - n}{r}$$

21. Write down the equations used in computing the path of a ray of light after refraction at a spherical surface. Apply them to compute the refracted ray in the following case:

$$r_1 = + 377\cdot85 \text{ mm.}, n = 1\cdot0, n' = 1\cdot5119.$$

The axial object point is situated 487 cm. in front of the surface; the incidence height of the ray is 12·7 cm. Show how to arrange the work in tabular form and calculate the first six quantities. Use four-figure logarithms.

22. Trace by trigonometrical computation the course, through the lens the

constants of which are given below, of a ray that crosses the axis at a point B 30 cm. in front of the first surface and is incident on the lens at an incident height of 2 cm. Find also the point where a paraxial ray incident through B crosses the axis after refraction.

$$r_1 = + 10 \text{ cm.}, r_2 = - 20 \text{ cm.}, d = 3 \text{ cm.}, n = 1 \cdot 624.$$

23. In general, the image formed by a simple spherical lens suffers from certain defects or aberrations. Enumerate these aberrations.

If the object consists of a point source of white light situated on the optical axis of the lens, with which aberration or aberrations will the image be affected when the aperture of the lens is (a) small; (b) large?

If the point source is moved away from the axis, what aberration or aberrations will be manifest when the aperture of the lens is small?

Give clear diagrams. (S.M.C.)

XIX - Lens Systems - General

240. Centred System of Spherical Surfaces. Paraxial Region

In dealing with the thin lens in Chapter VI and with lens systems and thick lenses in Chapter IX, we have considered only the case in which both object and image spaces are air. We may, however, have systems in which the object and image spaces have different refractive indices, the eye being an important example of this kind, and the general case of an optical system will now be considered.

Taking a system of centred surfaces as shown in Fig. 256 we can trace the path of the rays through the system graphically, or trigonometrically as in Chapter XVIII, or for pencils within the paraxial region we may apply the results, previously obtained for a single refracting surface (Chapter V), to each of the surfaces in turn.

Within the paraxial region, pencils from all points in the object plane remain homocentric after refraction at the first surface, and an image of the object is formed. This image is the object for the next surface, and as the pencils remain homocentric after refraction at the second surface, a second image is formed, which becomes the object for the third surface, and so on through the system.

An object perpendicular to the axis in the object space is imaged in a plane perpendicular to the axis by each surface in turn, and the final image in the image space is perpendicular to the axis and geometrically similar to the object.

In Fig. 256 the path of a paraxial ray $B_1 D_1$ from an object point B_1 on the axis is shown. Images are formed by each surface in turn at the points B'_1, B'_2, B'_3 and B'_4 respectively, where the ray crosses the axis, either actually or when produced back, in which case the image is virtual. If B_1Q_1 is a small object perpendicular to the axis, then the images formed by each surface will be perpendicular to the axis through

Fig. 256.—Centred system of spherical surfaces. Showing paths of rays from B_1 and additional construction for determining image sizes

B'_1, B'_2, etc. The image of Q_1 formed by the first surface is found by taking a ray from Q_1 through C_1 meeting the first image plane at Q'_1. A line from Q'_1 through C_2 fixes the position of Q'_2 and so on for each surface in turn.

241. Change of Vergence of Advancing Wave Front

Fig. 257 shows a wave front at A immediately after passing a refracting surface. The wave front is converging in a medium of refractive index n' towards a focus B' distant l' from A.

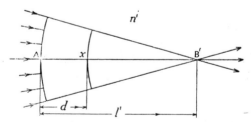

Fig. 257.—Change of vergence of advancing wave front

The *actual* vergence at A is therefore $\frac{1}{l'}$ and the reduced vergence is $\frac{n'}{l'} = L'$.

At some place X the actual vergence has increased to $\frac{1}{l' - d}$
The reduced vergence is therefore

$$\frac{1}{\dfrac{l'}{n'}-\dfrac{d}{n'}} = \frac{\dfrac{n'}{l'}}{1-\dfrac{d}{n'}\dfrac{n'}{l'}}$$

$$= \frac{L'}{1-\dfrac{d}{n'}L'} = L'_x \qquad \dots \qquad \dots \qquad \dots \quad (19.01)$$

If a second refracting surface exists at X, L'_x would be the value of L for the second surface.

242. Calculation for Paraxial Pencils through a Centred System

The focal power of each surface is given by

$$\left. \begin{aligned} \frac{n'_1-n_1}{r_1} &= F_1 \qquad \dots \qquad \dots \\[2mm] \frac{n'_2-n_2}{r_2} &= F_2 \qquad \dots \qquad \dots \\[2mm] \frac{n'_k-n_k}{r_k} &= F_k \qquad \dots \qquad \dots \end{aligned} \right\} \quad (19.02)$$

Each of these powers is added to the reduced vergence of the oncoming wave front in turn so that we have

$$\left. \begin{aligned} \text{First surface} \quad & L'_1 = L_1 + F_1 \\[2mm] & L_2 = \frac{L'_1}{1-\dfrac{d_2}{n_2}L'_1} \\[3mm] \text{Second surface} \quad & L'_2 = L_2 + F_2 \\[2mm] & L_3 = \frac{L'_2}{1-\dfrac{d_3}{n_3}L'_2} \\[3mm] \text{kth surface} \quad & L'_k = L_k + F_k \\[2mm] & l'_k = \frac{n'_k}{L'_k} \end{aligned} \right\} \quad (19.03)$$

Thus giving the position of the final image.

As the image formed by one surface is the object for the next, the size of each successive image and thence the size of the final image can be found as follows:

$$m_1 = \frac{h'_1}{h_1} = \frac{L_1}{L'_1}; \quad m_2 = \frac{h'_2}{h_2} = \frac{L_2}{L'_2}; \text{ etc.}$$

$$\text{and } h'_1 = h_2; \quad h'_2 = h_3; \quad \text{etc.}$$

Hence the total magnification

$$m = \frac{h'}{h} = \frac{L_1 \times L_2 \times \ldots \ldots L_k}{L'_1 \times L'_2 \times \ldots \ldots L'_k} = \prod_1^k \frac{L}{L'} \quad \ldots \text{ (19.04)}$$

where $h = h_1$, the height of the original object, and $h' = h'_k$, the height of the final image.

If the original object is at infinity and subtends an angle w, $L_1 = o$ and $h'_1 = f_1 \tan w$; then final $h' = h'_1 \times$ product of subsequent magnifications

$$h' = f_1 \tan w \times \prod_2^k \frac{L}{L'} \ldots \quad \ldots \quad \ldots \text{ (19.05)}$$

Considering the system as a whole, and making use of our definition of focal length in terms of the apparent size of the object (equation 5.11)

$$h' = f \tan w$$

where $f = $ first focal length of the system.

Hence

$$f = f_1 \times \prod_2^k \frac{L}{L'} \quad \ldots \quad \ldots \quad \ldots \text{ (19.06)}$$

By tracing the light through the system in the opposite direction we have

$$f' = f'_k \times \prod_{k-1}^1 \frac{L}{L'} \ldots \quad \ldots \quad \text{ (19.06A)}$$

Later it will be shown that

$$\frac{f'}{f} = -\frac{n'}{n}$$

where n and n' are the refractive indices of the original object and the final image spaces respectively.

For each surface
$$n_1 h_1 u_1 = n_2 h_2 u_2 = n_3 h_3 u_3 = \dots\dots\dots\dots$$
so for the whole system
$$n h u = n'h'u' \qquad \dots \qquad \dots \qquad \dots \qquad (19.07)$$
the Lagrange relation, $u = u_1$ being the angle that a ray from the axial object point makes with the axis in the object space and $u' = u'_k$, the angle the ray makes with the axis in the final image space.

243. Cardinal Points of a Centred System

Any system as a whole forms, of an infinitely distant object, either a real inverted image or a virtual erect image and is therefore either

Positive, convergent

or Negative, divergent

As we have previously seen with the thin lens combination or the thick lens in air, the result is as if the refraction took place completely at a plane distant f' back from F' for light in one direction and at a plane distant f back from F for light in the other direction. These two planes are the principal planes of the system and the points at which they intersect the axis, the principal points, P' and P. The distances f and f' can be found from equations (19.06) and (19.06A), and by taking L' and L in turn equal to zero we can find the distance f'_v of F' from the last surface.

We can therefore represent any lens system by its two focal planes through the principal foci F and F' and its two principal planes through the principal points P and P' where PF $= f$ and P'F' $= f'$. As was shown in Chapter IX, the principal planes are planes of unit magnification, and therefore any ray in the object space incident towards a point at a given distance above the axis in the first principal plane travels in the image space as though from a point in the second principal plane at the same distance from the axis.

244. Numerical Example. Schematic Eye

A useful numerical example of the above work is provided by the "schematic" eye. This is an optical system based on average values for the constants of the human eye and is intended closely to represent the optical effect of the eye. In the example, a simplified schematic eye is considered; this consists of three refracting surfaces, the first representing the cornea and the second and third, the two surfaces of the crystalline lens.

All lengths (except f and f') are given in metres, and powers and vergences are in dioptres.

Four figure logarithms have been used, and 10 has been added to the characteristics of the logarithms of numbers less than unity. The letter

	First Surface.		Second Surface.		Third Surface.	
	Number.	Logarithm.	Number.	Logarithm.	Number.	Logarithm.
Given data $\quad n$	1·0000	—	1·3333	0·1249	1·4160	0·1511
r	0·0078	7·8921	0·0100	8·0000	−0·0060	7·7782 n
n'	1·3333	0·1249	1·4160	0·1511	1·3333	0·1249
d	0·0036	7·5563	0·0036	7·5563		
$\dfrac{d}{n}$	0·0027	7·4314	0·00254	7·4052		
$n' - n$	0·333	9·5228	0·0827	8·9175	−0·0827	8·9175 n
$\dfrac{1}{r}$	128·2	2·1079	100·0	2·0000	−166·64	2·2218 n
F	42·73	1·6307	8·27	0·9175	13·78	1·1393
f (mm.)	−23·40	1·3693 n				
f' (mm.)	31·20	1·4942				
L	0	—	48·31	1·6840	66·10	1·8202
L'	42·73	1·6307	56·58	1·7527	79·88	1·9024
$1/L'$	0·0234	8·3693	0·01767	8·2473	0·01253	8·0976
$1/L' - d/n$	0·0207	8·3160	0·01513	8·1798	—	—
$L'/\!\left(1 - \dfrac{d}{n}\,L'\right)$	48·31	1·6840	66·10	1·8202	—	—
L/L'	—	—	0·8539	9·9313	0·8276	9·9178

n after a logarithm indicates that it is the logarithm of a negative quantity.

From the table, $L'_3 = + 79\cdot88$D., and therefore

$$l'_3 = \frac{1\cdot3333}{79\cdot88}m = +16\cdot69 \text{ mm.}$$

Hence the distance of F' from the cornea
$$= 16\cdot69 + d_2 + d_3 = + 23\cdot89 \text{ mm.}$$

$$f = f_1 \prod{}_2^3 \frac{L}{L'}$$

$$= -23\cdot40 \times 0\cdot8539 \times 0\cdot8276 = -16\cdot53 \text{ mm.}$$

$$f' = -\frac{n'}{n}f = 1\cdot3333 \times 16\cdot53 = +22\cdot04 \text{ mm.}$$

position of 2nd principal point $= 16\cdot69 - 22\cdot04$
$= -5\cdot35$ mm. from 3rd surface $= +1\cdot85$ mm. from cornea.

Focal power of eye, $F = \dfrac{1}{0\cdot01653} = \dfrac{1\cdot3333}{0\cdot02204} = +60\ 48$D

245. Special Case of Two Surfaces
Focal Length and Focal Power.

The focal length of a combination of two surfaces is given by

$$f = f_1 \prod{}_2^2 \frac{L}{L'} = f_1 \times \frac{L_2}{L'_2}$$

$$\frac{1}{f} = \frac{1}{f_1}\cdot\frac{L_2+F_2}{L_2} = \frac{1}{f_1}\left[1+F_2\frac{(f'_1-d_2)}{n_2}\right] = -\frac{F_1}{n_1}\left[1+\frac{F_2}{F_1}-\frac{d_2}{n_2}F_2\right]$$

$$\text{or } -\frac{n_1}{f} = F_1+F_2-\frac{d_2}{n_2}F_1F_2 = F \quad \ldots \quad \ldots \text{ (19.08)}$$

giving the general expression for the power of a combination of two surfaces forming a lens, the media in front and behind having refractive indices n_1 and n_3 respectively.

By tracing the light in the opposite direction we obtain

$$\frac{n_3}{f'} = F_1+F_2-\frac{d_2}{n_2}F_1F_2 = F \quad \ldots \quad \text{(19.08A)}$$

246. Back Vertex Power and Positions of Prinicipal Points

When the light is incident from an infinitely distant object the reduced vergence immediately on leaving the last surface is the **back vertex power,** B.V.P. of the system. This value is of particular importance in the case of the spectacle lens.

$$\left. \begin{aligned} \text{B.V.P.} &= \frac{F_1}{1 - \frac{d_2}{n_2}F_1} + F_2 = \frac{F_1 + F_2 - \frac{d_2}{n_2}F_1 F_2}{1 - \frac{d_2}{n_2}F_1} \\ &= \frac{F}{1 - \frac{d_2}{n_2}F_1} = \frac{n_3}{f'_V} \end{aligned} \right\} \quad \dots (19.09)$$

The distance from the second surface to the second principal point is

$$f'_V - f' = \frac{n_3}{F_1 + F_2 - \frac{d_2}{n_2}F_1 F_2}\left(1 - \frac{d_2}{n_2}F_1 - 1\right) = -n_3\frac{d_2}{n_2}\frac{F_1}{F}$$

Calling this distance e' and the distance of the first principal point from the first surface, e, we have

$$e' = -n_3\frac{d_2}{n_2}\frac{F_1}{F} \text{ or } \frac{e'}{n_3} = -\frac{d_2}{n_2}\frac{F_1}{F} \quad \dots \quad \dots (19.10)$$

and similarly

$$e = n_1\frac{d_2}{n_2}\frac{F_2}{F} \text{ or } \frac{e}{n_1} = \frac{d_2}{n_2}\frac{F_2}{F} \quad \dots (19.10A)$$

It will be seen that all these expressions are similar to those obtained in Chapter IX for a thick lens in air, in which case n_1 and $n_3 = 1$.

247. Conjugate Relations of Centred System

Having once determined the cardinal points of a given lens system we may make use of their properties to find the position and size of the image of *any* object formed by that system. In this way the rather tedious process of tracing the light through each surface of the system is avoided, and any further calculations with the given system are greatly simplified. If was for this purpose that the theory, originally due to Gauss, was developed, and with a knowledge of the cardinal points, any further work with the system is concerned only with the object and image spaces.

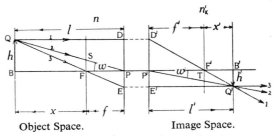

Fig. 258.–Cardinal points of centred system

Thus if the positions of the cardinal points are known, we can easily find the position of the image of a given object by construction. In Fig. 258 F and F′ are the principal foci and P and P′ the principal points of a system. Then a ray QD becomes D′F′ after refraction, and a ray QF after refraction becomes E′Q′ parallel to the axis. These refracted rays intersect at Q′, the image of Q, and as we are dealing with the paraxial region, a perpendicular to the axis through Q′ gives the point B′, the image of B.

Relations connecting the object and image, similar to those for a single surface or thin lens, can easily be found from the geometry of the figure.

Since the image of a point S in the first focal plane is at infinity and a ray from S parallel to the axis when refracted must pass through F′, then this refracted ray and P′T must be parallel and hence F′T = − FS. Incidentally this provides a construction for finding the path of the refracted ray corresponding to any incident ray, similarly to that already described for a thin lens.

$$FS = f \tan w = fw \text{ and } F T = f' \tan w' = f'w'$$

$$\text{Therefore} \frac{f'}{f} = -\frac{w}{w'}, \text{ since } F'T = -FS$$

For any system $n h u = n'h'u'$, and therefore for the principal points, since $h' = h$, we have $nu = n'u'$, i.e., $nw = n'w'$, or an incident ray QP leaves the system as P′Q′ as if it had suffered one refraction at a surface separating n and n'

$$\text{hence} \frac{f'}{f} = -\frac{n'}{n} \quad \ldots \quad \ldots \quad \ldots \quad (19.11)$$

$$m = \frac{h'}{h} = \frac{B'Q'}{BQ} = \frac{B'Q'}{P'D'} = \frac{F'B'}{F'P'} = -\frac{x'}{f'} \left.\begin{array}{c} \\ \\ \\ \\ \end{array}\right\} \quad \ldots \quad (19.12)$$

$$\text{also } m = \frac{B'Q'}{BQ} = \frac{PE}{BQ} = \frac{FP}{FB} = -\frac{f}{x}$$

and $xx = ff'$ (Newton's relation) ... (19.13)

$$x = -f + PB \text{ and } x' = -f' + P'B'$$

$$(-f + PB)(-f' + P'B') = ff'$$

$$PB \cdot P'B' - PBf' - P'B'f = 0$$

or

$$\frac{f'}{P'B'} + \frac{f}{PB} = 1$$

$$\frac{n'}{P'B'} \cdot \frac{f'}{n'} - \frac{n}{PB} \left(-\frac{f}{n} \right) = 1$$

$$\frac{n'}{P'B'} - \frac{n}{PB} = \frac{n'}{f'} = -\frac{n}{f} = F$$

Thus it will be seen that if the distances of object and image are measured from the respective principal points, these distances are connected with the power of the system by the same simple expression as was previously obtained for a surface or thin lens, i.e.,

$$\frac{n}{l'} - \frac{n}{l} = F \text{ or } L' = L + F \qquad \dots \qquad \dots \quad (19.14)$$

Magnification (Lateral).

$$m = \frac{h'}{h} = \frac{B'Q'}{BQ} = \frac{l'w'}{lw}$$

$$= \frac{l'}{n'} \bigg/ \frac{l}{n} = \frac{L}{L'} \qquad \dots \qquad \dots \qquad \dots \quad (19.15)$$

Magnification is given by the relation between the *reduced* distances of object and image in the same way as with a single surface.

248. Nodal Points

Two further points, as was mentioned in Chapter IX, are usually added to the focal and principal points under the heading of Cardinal Points; these are the **nodal points**. Of all the rays from Q (Fig. 259) which reunite in a conjugate image point Q' there is one pair of conjugate rays for which $u = u'$, that is, the refracted ray is unchanged in direction although displaced. These conjugate rays cut the axis at two special conjugate points, the nodal points.

Fig. 260 shows a parallel pencil from a distant object point Q incident on a system. The ray FD is refracted parallel to the axis and, since the image is formed in the second focal plane, it is situated at Q'. The second nodal point N′ is located by drawing the line Q′N′T′ through Q′ parallel to FD, and as this is a ray that leaves the second principal plane at a distance P′T′ from the axis, the conjugate incident ray is a ray parallel

Fig. 259.—Nodal Points

Fig. 260.—Centred system. Positions of nodal points

to FD meeting the first principal plane at a distance PT = P′T′ from the axis. This incident ray RT locates the first nodal point N.

$$NN' = TT' = PP'$$

or the nodal points have the same separation as the principal points

$$T'D' = TD = RF$$

Therefore $\qquad FN = D'Q' = P'F' = f' \left.\rule{0pt}{10pt}\right\}$

similarly $\qquad N'F' = FP = -f \quad\left.\rule{0pt}{10pt}\right.$ $\qquad$ ... $\qquad$ ... (19.16)

Thus the distance between the *first* focal and nodal points is equal to the *second* focal length, and the distance between the *second* focal and nodal points is equal to the *first* focal length, the distance being measured *from* the focal *to* the nodal point, that is, in the opposite directions to the focal lengths.

Relative to the principal points the nodal points are displaced towards the medium of higher refractive index.

In a system for which $n' = n$ (such as a thick lens in air) the focal lengths are numerically equal and it is easily seen that the principal and nodal points then coincide; the combined points are often called the **equivalent points.**

439

Nodal points are of considerable importance in connection with image size, as this is determined by the ray emerging from the second nodal point. Frequent use of this fact is made in work on the optical system of the eye, where the size of the retinal image of any object is quickly found when the distance of the retina from the second nodal point is known.

In a single refracting surface the nodal points coincide at the centre of curvature C, while the principal points coincide at the vertex of the surface A.

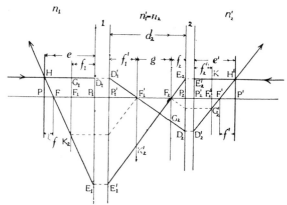

Fig. 261.—Combination of two systems

249. Combination of Two Centred Systems

From the above expressions we can find analytically the positions of the cardinal points of any system consisting of two surfaces. It is now necessary to consider how we can deal with the combination of two such systems.

Fig. 261 represents two systems 1 and 2, the first having its principal points at P_1 and P'_1 and its focal points at F_1 and F'_1; the second system has its principal points at P_2 and P'_2 and its focal points at F_2 and F'_2. Two rays, in opposite directions, are traced graphically through the combination, using a similar construction to that in Fig. 76, Chapter VI. It is clear that F and F' are the principal foci of the compound system, and as the two rays intersect at H and H' at equal distances PH and P'H' above the axis, PH and P'H' are the principal planes. Then PF $= f$ and P'F' $= f'$, the focal lengths of the combined system.

As F and F_2 are conjugate points with respect to system 1, from expression (19.13)

$$F_1F \times F'_1F_2 = f_1f'_1$$

Similarly F'_1 and F' are conjugate points with respect to system 2 and

$$F_2F'_1 \times F'_2F' = f_2f'_2$$

then

$$F_1F = \frac{f_1f'_1}{g}$$

and

$$F'_2F' = -\frac{f_2f'_2}{g}$$

where

$$F'_1F_2 = g$$

In the similar triangles FPH and FP_1E_1

$$\frac{FP}{FP_1} = \frac{PH}{P_1E_1}$$

and as

$$PH = P_2E_2 \text{ and } P_1E_1 = P'_1E'_1$$

$$\frac{FP}{FP_1} = \frac{P_2E_2}{P'_1E'_1}$$

In the similar triangles $F_2P_2E_2$ and $F_2P'_1E'_1$

$$\frac{P_2E_2}{P'_1E'_1} = \frac{F_2P_2}{F_2P'_1}$$

and hence

$$FP = \frac{F_2P_2}{F_2P'_1} \cdot FP_1$$

$$FP_1 = FF_1 + F_1P_1 = \frac{-f_1f'_1}{g} - f_1 = \frac{-f_1}{g}(f'_1 + g)$$

and

$$F_2P'_1 = F_2F'_1 + F'_1P'_1 = -g - f'_1$$

Putting

$$FP = -f \text{ and } F_2P_2 = -f_2$$

we obtain

$$f = \frac{f_1f_2}{g}$$

Similarly we find

$$f' = -\frac{f'_1 f'_2}{g}$$

From the last two equations

$$F = -\frac{n_1}{f} = \frac{n'_2}{f'} = -\frac{n'_2 g}{f'_1 f'_2}$$

$$= -\frac{g}{n_2} \cdot \frac{n_2}{f'_1} \cdot \frac{n'_2}{f'_2} \quad \ldots \quad \ldots \quad \ldots \quad (19.17)$$

$$= -\frac{g}{n_2} \cdot F_1 \cdot F_2 \quad \ldots \quad \ldots \quad (19.17A)$$

For two systems separated by air, $n_2 = 1$ and $F = -gF_1F_2$. The microscope is a system of this type where g is the optical tube-length.

When $g = 0$, $F = 0$ and we have a system of zero power—an afocal or telescope system (compare § 110).

$$P'_1 P_2 = d_2 = f'_1 + g - f_2$$
$$g = d_2 - f'_1 + f_2$$

substituting this value of g in expression (19.17A) gives us

$$F = F_1 + F_2 - \frac{d_2}{n_2} F_1 F_2 \ldots \quad \ldots \quad \ldots \quad (19.18)$$

From the geometry of the figure we also obtain

$$P_1 P = e = n_1 \frac{d_2}{n_2} \frac{F_2}{F} \quad \ldots \quad \ldots \quad \ldots \quad (19.19)$$

and

$$P'_2 P' = e' = -n'_2 \frac{d_2}{n_2} \frac{F_1}{F} \quad \ldots \quad (19.19A)$$

It will be seen that the above expressions for the combination of two systems are the same as those for the combination of two surfaces.

The focal power and the positions of the cardinal points of *any* system of centred surfaces may be found by the above method. Each pair of adjacent surfaces may be combined to form a system, the systems so formed are then combined again in pairs, and so on until the cardinal points of the complete system are found. Thus, with a system of six surfaces, the cardinal points are first found of each of the three systems formed by combining surfaces 1 and 2, 3 and 4, and 5 and 6. The cardinal points of the system formed by the combination of the first two of these systems are found, and this system is then combined with the third.

As an example we will consider the schematic eye which we have previously calculated. Here it will be convenient to find the power and principal points of the system formed by the second and third surfaces, representing the crystalline lens, and combine this with the first surface, The work is shown in § 244.

It will be seen from the table that the first principal point is + 1·551 mm. from the cornea. The second principal point is − 4·060 mm. from the second principal point of the crystalline lens and this is − 1·288 mm. from the third surface; the second principal point is therefore 7·2 − 1·288 − 4·060 = + 1·85 mm. from the cornea. The first principal focus is − 16·53 + 1·55 = − 14·98 mm. from the cornea, and the second principal focus 1·85 + 22·04 = + 23·89 mm. from the cornea. The distances from the cornea of the first and second nodal points respectively are − 14·98 + 22·04 = + 7·06 mm. and 23·89—16·53 = + 7·36 mm.

250. Lens Mirror or Thick Mirror

These terms are applied to any system of surfaces or lenses, the last surface of which acts as a mirror, the light being thereby reflected back through the system. Such systems occur in the Mangin mirror frequently

Fig. 262.–Lens mirror or thick mirror

used as the reflector in various projector lamps, in the shaving glass formed by silvering the back surface of a double convex lens and in the various surfaces of the eye media, the reflections from which give rise to the Purkinje images.

In Fig. 262 P_1 and P'_1 are the principal points of a system in front of a convex spherical reflecting surface at A_2 with radius of curvature A_2C_2. A ray TD_1 incident parallel to the axis will, after refraction and reflection, travel along the path $D_2E'_1 E_1S$, and the total effect of the complete system will be the same as if the light had undergone a simple

	First Surface No.	First Surface log.	Second Surface No.	Second Surface log.	Third Surface No.	Third Surface log.	Complete System No.	Complete System log.
Given Data								
n	1·0000	—	1·3333	0·1249	1·4160	0·1511		
r	0·0078	7·8921	0·0100	8·0000	−0·0060	7·7782 n		
n'	1·3333	0·1249	1·4160	0·1511	1·3333	0·1249		
d	0·0036	7·5563	0·0036	7·5563	—	—		
$\frac{d}{n}$	0·0027	7·4314	0·002543	7·4052	—	—		
$n'-n$	0·3333	9·5228	0·0827	8·9175	−0·0827	8·9175 n		
$\frac{1}{r}$	128·2	2·1079	100·0	2·0000	−166·67	2·2218 n		
F	42·73	1·6307	8·27	0·9175	13·78	1·1393		
f (mm.)	−23·40	1·3693 n	−161·2	2·2074 n	−102·8	2·0118 n		
f' (mm.)	31·20	1·4942	171·2	2·2336	96·74	1·9856		

Combining Second and Third Surfaces.

	Combining Second and Third Surfaces No.	log.	Complete System No.	Complete System log.
F_1+F_2	22·05	—	64·49	—
d	0·002543	7·4052	0·00431	7·6345
$\frac{d}{n}F_1F_2$	0·2897	9·4620	4·007	0·6029
F	21·76	1·3377	60·48	1·7816
f (mm.)	−61·26	1·7872 n	−16·53	1·2184 n
f' (mm.)	61·26	1·7872	22·04	1·3433
e	0·002146	7·3317	0·001551	7·1906
e'	−0·001288	7·1099 n	−0·00406	7·6085 n

reflection at a mirror of suitable curvature at a position D'_2. $A'_2D'_2$ represents therefore the *equivalent mirror* which can replace the complete lens-mirror system.

Position of Equivalent Mirror.—Light from any point D_2 on the mirror surface will, after passing through the system in front of the mirror, emerge as though it had come from D'_2; that is, D'_2 is the image of D_2 as formed by the system. But D'_2 is a point on the equivalent mirror corresponding to any point D_2 on the actual mirror, hence the equivalent mirror occupies the position of the image of the actual mirror as formed by the system before the mirror.

Position of the Centre of Curvature of Equivalent Mirror.—A ray QH_1 incident in such a direction that after refraction at the lens system it meets the mirror normally at D_2 will be reflected and refracted back along its original path. Such a ray must be incident towards the centre of curvature C' of the equivalent mirror. Hence this ray is, after reflection coming apparently from C_2 the centre of the mirror, and after emergence from the system is coming apparently from C', then C', the centre of curvature of the equivalent mirror, is the image of C_2 as formed by the system in front of the mirror.

Example.—To find the position and radius of curvature of the equivalent mirror corresponding to the anterior surface of the crystalline lens in the simplified schematic eye (§ 244). The reflection from this surface gives rise to the third image of Purkinje.

Position of Equivalent Mirror.—Taking anterior surface of crystalline lens as object, the light travelling from crystalline lens to cornea,

$$l = -0.0036 \text{ m.} \quad L = -\frac{1.333}{0.0036} = -370.4 \text{ D}$$

$$n = 1.333$$

$$F = +42.73 \text{ D (power of cornea)}$$

$$n' = 1$$

$$l' = -3.052 \text{ mm.}$$

Therefore the equivalent mirror is situated 3.05 mm. behind the cornea.

Position of Centre of Curvature of Equivalent Mirror.—Taking the centre of curvature of the anterior surface of the crystalline lens as object,

$$l = -0.0136 \text{ m.} \quad L = -\frac{1.333}{0.0136} = -98.06 \text{ D}$$

$$L' = -98 \cdot 06 + 42 \cdot 73 = -55 \cdot 33 \text{ D}$$

$$l = -18 \cdot 08 \text{ mm.}$$

Therefore the centre of curvature of the equivalent mirror is $18 \cdot 08$ mm. behind the cornea, and the radius of curvature is $18 \cdot 08 - 3 \cdot 05 = +15 \cdot 03$ mm.

If a lamp is placed, say 30 cm. in front of the cornea, the position and magnification of the third image of Purkinje may now be found as follows:

Object distance from equivalent mirror,

$$l = -30 \cdot 3 \text{ cm.}$$

Radius of equivalent mirror $r = +1 \cdot 503$ cm.

$$L' = L + F = L - 2R$$

$$= \frac{100}{-30 \cdot 3} - \frac{200}{1 \cdot 503} = -136 \cdot 4 \text{ D}$$

Therefore $l' = +0 \cdot 733$ cm.

Hence, the image is situated $3 \cdot 05 + 7 \cdot 33 = 10 \cdot 38$ mm. behind the cornea.

The magnification of the image

$$m = \frac{7 \cdot 33}{303} = 0 \cdot 0242 \text{ or practically } \tfrac{1}{40}.$$

CHAPTER XIX   EXERCISES

1. An equi-convex lens having curves of 10 cm. radius, a central thickness of 2 cm. and refractive index $1 \cdot 61$ closes one end of a long tube filled with water. An object 5 cm. long is placed 60 cm. in front of the lens; find the position and size of the image.

2. Explain the terms: equivalent focal length, principal points and nodal points, and illustrate each by a diagram of the eye.

3. A $+ 10$D equi-convex lens of glass of $n = 1 \cdot 62$ is mounted in the centre of a tank 20 cm. long and having thin plane glass ends; the tank is filled with water ($n = 1 \cdot 33$). Find the power of the system and the position of the focus when parallel light is incident on one end of the tank.

4. What are the properties of the principal and nodal points of a lens?

A double convex lens of glass of refractive index $1 \cdot 53$ having curves of 5D and 8D respectively and thickness of 4 cm. has its second surface immersed in water ($n = 1 \cdot 33$). Find the focal lengths and positions of the principal and nodal points.

5. A thin + 10D equi-convex lens of refractive index 1·53 forms one end of a long tube filled with water of refractive index 1·33; find the positions of the principal foci and the equivalent focal length of the system.

6. Find the focal power and positions of the nodal and principal points of an aphakic eye, radius of cornea = 8 mm.; refractive index = 1·336 .What correcting lens is necessary to focus distant objects on the retina, the lens being placed 15 mm. from the cornea and the axial length of the eye being 24 mm.?

7. A long tank filled with water has an equi-convex lens of $n = 1·58$, thickness $\frac{3}{4}$ inch and radii of curvature of 2 inches fitted in one end. Find the position of the principal, nodal and focal points.

8. A small object is situated on the second surface of the crystalline lens; find, using the data for the schematic eye, the position and magnification of the image of this object formed by the system in front of it.

9. Show that in a concentric lens, i.e., one in which the two surfaces have a common centre of curvature, the principal points coincide at the centre of curvature.

10. In Gullstrand's Schematic Eye, the cornea and the aqueous humour have the following constants:

Thickness of cornea	... ...	0·5 mm.
Radius of anterior surface	...	+7·7 mm.
Radius of posterior surface	...	+6·8 mm.
Refractive index of cornea	...	1·376
Refractive index of aqueous humour		1·336

Find the two focal lengths and the power of the cornea.

11. A hollow globe of glass ($n = 1·52$) has an external diameter of 10 inches and an internal diameter of $9\frac{1}{2}$ inches; it is filled with water. Find the position at which a narrow parallel pencil of light incident on the globe will be focused. What will be the positions of the principal points of the system?

12. Calculate the equivalent focal length of a lens system consisting of three thin positive lenses of 4 inches, 5 inches and 6 inches focal length respectively, each separation being 1 inch.

13. An object 5 cm. high is placed 25 cm. in front of the cornea of the schematic eye (§ 244). Find by tracing through the various surfaces in turn the position and size of the image formed by the system. Compare the result with that obtained by making use of the equivalent focal power and the positions of the principal and nodal points. (Page 444).

14. Find the equivalent focal length and the positions of the principal points of the Rapid Landscape photographic lens having the following constants:

$$r_1 = -120·8 \text{ mm.} \qquad n_1 = 1$$
$$d_2 = 6 \text{ mm.}$$
$$r_2 = -34·6 \text{ mm.} \qquad n_2 = 1·521$$
$$d_3 = 2 \text{ mm.}$$
$$r_3 = -96·2 \text{ mm.} \qquad n_3 = 1·581$$
$$d_4 = 3 \text{ mm.}$$
$$r_4 = -51·2 \text{ mm.} \qquad n_4 = 1·514$$
$$n_5 = 1$$

15. Find from the following data (all distances in mm.) the equivalent focal length and the positions of the principal points of the Cooke Series IV photographic objective:

$$r_1 = + 19 \cdot 44 \quad d_2 = \quad 4 \cdot 29 \quad n_2 = 1 \cdot 6110$$
$$r_2 = - 128 \cdot 3 \quad d_3 = \quad 1 \cdot 63 \quad n_3 = 1 \cdot 000$$
$$r_3 = - 57 \cdot 85 \quad d_4 = \quad 0 \cdot 73 \quad n_4 = 1 \cdot 5754$$
$$r_4 = + 18 \cdot 19 \quad d_5 = 12 \cdot 9 \quad n_5 = 1 \cdot 000$$
$$r_5 = + 311 \cdot 3 \quad d_6 = \quad 3 \cdot 03 \quad n_6 = 1 \cdot 6110$$
$$r_6 = - 66 \cdot 4$$

16. Having given the following data of Helmholtz's schematic eye at rest, find:

(a) Power and focal lengths of crystalline lens.

(b) Positions of principal points of crystalline lens.

(c) Power and focal lengths of cornea.

(d) Power of whole schematic eye.

Data:

Radius of cornea at vertex	+ 7·829 mm.
Radius of anterior surface of crystalline lens	+10·0 ,,
Radius of posterior surface of crystalline lens	− 6·0 ,,
Depth of anterior chamber	3·6 ,,
Thickness of crystalline lens	3·6 ,,
Refractive index of aqueous humour	1·3365
,, ,, ,, crystalline lens	1·4371
,, ,, ,, vitreous humour	1·3365

17. An equi-convex lens has surfaces of 15 cm. radius of curvature and a central thickness of 2 cm. and $n = 1 \cdot 51$. Part of the light which enters the lens from a distant luminous point is reflected at the second surface and emerges again from the first. Where will this reflected light focus?

18. What will be the focal power and position of the equivalent thin mirror corresponding to a meniscus lens having curvatures $r_1 = - 15$ cm., $r_2 = - 25$ cm., central thickness 2 cm., $n = 1 \cdot 53$, the second surface of the lens being silvered?

19. Using the data for the schematic eye given in § 244, find the position and size of the image formed by reflection at the second surface of the crystalline lens (the fourth image of Purkinje), the object being 5 cm. diameter and situated 30 cm. from the cornea.

Appendix

I.—Reflection Coefficients

Material	Percentage of Incident Light reflected at 30° to Normal.
Magnesium Oxide	96
Plaster of Paris	91
Matt White Celluloid	80—85
White Blotting Paper	80—85
Ground Opal Glass	76
Foolscap Paper	70
Black Cloth	12
Black Velvet	4

II.—Principal Fraunhofer Lines

Line.	Position in Spectrum.	Element to which line corresponds.	Wave-length nm
A	Extreme red	O	759·4
B	Red	O	686·7
C	,,	H	656·3
d	Yellow	He	587·6
D_1	Yellow	Na	589·6
D_2	,,	,,	589·0
E	Green	Fe & Ca	527·0
b	,,	Mg	518·4
F	Blue	H	486·1
G	Violet	H	434·0
G	,,	Fe & Ca	430·8
H	,,	Ca	396·9
K	,,	,,	393·4

III.—Optical Glass Types (Chance)

Wave-Length Table (Angstroms) — Partial and Relative Partial Dispersions. Wavelengths: A' 7682 K, C 6563 H, D 5893 Na, d 5876 He, e 5461 H, F 4861 H, g 4358 Hg, G' 4340 H, h 4047 Hg.

Glass Type	Type / Refractive index n_d	V ($\frac{n_d - 1}{n_F - n_C}$)	Mean Dispersion C − F	A' − C	C − d	d − e	e − F	F − g	g − h	n_C	n_D	n_G
510644	Boro-silicate Crown, 1·50970	64·44	0·00791	0·00279 / 0·360	0·00243 / 0·307	0·00188 / 0·238	0·00360 / 0·455	0·00423 / 0·535	0·00348 / 0·440	1·50727	1·50962	1·51959
519604	Hard Crown, 1·51899	60·42	0·00859	0·00298 / 0·351	0·00262 / 0·305	0·00205 / 0·239	0·00392 / 0·456	0·00465 / 0·541	0·00386 / 0·449	1·51637	1·51890	1·52981
541595	Light Barium Crown, 1·54065	59·54	0·00908	0·00310 / 0·341	0·00277 / 0·305	0·00217 / 0·239	0·00414 / 0·456	0·00493 / 0·543	0·00406 / 0·447	1·53788	1·54056	1·55210
572577	Medium Barium Crown, 1·57220	57·74	0·00991	0·00341 / 0·344	0·00301 / 0·304	0·00236 / 0·238	0·00454 / 0·458	0·00540 / 0·545	0·00449 / 0·453	1·56919	1·57210	1·58473
612585	Dense Barium Crown, 1·61230	58·54	0·01046	0·00357 / 0·341	0·00318 / 0·304	0·00249 / 0·238	0·00479 / 0·458	0·00569 / 0·544	0·00472 / 0·451	1·60912	1·61219	1·62551
575426	Light Flint, 1·57453	42·62	0·01348	0·00441 / 0·327	0·00400 / 0·297	0·00320 / 0·237	0·00628 / 0·466	0·00770 / 0·571	0·00660 / 0·490	1·57053	1·57439	1·59204
613369	Dense Flint, 1·61323	36·92	0·01661	0·00528 / 0·318	0·00489 / 0·294	0·00393 / 0·237	0·00779 / 0·469	0·00965 / 0·581	0·00836 / 0·503	1·60834	1·61306	1·63502
653335	Extra Dense Flint, 1·65348	33·48	0·01952	0·00615 / 0·315	0·00570 / 0·292	0·00461 / 0·236	0·00921 / 0·472	0·01150 / 0·589	0·01006 / 0·515	1·64778	1·65328	1·67930
748278	Double Extra Dense Flint, 1·74842	27·85	0·02687	0·00826 / 0·307	0·00775 / 0·288	0·00635 / 0·236	0·01277 / 0·475	0·01620 / 0·603	0·01438 / 0·535	1·74067	1·74815	1·78445

IV. Chief Spectral Lines used in Optics (Wavelengths in nanometres)

Colour	Hydrogen	Helium	Cadmium	Mercury	Sodium
Red	C 656·3	–	C′ 643·9	–	–
Yellow	–	d 587·6	–	–	D₁ 589·6 D₂ 589·0
Green	–	–	–	e 546·1	–
Blue	F 486·1	–	F′ 480·0	g 435·8	–
Violet	–	–	–	h 404·7	–

V.–Refractive Indices of Some Solids and Liquids

	n_D	v
Diamond	2·4173	55·4
Iceland Spar (ordinary ray)	1·6584	48·9
,, ,, (extraordinary ray) ...	1·4864	79·8
Quartz (ordinary ray)	1·5442	69·95
,, (extraordinary ray)	1·5533	69·0
Quartz (fused)	1·4585	67·5
Rock Salt	1·5443	42·8
Canada Balsam	1·526	41·5
Fluorite	1·4338	97·3
Ice (ordinary ray)	1·308	
,, (extraordinary ray)	1·313	
Polymethyl methacrylate (Perspex) ...	1·4900	53·7
Polystyrene	1·5900	31·0
Monobromnaphthaline (20° C.)	1·6582	20·3
Carbon Disulphide (15° C.)	1·6303	18·3
Oil of Cassia (10° C.)	1·6104	15·9
Cinnamic Ether	1·5607	11·0
Methyl Salicylate (21° C.)	1·5319	24·7
Cedar Wood Oil	1·510	
,, ,, ,, (hardened)	1·520	
Glycerine (20° C.)	1·4730	60·6
Turpentine (11° C.)...	1·4744	46·5
Water (19° C.)	1·3336	54·7

Answers to Exercises

CHAPTER I

1. 0·000667, 0·000527, 0·000431 mm.
3. 9°31′; 240 ft.
5. 300 sq. cm.
6. 5·59, 0·349, 0·785 sq. cm.
7. Circular patch of full illumination 13 cm. diameter, surrounding ring of partial illumination 8·33 cm. wide.
8. 2916 sq. cm.; full illumination 648 sq. cm., partial illumination 4536 sq. cm.
9. 52·2 ft.
10. 8 ft. 7½ in.
11. 8⅝ ft. diameter; umbra 3⅔ ft. diameter, penumbra 13⅝ ft. diameter.
13. 5·45 in. from image.
14. 13·5 in.
15. 53·7 in.
17. 27 in. diameter; penumbra.
19. Umbra 5766 miles, penumbra 10234 miles.
21. 144 ft.; 20 in.

CHAPTER II

7. (a) 225·5, (b) 206·8, (c) 181·8 million metres per sec.
8. (a) 1·205, (b) 0·8297.
9. (a) 1·128, (b) 1·819, (c) 0·5496, (d) 0·6060, (e) 1·10.
16. $i'_1 = 28° 45′, i'_2 = 32° 7′$.
17. 122° 52′.

CHAPTER III

3. 100 cm. × 75 cm.
4. 12 ft.
5. 55 cm. × 36·7 cm.; 133·2 cm. above the ground.
7. 2·087 ft. × 1·391 ft.
8. 178° 13′.

452

9. 5 ft. 10 in. × 4 ft. 2 in.; lower edge 4 ft .4½ in. above the ground.
10. 1 ft. 10 in. × 8 in.
11. 12 cm.; two portions 4 cm. wide, separated by 4 cm.
12. 1° 26′.
22. 90°; 15° W. of N.
23. From M_1, 40, 160, 240 cm.
    From M_2, 60, 140, 260 cm.
24. 35°.
28. 340·3 yds.
30. 1⅓°.

CHAPTER IV

1. $i' = 27°\ 32'$.
2. $i_C = 52°\ 48'$.
3. (a) $i'_1 = i'_2 = i'_3 = 0°$;
   (b) $i'_1 = 32°\ 2'$, $i'_2 = 25°\ 53'$, $i'_3 = 45°$.
4. $i'_1 = 25°\ 53'$; displacement 1·82 cm; angle of emergence in oil 28° 20′.
8. 0·29 in.
11. (a) 41° 4′, (b) 61° 7′, (c)74° 54′.
12. 3·02 in.
13. 4 in.
15. 1 ft. 9 in. from side nearer eye.
17. Blue light totally reflected at 40° 45′ to normal.
    Red light emerges at 80° 42′ to normal.
20. $i'_1 = 32°\ 7' = i_2 = r_2$, $i_3 = 57°\ 53'$, $i'_3 = 47°\ 2' = i_4$, light totally internally reflected in glass side.
22. (a) $i'_2 = 24°\ 28'$, $d = 18°\ 56'$ (b) $i'_2 = 0°$, $d = 23°\ 8'$.
    (c) $i'_2 = 53°\ 8'$, $d = 23°\ 8'$.
23. $i'_1 = 24°\ 51'$, $i_2 = 35°\ 9'$, $i'_2 = 61°\ 45'$, $d = 41°\ 45'$.
    Minimum $d = 39°\ 49'$.
24. 1·556.
26. 1·618.
27. 9° 12′.
28. $i'_1 = 33°\ 19'$, $i_2 = 41°\ 41'$, $r_2 = 41°\ 41'$, $i_3 = 26°\ 41'$,
    $i'_3 = 47°\ 49'$, $d = 107°\ 11'$.
29. $i'_1 = 19°\ 5'$, $i_2 = 40°\ 55'$, $r_2 = 40°\ 55'$, $i_3 = 19°\ 5'$, $i'_3 = 30°$,
    $d = 120°$, i_1 for minimum deviation $= 49°\ 55'$.
30. 36·6 cm.
31. 99° 49′.
32. 65° 4′.
33. Images 1 ft. behind prism, 0·64 in. apart.
34. 15° 53′; using approximate formula 16° 45′.
35. 2′ 18;.
36. (a) Not smaller than 6° 24′, (b) Not greater than 24° 19′ on other side of normal.
37. (a) 16° 49′, (b) 1° 42′, (c) − 18° 40′.
38. $d = 2·19$ prism dioptres, $a = 2·5°$; $n = 1·5$.
39. 1·577.
40. 1·5811.
41. 1·568.
42. 22° 23′.

CHAPTER V

2. $+ 8.726D.: 1.18\%$.
3. 0.05 in.
4. 2.88 mm.
5. 1.528.
6. 1.543.
7. 3.91 mm.; 6.21 mm. + allowance for working.
8. 32.6 mm.
9. $+ 5D.; + 5.26D.; + 1000D.; - 100D.$
10. $- 1\frac{1}{3}D.; l' = + 37.5$ cm.
11. $- 14.3$ cm. from lens; $+ 10.33D.$
15. Virtual erect image $- 37.38$ cm. from surface, 1.64 cm. high.
    $F - 38.46$ cm., $F' + 58.45$ cm. from surface.
17. 0.76 mm.
18. 1.504; 75.1 mm.
19. Centre; 3.02 cm. from front surface.
20. $- 1.24$ in. from upper surface; $m = 1.26$.
21. $+ 22.22$ mm.; $- 22.5$ cm.
22. $l' = - 3.05$ mm.; 4.52 mm. diameter.
23. (a) 20 cm., (b) 80 cm. from curved surface.
24. $+ 53.33D.; + 1.875$ cm. from plane surface.
25. 0.29 mm.

CHAPTER VI

1. 74.2 cm.; 12.37 cm.
3. 16.5 cm; 8.25 cm.
4. 12.4 cm.
5. 2 in.; $+ 3.33D.$; 12 in.
7. 60 cm.
8. $- 6.06$ cm. virtual; $- 11.76$ cm. real.
9. 41.84 cm.
10. (a) $+ 25$ cm. $+ 4D.$; (b) $+ 28.57$ cm. $+ 3.5D.$; (c) $+ 33.33$ cm. $+ 3D.$;
    (d) $+ 50$ cm. $+ 2D.$; (e) ∞, 0D.; (f) $- 16.67$ cm. $- 6D.$
11. -17.65 cm.; 1.47 mm. erect.
13. $+ 20$ cm., $+ 5D.$
14. 6.98 in. or 29,02 in. from lamp; $m = - 4$ or $- \frac{1}{4}$ (approx.).
15. $+ 15.55$ cm.
17. 40 mm.
18. $- 14$ in.
19. $+ 7.58$ in.
21. $+ 229.5$ in. from lens; 4.5 in.
22. 16 cm. from object; $f' = + 12.8$ cm.
23. 8 in. from object; $+ 6.57D.$
24. $+ 12.25$ cm.
25. (a) $+ 50$ cm., 2 mm.; (b) $+ 36.4$ cm., 0.9 mm.
26. $+ 7.41$ in.
27. Image 16 in. long, one end 14 in. from lens; image 13.34 in. long, 16.67 in.
    from lens.
28. $- 6.22$ in. from lens.
29. $- 80$ in.; $m = + 5$.

31. $- 60$ cm.
32. $+ 4$ cm.; $+ 24$ cm.
33. 10 cm.; $+ 3$D.
34. 185·4 cm.
35. 10·7 cm.
36. 2·65 cm.
37. 91·45 mm.
38. 30 cm. from 1st lens ($m = + \frac{1}{2}$).
    110 cm. from 1st lens ($m = - \frac{1}{2}$).
39. 4·37 cm.; 15·63 cm. from the 1st lens.
40. $+ 100$ cm.
42. $- 1·2$ in.; $- 32·8$D.
43. 0·34 cm.
44. Yellow 288·2 mm.; $+ 3·47$D.
    Red   289·6 mm.; $+ 3·45$D.
    Blue 284·9 mm.; $+ 3·51$D.
46. 67·08 cm.
47. 2·17 cm.
49. $- 60$ cm.
50. $+ 138·6$ cm. from lens.
51. $+ 4·75$D.; 30 mm.
55. $+ 6·4$ in.
57. $- 8$D.

CHAPTER VII

3. $- 9·95$ cm. from mirror.
4. $- 10·53$ cm.; $- 22·22$ cm.
6. 34·2 sq. cm.; 45·8 cm. or 12·5 cm.
9. $\frac{5}{8}r$; $\frac{3}{8}r$.
10. $- 3·15$D.; $- 21·25$ in.; $- 8·85$ in.
11. $- 8·53$ cm. from mirror.
12. $(a) + 3·33$ cm.; 1 cm.; $(b) + 15$ cm.; 9 cm.
14. $l = - 3·86$ in.
15. $l = + 10·34$ cm.; 0·86 mm.; virtual.
16. $- 18$ in.
17. $+ 5·75$D.; $- 2·875$D.; $- 34·78$ cm.
18. 20 cm.; 40 cm.
19. 0·6 ft. from mirror.
20. $m = - 0·6$.
21. $l = - 3·5$ in.
22. 5·02 in. by 4·18 in.; 8·37 in. from mirror.
24. $- 46·17$ cm.; 3·34 mm.
25. 48 ft. 9 in. by 29 ft. 3 in.
26. 1·18 mm. long; 3·92 mm. behind cornea.
27. $- 53,33$ cm.
28. 2 ft. from flame; $f = - 1\frac{1}{2}$ ft.
29. 9·78 cm.; 14·22 cm.; 25·78 cm.
32. 591 cm.
33. 70 cm.
34. 12·73 cm. in front of concave mirror; 0·273 cm. high; inverted.

455

OPTICS

35. − 13·85 cm.
36. + 10D.
37. Real inverted image 60 cm. in front of 2nd mirror; 1 cm. high.
39. + 3·97 mm.; 4·96 mm.
40. Virtual image 60 cm. from lens; 15 cm. high.
42. − 8 cm.; − 3·87 cm.
43. − 62·86 cm.; − 19·14 cm.
44. $r_2 = -15$ cm.; $r_1 = +7·5$ cm.
45. 1·415.
46. + 18·34 cm.
47. $n = 1·5$; $r = 24$ cm.
49. − 3·09 cm. from front surface; − 5·33 from front surface.
50. Light focuses 20·58 cm. in front of lens.
52. $\frac{a}{b} = 7$
53. AB = 18·21 cm.
55. 36°.

CHAPTER VIII

1. 5·23D.; 0; 1·31D.
2. 1·97 mm.; 3·42 mm.
4. + 4·0D.S./− 6·0D.C., ax. 170° or − 2·0D.S./+ 6·0D.C., ax. 80°.
5. 13·82 cm.; 22·62 cm.
6. (a) + 33·3 cm. from lens; 6·66 cm. long, vertical.
   (b) Vertical line 3·33 cm. long; + 16·67 cm. from lens; horizontal line 20 cm. long. + 100 cm. from lens.
   Circle of least confusion 2·86 cm. diameter; + 28·57 cm. from lens.
7. Horizontal line 3·77 cm. long; + 21·4 cm. from lens.
   Vertical line 15·01 cm. long; + 85·7 cm. from lens.
8. Horizontal line 1·97 cm. long, + 21·88 cm. from lens; vertical line 3·88 cm long, + 43·08 cm. from lens; circle of least confusion 1·31 cm. diameter, + 29·1 cm. from lens.
9. Circular, 1·5 cm. diameter; + 2·0D.S./− 4·0D.C., ax. H.
10. (a) 1·8 cm. horiz., 2·7 cm. vert.; (b) 0·9 cm. horiz., 0·9 cm. vert.; (c) 3·6 cm. horiz., 0·9 cm. vert.
11. (a) Line 45 mm. long; (b) Patch 22·5 mm. square.
12. + 3·0D.C. ax. V./+ 5·0D.C. ax. H.; toric.
13. 52·3 mm.; 87·2 mm.
14. 0·3 in.
15. H line − 27·64 cm. or − 72·36 cm.
   V line − 50 cm.
16. − 50 cm.; infinity.
17. $l = -50$ cm. Circle 62·6 mm. diameter.

CHAPTER IX

1. + 4·65D.; + 21·5 cm.
2. Virtual image − 60 cm. from 2nd lens.
3. + 15·64D.
4. − 4·21D.
8. 2·92 cm.; 9·17 cm. from 1st lens.

9. $+ 5\cdot39$ cm. from 2nd lens, $2\cdot71$ cm. high; $- 3\cdot85$ cm. from 1st lens.
11. $- 14\cdot5$ cm. from 2nd lens; virtual vertical line.
12. $\frac{4}{5}$ cm. high; 2 cm. from 2nd lens.
13. $5\cdot11$ mm.; $+ 8\cdot1$ D.
15. $+ 46,73$D.C. ax. V./$- 12\cdot12$D.C. ax. H.
16. Inverted image $3\cdot37$ cm. high, $+ 11\cdot54$ cm. from 2nd lens;
     $f' = + 19\cdot23$ cm. Image of same size, $+ 15\cdot39$ cm. from $+ 4$D. lens; f'
     as before.
17. $+ 6$ cm.
18. $f' = + 4\cdot57$ cm.; $f_V = - 1\cdot91$ cm.; $f'_V = + 2\cdot57$ cm.;
     $e = + 2\cdot66$ cm., $e' = - 2\cdot0$ cm.
19. $f' = + 41\cdot67$ cm.; image $+ 68\cdot73$ cm. from 2nd lens. $m = - 1\cdot25$.
20. $F = + 6$D.; $e = - 6\cdot67$ cm.; $e' = - 8\cdot33$ cm. Telephoto system.
21. $3\cdot513$ inches.
22. $f_V = - \frac{1}{2}$ in., $f'_V = + \frac{1}{2}$ in., $e = + 1$ in., $e' = - 1$ in.
23. $F = - 3$D.; $f_V = - 33\cdot3$ cm.; $f'_V = + 133\cdot3$ cm.; $e = - 66\cdot6$ cm.
     $e' = + 166\cdot6$ cm.
24. $+ 20$ cm. from 2nd lens; $8\cdot81$ cm. high.
25. $f' = + 17\cdot2$ cm.; $e = + 5\cdot2$ cm.; $e' = - 6\cdot9$ cm.; image $+ 18\cdot2$ cm. from
     2nd lens; $1\cdot38$ cm. high.
26. $F = + 1\cdot84$D.; $e = - 34\cdot8$ cm.; $e' = - 26\cdot1$ cm.; virtual image $- 47\cdot2$ cm.
     from 2nd lens; $4\cdot17$ cm. high.
29. $f' = 1\cdot429$ in., $e = + 2\cdot5$ in.; $e' = - 1$ in.
32. $+ 33\cdot33$ cm.
34. $+ 14\cdot55$D.
35. $+ 9\cdot87$D., $+ 11\cdot12$D., $+ 8\cdot99$ cm. from 2nd surface.
36. $+ 9\cdot64$ in. from 2nd surface; $1\cdot33$ in. long.
37. Afocal.
38. $- 2\cdot25$ in.
39. $1\cdot55$; $9\cdot09$D.
40. $+ 1\cdot925$ in.; principal points at centre of sphere.
42. $+ 7\cdot14$ in.
45. $F = + 6$D.; $h' = \cdot834$ cm.
     f'_V changes from $+ 8\cdot33$ cm. to $+ 23\cdot3$ cm.; h' is unchanged.
47. PP' $= - 5$ mm.; (a) $+ 20$D, (b) 20 mm.
48. $+ 0\cdot268$D.
49. $e = + 8$ mm.; $e' = - 5$ mm.

CHAPTER X

3. $2° 3'$.
4. $- 7\cdot78$ cm. from the lens; $1\cdot8$ cm.
5. (a) $3\cdot75$ (b) $4\cdot0$.
6. $2\cdot86$.
8. 72.
9. 10; $0\cdot08$ in.
10 $-2\cdot66$ in. from objective.
11. $- 1\cdot66$ cm. from objective; 75; $3\cdot33$ mm.
13. $0\cdot014$ mm. away from object; 146.
15. $+ 2\frac{1}{8}$ in.
16. (a) $10\frac{2}{3}$ in. and $1\frac{1}{3}$ in.; (b) $13\frac{5}{7}$ in. and $- 1\frac{5}{7}$ in.
19. $- 36$ mm., 138 mm.

OPTICS

20. Inverted and magnified 4 times in horizontal meridian.
22. 1·1 in.
24. 10·7; 15·04 in. (assuming 40 in. from objective).
25. (a) Eyepiece moved inwards 7·58 mm. $M = 3·8$.
    (b) Eyepiece moved outwards 2·57 mm. $M = 5·6$.
26. Principal points at infinity, telescopic system, angluar magnification $- 4$; $+ 6·25$ cm. from 20D. surface, $m = - \frac{1}{4}$.
27. $f' = + 1·43$ in., $f_V = + 1·07$ in., $f'_V = + 0·428$ in.; $M = 14$.
30. $f' = + 53·30$ cm., $e = - 100·0$ cm., $e' = - 40·0$ cm., $f'_V = + 13·3$ cm., $h' = 1·86$ cm.
32. $f/5·5, f/8, f/16$; 1 sec., 8 sec.
33. 4·86 metres $\times$ 3·89 metres.

CHAPTER XI

1. (a) 1·44 lm./ft.², (b) 15·5 lm./m.²
2. 28·8 c.p.
3. 3·08 $\times$ 10¹⁶.
6. 1 lm./ft.² = 0·00108 lm./cm.² = 10·8 lm./m.²; 0·347 lm./ft.²; 3·75 lm./m.²
7. 0·18 lm./ft.²
10. 59° 48′.
11. 4·67 lm./m.²
12. 4·73 ft.
13. 8·73 lumens; 7·66 lm./ft.²
14. (a) 0·44 lm./ft.² (b) 0·096 lm./ft.²
15. 1·41 lumens; 260 lm./ft.²
16. 1/120 lm./cm.²; no change.
17. 0·4 lm./ft.²; 1·28 lm./ft.²
18. (a) 0·78 lm./ft.²; (b) 0·122 lm./ft.²
19. 6·44 lm./ft.²
20. 9·39 lm./ft.²; 4·8 lm./ft.²
21. 19·4 lm./ft.²; 17·6 lm./ft.²
22. 4·47 ft.; 0·88 lm./ft.²
23. (a) 0·28 lm./ft.², (b) 171 c.p.
24. 0·205 lm./ft.²
25. 5·69 lm./ft.²
26. 1·307 lm./ft.²; 340 c.p.
27. 3·2, 2·4, 0·267; 1:1·78:144.
28. 1·72 lm./ft.²
29. 17·32 candles; 80; 45.
32. 22·5 c.p.
34. 6·37 cd. per sq. mm.; 0·8 lm./mm.²
35. 110 cm. and 1090 cm. from 30 c.p. lamp.
36. D = 0·699, T = 4%, D = 1·398
38. 8·2%.
39. 96·6%.
40. 1 metre from either lamp.
41. 108·5 cm. from brighter lamp.
42. 6 ft.
43. (a) 50 cd. (b) 25 cd.; 0·0628 lumens.
44. 3⅕ lm./mm.²

458

45. (a) 0·0014 lm. (b) 0·0053 lm./ft.²
46. 11·1%; 0·95.

CHAPTER XII

2. 7595A, 759·5 nm, 0·7595μ, 2·99 × 10⁻⁵ in.
   5894A, 589·4 nm, 0·5894μ, 2·32 × 10⁻⁵ in.
   4862A, 486·2 nm, 0·4862μ, 1·91 × 10⁻⁵ in.
3. A stationary, above.
   B upwards, level.
   C downwards, below.
9. − 8.
13. 186,700 miles per second.
14. 16⅔ times per second.
15. 3·08 × 10⁸ metres per second.
16. 5 × 10¹³ miles.

CHAPTER XIII

4. 0·01646; Hard Crown.
8. 1° 41′.
10. $R_1 - R_2 = - 13·0D$.
11. 0·0821D.; − 2·96D.
12. $r_1 = + 10·04$ cm., $r_2 = - 10·04$ cm., $r_3 = - 10·04$ cm., $r_4 = + 294·1$ cm.
13. $r_1 = + 8·67$ cm., $r_2 = - 13·0$ cm., $r_3 = - 12·2$ cm., $r_4 = + 68·7$ cm.
14. $r_1 = + 8·74$ cm., r_2 and $r_3 = - 10·2$ cm., $r_4 =$ infinity.
15. 0·365 mm.; 0·036 mm.
16.

	D	F	C
F	+ 4·253D.	+ 4·312D.	+ 4·235D.
f'	+ 235·1 mm.	+ 231·9 mm.	+ 236·2 mm.
e	− 18·53 mm.	− 18·41 mm.	− 18·55 mm.
e'	− 29·36 mm.	− 29·17 mm.	− 29·39 mm.
h'	41·46 mm.	40·89 mm.	41·64 mm.

19. 4°; 1·7°.
20. 2·16′.
21. Crown 3° 56′; flint 1° 58′.
22. Crown 2·77°; flint 1·4°.
23. $a = 5°$, $d = 2° 7·5′$.
28. Red to yellow 30·5′, yellow to blue 1° 15·5′.
33. Crown 0·0157; 0·375; 0·625; 1·125.
    Flint 0·0357; 0·296; 0·704; 1·333.
36. B = 2A.

CHAPTER XIV

6. 0·514 mm.
7. 0·00064 mm.
9. 0·000658 mm.
10. 0·0053 mm.
11. 9·96 mm
12. 0·105D.
14. 12·25 metres.

16.   1·57 mm.; 4·96 mm.
20.   1·18 mm.
21. 6×10^{-4} mm.
22.   0·145 mm.
24.   0·006 mm.
30.   273 nm, 537·7 nm, 472·8 nm
31.   90·6 nm
32.   89·28 nm, 1%, 164%, 1·3%

CHAPTER XV

1.   1·54 mm. radius; intensity approximately zero.
5.   1·55, 2·19, 2·68, 3·1 mm.
8.   0·93 mm.
9.   3·7 sec.
11.   0·008 mm.
12.   54 mm.
14.   0·99 in.
15. $w = 1·85$ sec.; 0·56 in.
16.   0·00325 mm.
18.   9·39 cm.
19.   11·28 mm.
21.   Radius of 1st dark ring 0·000298 in.
      $w = 0·198$ sec.; magnification 454.
24.   1′ 15″;.
25.   0·0133, 0·1975, 0·6893 in.
28.   9·6 mm.

CHAPTER XVI

10.   90°.
12.   1° 2′.
14. π.
16.   (a) 0·0328 mm. (b) 0·0164 mm.
17.   1, 0·75, 0·5, 0·25, 0.
18.   0·00446 k mm., where k is any odd integer.
20.   0·175 g. per c.c.
22. $a_2 = 29° 29′$; $d = 5° 1′$.
23.   0·03 mm.

CHAPTER XVII

1.   3·052 mm. behind cornea; $m = 1·13$.
2.   13·3 cm. in front of lens; 10 mm. diameter.
3.   Entrance pupil 4·29 cm. behind 1st lens, 2·86 cm. diameter; exit pupil 2·31 cm. in front of the 2nd lens, 2·31 cm. diameter; entrance port is aperture of 1st lens; exit port 7·5 cm. in front of 2nd lens; $m = 1·5$.
4.   Field of full illumination 1° 52′; total field 68° 36′.
6.   28·5 mm.; 16·5 cm.
8.   (a) 14° 14′, 3° 34′; (b) 7° 8′, 1° 47′; (c) 21° 15′, 5° 19′.
10.   Objective $f' = 8$ in., diameter 1·33 in.
      Eyepiece, field lens $f' = 1·33$ in., diameter 1·6 in.
      eye lens $f' = 1·33$ in. diameter 0·8 in.

12. 23·15 metres.
13. 11 ft. 5 in. to 8 ft. 10½ in.
14. $\frac{1}{7·5}$ sec., 0·032 in.
16. 13·04 cm. to 42·86 cm. from the lens.

CHAPTER XVIII

4. $l = + 25$ cm., $l' = + 16·67$ cm., $m = 0·44$.
5. $l = + 5·2$ in., $l' = + 3·25$ in.
6. $m = \left(\dfrac{n}{n'}\right)^2$
11. $t' = + 8·57$ cm.; $s' = + 11·57$ cm.
12. $l' = + 14·4$ in.
21. $l' = + 128·54$ cm.
22. $l' = + 15·39$ cm. (marginal): $+ 15·87$ cm. (paraxial).

CHAPTER XIX

1. $l'_2 = + 17·8$ cm.; $h' = - 1·18$ cm.
3. $F = + 4·68$D.; $+ 23·85$ cm. from lens.
4. $f = - 24·15$ cm., $f' = + 32·13$ cm., $e = + 1$ cm., $e' = - 2·23$ cm., $N + 4·98$ cm. and $N' + 5·75$ cm. from 2nd surface.
5. $f = - 14·51$ cm., $f' = + 19·34$ cm.
6. $F = + 42$D.; principal points at surface, nodal points at centre of curvature. $F = + 11·34$D.
7. $f_V = - 2·365$ in., $f'_V = + 2·892$ in., $e = + 0·149$ in., $e' = - 0·46$ in. $N + 0·237$ in., $N' + 0·378$ in. from 2nd surface.
8. 6·84 mm. from cornea; $m = 1·32$.
10. $F = + 43·05$D.; $f = - 23·22$ mm., $f' = + 31·03$ mm.
11. $+ 5·28$ in from back surface; principal points at centre of curvature.
12. 2·18 in.
13. Image $+ 18·25$ mm. from 3rd surface; 3·52 mm. high.
14. $f' = + 198·8$ mm. 1st principal point $+ 13·8$ mm. from 1st surface,
    2nd principal point $+ 7·0$ mm. from 2nd surface.
15. $f' = + 101·6$ mm.; 1st principal point $+ 6·3$ mm. from 1st surface.
    2nd principal point $- 16·8$ mm. from 2nd surface.
16. (a) $+ 26·39$D.: $f = - 50·63$ mm.; $f' = + 50·63$ mm.
    (b) $e = + 2·126$ mm.; $e' = - 1·276$ mm.
    (c) $+ 42·97$D.; $f = - 23·27$ mm., $f' = + 31·1$ mm.
    (d) $+ 64·5$D.
17. 2·65 cm. in front of 1st surface.
18. $+ 6$D.; 1·25 cm. behind 1st surface.
19. 3·74 mm. behind cornea: 0·49 mm. diameter.

Index